Functional Materials Based on Metal Hydrides

Functional Materials Based on Metal Hydrides

Special Issue Editors

Torben R. Jensen
Hai-Wen Li
Min Zhu
Craig Buckley

MDPI • Basel • Beijing • Wuhan • Barcelona • Belgrade

MDPI

Special Issue Editors

Torben R. Jensen
Aarhus University
Denmark

Hai-Wen Li
Kyushu University Platform of Inter/Transdisciplinary Energy Research (Q-PIT)
Japan

Min Zhu
South China University of Technology
China

Craig Buckley
Curtin University of Technology
Australia

Editorial Office
MDPI
St. Alban-Anlage 66
Basel, Switzerland

This is a reprint of articles from the Special Issue published online in the open access journal *Inorganics* (ISSN 2304-6740) from 2017 to 2018 (available at: https://www.mdpi.com/journal/inorganics/special_issues/metal_hydrides)

For citation purposes, cite each article independently as indicated on the article page online and as indicated below:

LastName, A.A.; LastName, B.B.; LastName, C.C. Article Title. *Journal Name* **Year**, *Article Number*, Page Range.

ISBN 978-3-03897-282-2 (Pbk)
ISBN 978-3-03897-283-9 (PDF)

Contents

About the Special Issue Editors

Torben R. Jensen, Professor D.Sc., has investigated the synthesis and characterization of new inorganic compounds over the past two decades and has developed sample environments for in situ powder diffraction at variable temperature and pressures. His main interest is to create new materials for energy storage in batteries and or hydrogen in the solid state. He has been an expert for the International Energy Agency (IEA) Hydrogen implementation program since 2006. Torben has published about 250 scientific papers in peer-reviewed journals.

Hai-Wen Li, Associate Professor, mainly focuses on the development of advanced materials and systems for efficient hydrogen storage and transportation, as well as the development of hydrides for other energy-related functions such as electrochemical properties.

Min Zhu, Professor, has research interests that include hydrogen storage alloys, electrode materials for lithium-ion batteries, and mechanical alloying. He has published more than 250 papers in peer-reviewed journals. He has also authorized more than 20 patents, and some of them have been applied in industry. He received the "hydrogen and energy award" in 2016 and served as a steering committee member of the international symposium for metal-hydrogen systems.

Craig Buckley, Professor of Physics, is Head of the Hydrogen Storage Research Group and Deputy Director of the Fuels & Energy Technology Institute at Curtin University, Australia. Since 1994, he has held various research positions in the U.K, USA, and Australia prior to being awarded tenure at Curtin University. He is a Fellow of the Australian Institute of Physics and a member of several National and International science and advisory committees including the Australian executive committee member for the International Energy Agency (IEA) Hydrogen Technology Collaboration Program, and an Australian expert on the IEA Hydrogen Task 32: Hydrogen-Based Energy Storage. His research focusses on hydrogen storage materials and the investigation of the structural properties of a wide range of materials. Craig has published 145 scientific papers in peer-reviewed journals.

inorganics

MDPI

Editorial

Functional Materials Based on Metal Hydrides

Hai-Wen Li [1], Min Zhu [2], Craig Buckley [3] and Torben R. Jensen [4,*]

[1] Kyushu University Platform of Inter/Transdisciplinary Energy Research, International Research Center for Hydrogen Energy and WPI International Institute for Carbon-Neutral Energy Research, Kyushu University, Fukuoka 819-0395, Japan; li.haiwen.305@m.kyushu-u.ac.jp

[2] School of Materials Science and Engineering, Key Laboratory of Advanced Energy Storage Materials of Guangdong Province, South China University of Technology, Guangzhou 510641, China; memzhu@scut.edu.cn

[3] Department of Physics and Astronomy, Faculty of Science & Engineering, Curtin University of Technology, GPO Box U 1987, Perth 6845, WA, Australia; c.buckley@curtin.edu.au

[4] Center for Materials Crystallography, Interdisciplinary Nanoscience Center and Department of Chemistry, Aarhus University, DK-8000 Aarhus C, Denmark

* Correspondence: trj@chem.au.dk

Received: 28 August 2018; Accepted: 30 August 2018; Published: 4 September 2018

Abstract: Storage of renewable energy remains a key obstacle for the implementation of a carbon free energy system. There is an urgent need to develop a variety of energy storage systems with varying performance, covering both long-term/large-scale and high gravimetric and volumetric densities for stationary and mobile applications. Novel materials with extraordinary properties have the potential to form the basis for technological paradigm shifts. Here, we present metal hydrides as a diverse class of materials with fascinating structures, compositions and properties. These materials can potentially form the basis for novel energy storage technologies as batteries and for hydrogen storage.

Our extreme and growing energy consumption, based on fossil fuels, has significantly increased the levels of carbon dioxide, which may lead to irreversible global climate change. The planet has plentiful renewable energy, e.g., sun and wind, but the fluctuations over time and geography call for a range of new ideas, and possibly novel technologies. The most difficult challenge appears to be the development of efficient and reliable storage of renewable energy [1]. Hydrogen has long been considered a potential means of energy storage; however, storage of hydrogen is also challenging [2]. Therefore, a wide range of hydrogen-containing materials with energy-related functions has been discovered over the past few decades [3,4]. The chemistry of hydrogen is diverse, and so also are the new hydrides that have been discovered, not only in terms of structure and composition, but also in terms of their properties [5,6]. This has led to a wide range of possible applications for metal hydrides that permeate beyond solid-state hydrogen storage [7–11]. A variety of new hydrides, proposed as battery materials, have been discovered [3]. Their properties can be exploited as fast ion conductors or as conversion-type electrodes with much higher potential energy capacities compared to materials currently used in commercial batteries [3,7,12]. Solar heat storage is also an area of great potential for metal hydrides, in principle offering orders of magnitude higher energy densities than existing materials [3,13–15].

This Special Issue of *Inorganics*, entitled "Functional Materials Based on Metal Hydrides", is dedicated to the wide range of emerging energy-related inorganic hydrogen-containing materials. We have collected thirteen publications, which clearly document that metal hydrides are a diverse class of materials with a range of properties, here denoted as "multi-functional materials". We hope you will enjoy the breadth of science presented in this open-access Special Issue that highlights the many different potential applications of metal hydrides.

An excellent overview of hydride composition and properties is provided by C. Pistidda and co-workers, which also includes important historical landmarks for this class of materials from the past century [16]. In this review, a detailed account of selected aspects of the most relevant complex hydrides is reported. Owing to their high gravimetric hydrogen capacity, tetrahydroborates are extremely appealing for hydrogen storage. However, these materials have some limitations for their use for practical applications, as highlighted in [16], and as also discussed in several research papers in this special issue. The energy density of the reported materials in this special issue are compared in Table 1, which exhibit great potential of hydrides for high capacity energy storage.

Table 1. Comparison of volumetric and gravimetric energy density in various materials.

Materials	Available Gravimetric Hydrogen Density (wt %)	Available Volumetric Hydrogen Density (g/L)	Gravimetric Energy Density (kWh/kg) *	Volumetric Energy Density (kWh/L) *	Ref.
Gaseous H_2 (70 MPa)	5.7	40	39	1.6	[1]
$2LiBH_4-MgH_2$	11.8	96	5.5	3.1	[5]
$2LiBH_4-Al/TiF_3$	9.3	125	3.6	2.4	[17]
$LiAlH_4-h$-BN	7.7	71	3.0	2.8	[18]
$NaBH_4$	10.8	112	4.2	4.3	[19]
$3LiBH_4-Er(BH_4)_3-3LiH$	9.0	147	3.5	2.3	[20]
$LiBH_4-Mg_2FeH_6$	6.0	151	2.3	4.1	[21]
MgH_2	7.6	110	3.0	4.4	[22,23]
$LiAlH_4$	7.9	73	3.1	2.9	[24]
$Ti_1V_{0.9}Cr_{1.1}/4$ wt %Zr	3.5	210	1.4	4.3	[25]
MgH_2-TiO_2-EG	None	None	0.05 **	0.07 **	[26]
Al anode/$LiBH_4$ electrolyte	None	None	0.3 **	0.8 **	[27]

* The energy density was calculated based on 140.4 MJ/kg H_2 at standard condition; ** The energy density was calculated only based on the weight of active materials of electrodes without consideration of the total weight of cell.

An interesting research paper by Chong, Autrey and Jensen investigate one of the most promising hydrogen storage materials, magnesium borohydride, $Mg(BH_4)_2$, with a high gravimetric hydrogen density of 14.7 wt % H_2 [28]. Their recent research reveals that tetrahydofuran complexes of magnesium borohydride are significantly more prone to release and uptake of hydrogen under more moderate conditions. Hydrogen is released at $T < 200$ °C via the intermediate $Mg(B_{10}H_{10})$. These studies reveal new methods to optimize the hydrogen release and uptake reaction through modification of ligand coordination to magnesium [28].

The properties of complex hydrides, such as lithium borohydride, $LiBH_4$, can also be improved creating a reactive hydride composite (RHC) [16] using aluminium, i.e., $LiBH_4-Al$ [29]. Here, Carrillo-Bucio, Tena-García and Suárez-Alcántara investigate the dehydrogenation reactions of the mechanochemically treated $2LiBH_4-Al$ composite; also with additives, TiF_3 or CeO_2 [17]. A reduction in the dehydrogenation temperature and an increase in the reaction kinetics compared to $LiBH_4$ is observed. The dehydrogenation reactions were observed to take place in two main steps, with onsets at 100 °C and 200–300 °C, with a maximum release of 9.3 wt % H_2 for $2LiBH_4-Al/TiF_3$ [17]. However, aluminium-based complex hydrides were the first complex hydrides to be successfully catalysed. Nakagawa, Isobe, Ohki and Hashimoto recently investigated new types of additive effects of ball milled lithium alanate with hexagonal boron nitride, $LiAlH_4-h$-BN [18]. The role of h-BN on the desorption process of $LiAlH_4$ was discussed in comparison with the desorption properties of $LiAlH_4-X$ (X = graphite, LiCl or LiI) [18].

The paper by Ouyang, Zhong, Li and Zhu, discuss the use of another additive, namely water, to promote hydrogen production from $NaBH_4$ by hydrolysis [19]. The reaction of $NaBH_4$ and water readily produces hydrogen, but also a stable oxide, $NaBO_2$, as a by-product. The paper discusses the possible regeneration mechanisms of $NaBO_2$ using magnesium [19].

Inorganics **2018**, *6*, 91

Michael Heere and co-workers present an interesting paper concerning hydrogen release and uptake of a new type of reactive hydride composite based on a rare earth metal borohydride, erbium borohydride, $Er(BH_4)_3$, combined with $LiBH_4$ and/or LiH, $3LiBH_4-Er(BH_4)_3-3LiH$ with a capacity of 9 wt % H_2 [20]. Sieverts-type measurements reveal a three-step desorption-absorption cycle with release of 4.2, 3.7 and 3.5 wt % H_2 for the first, second and third cycle. This work adds a new example to the growing interest in utilisation of rare earth metals and their unique properties [30,31].

Chaudhary, Dornheim, Orimo and co-workers investigated a transition metal containing a reactive hydride composite $(1 - x)LiBH_4–xMg_2FeH_6$, and measured excellent pressure–composition–isothermal (PCT) data [21]. The results indicate that the same stoichiometric reaction ($x = 0.5$) occurred in all investigated samples with optimal hydrogen storage and reversibility properties. This is much lower than those required for the partial rehydrogenation of $LiBH_4$. Moreover, the $x = 0.5$ composite could be reversibly hydrogenated for more than four cycles without degradation of the H_2 capacity. The authors conclude that magnesium hydride plays a vital role as an intermediate in the hydrogen release and uptake reactions [21].

Magnesium hydride remains highly relevant as a hydrogen storage material, and has been investigated in great detail previously [32,33]. Priscilla Huen, Mark Paskevicius, Dorthe B. Ravnsbæk and co-workers investigated the direct synthesis of nano-particulate MgH_2 in a nanoporous carbon scaffold [22]. A solvent-based approach using dibutyl magnesium, $MgBu_2$, infiltrated into four different carbon aerogels with different porosities is investigated. Three independent infiltrations of $MgBu_2$, each with individual hydrogenations, are conducted for each scaffold. Systematic experimental work shows that butane is release over many cycles, inferring that the dibutyl magnesium precursor is not completely reduced. The large difference in molar volume of MgH_2 and $MgBu_2$ is highlighted as a drawback for this approach [22].

Nano-particles are known to have different properties as compared to the bulk state, and this can also be used to improve metal hydride properties, including hydrogen storage [34]. Lei Wang and Kondo-Francois Aguey-Zinsou report on a novel approach to produce nanoscale $LiAlH_4$ via a bottom-up synthesis [24]. Upon further coating of these nanoparticles with Ti, the composite nanomaterial was found to decompose at 120 °C in one single and extremely sharp exothermic event with instant hydrogen release. This finding implies a significant thermodynamic alteration of the hydrogen properties of $LiAlH_4$ induced by the synergetic effects of the Ti catalytic coating and nanosizing effects [24].

Nicola Patelli, Marco Calizzi, and Luca Pasquini analysed the effect of the interfacial free energy on the thermodynamics of hydrogen sorption in nano-scaled materials [23]. When the enthalpy and entropy terms are the same for all interfaces, as in an isotropic bi-phasic system, a compensation temperature is obtained, which does not depend on the system size nor on the relative phase abundance. They also consider the possible effect of elastic strains on the stability of the hydride phase and on hysteresis. They compare a simple model with experimental data obtained on two different systems: (1) bi-phasic nanocomposites where ultrafine TiH_2 crystallite are dispersed within a Mg nanoparticle and (2) Mg nanodots encapsulated by different phases [23].

Most frequently, nano-materials for hydrogen storage are produced by mechano-chemistry [35]. However, Sleiman and Huot [25] investigated the effect of arc melting $Ti_1V_{0.9}Cr_{1.1}$ with zirconium additives on the microstructure and hydrogen storage properties of main BCC alloy and secondary Laves phase alloys C15 and C14. Small amounts of Zr produced fast kinetics and high hydrogen storage capacity [25].

Metal hydrides have been receiving increasing interest as battery materials, both as electrodes and ion conductors [3,5,12]. This is clearly demonstrated by Yang, Wang, Ouyang, Liu and Zhu [26] who prepared MgH_2-based composites with expanded graphite (EG) and TiO_2 by a plasma-assisted milling process as composite electrodes. A stable discharge capacity of 305.5 mAh/g could be achieved after 100 cycles for the 20 h-milled $MgH_2–TiO_2–EG$-20 h composite electrode and the reversibility of the conversion reaction of MgH_2 could be greatly enhanced. This improvement in cyclic performance is attributed mainly to the composite microstructure by the specific plasma-assisted milling process, and the additives TiO_2 and graphite that could effectively ease the volume change during the de-/lithiation process, as well as inhibit the particle agglomeration [26].

Weeks, Tinkey, Ward, Lascola, Zidan, and Teprovich [27] analyse and compare the physical and electrochemical properties of an all solid-state cell utilizing $LiBH_4$ as the electrolyte and aluminium as the active anode material. An initial capacity of 895 mAh/g was observed and is close to the theoretical capacity of aluminium due to the formation of a LiAl (1:1) alloy. This work is the first example of reversible lithiation of aluminium in a solid-state cell and further emphasizes the robust nature of the $LiBH_4$ electrolyte. This demonstrates the possibility of utilising other high-capacity anode materials with a $LiBH_4$-based solid electrolyte in all-solid-state batteries [27].

We hope you will enjoy the breadth of science offered in this open-access Special Issue of *Inorganics*, now organised as a book. Many current frontier research challenges are discussed in the included articles. We hope this unique collection of science can create inspiration for the design and synthesis of other novel "energy-materials". This research field emphatically contributes towards a cleaner and carbon-free future.

References

1. Møller, K.T.; Jensen, T.R.; Akiba, E.; Li, H.-W. Hydrogen—A sustainable energy carrier. *Prog. Nat. Sci. Mater.* **2017**, *27*, 34–40. [CrossRef]
2. Ley, M.; Jepsen, L.; Lee, Y.-S.; Cho, Y.; Colbe, J.; Dornheim, M.; Rokni, M.; Jensen, J.; Sloth, M.; Filinchuk, Y.; et al. Complex hydrides for hydrogen storage—New perspectives. *Mater. Today* **2014**, *17*, 122–128. [CrossRef]
3. Møller, K.T.; Sheppard, D.; Ravnsbæk, D.; Buckley, C.; Akiba, E.; Li, H.-W.; Jensen, T.R. Complex metal hydrides for hydrogen, thermal and electrochemical energy storage. *Energies* **2017**, *10*, 1645. [CrossRef]
4. Yu, X.; Tang, Z.; Sun, D.; Ouyang, L.; Zhu, M. Recent advances and remaining challenges of nanostructured materials for hydrogen storage applications. *Prog. Mater. Sci.* **2017**, *88*, 1–48. [CrossRef]
5. Paskevicius, M.; Jepsen, L.; Schouwink, P.; Černý, R.; Ravnsbæk, D.; Filinchuk, Y.; Dornheim, M.; Besenbacher, F.; Jensen, T.R. Metal Borohydrides and derivatives—Synthesis, structure and properties. *Chem. Soc. Rev.* **2017**, *46*, 1565–1634. [CrossRef] [PubMed]
6. Jepsen, L.; Ley, M.; Lee, Y.; Cho, Y.; Dornheim, M.; Jensen, J.; Filinchuk, Y.; Jørgensen, J.; Besenbacher, F.; Jensen, T.R. Boron-nitrogen based hydrides and reactive composites for hydrogen storage. *Mater. Today* **2014**, *17*, 129–135. [CrossRef]
7. Paskevicius, M.; Hansen, B.; Jørgensen, M.; Richter, B.; Jensen, T.R. Multifunctionality of silver closo-boranes. *Nat. Commun.* **2017**, *8*, 15136. [CrossRef] [PubMed]
8. Payandeh GharibDoust, S.; Dorthe, B.; Černý, R.; Jensen, T.R. Synthesis, structure and properties of bimetallic sodium rare earth (RE) borohydrides, NaRE(BH₄)₄, RE = Ce, Pr, Er or Gd. *Dalton Trans.* **2017**, *46*, 13421–13431. [CrossRef] [PubMed]
9. Schouwink, P.; Ley, M.; Tissot, A.; Hagemann, H.; Jensen, T.R.; Smrčok, L.; Černý, R. Structure and properties of complex hydride perovskite material. *Nat. Commun.* **2014**, *5*, 5706. [CrossRef] [PubMed]
10. Paskevicius, M.; Ley, M.; Sheppard, D.; Jensen, T.R.; Buckley, C. Eutectic melting in metal borohydrides. *Phys. Chem. Chem. Phys.* **2013**, *15*, 19774–19789. [CrossRef] [PubMed]
11. He, L.; Li, H.-W.; Nakajima, H.; Tumanov, N.; Filinchuk, Y.; Hwang, S.-J.; Sharma, M.; Hagemann, H.; Akiba, E. Synthesis of a bimetallic dodecaborate $LiNaB_{12}H_{12}$ with outstanding superionic conductivity. *Chem. Mater.* **2015**, *27*, 5483–5486. [CrossRef]
12. Hansen, B.; Paskevicius, M.; Li, H.-W.; Akiba, E.; Jensen, T.R. Metal boranes: Progress and applications. *Coord. Chem. Rev.* **2016**, *323*, 60–70. [CrossRef]
13. Harries, D.; Paskevicius, M.; Sheppard, D.A.; Price, T.; Buckley, C.E. Concentrating solar thermal heat storage using metal hydrides. *Proc. IEEE* **2012**, *100*, 539–549. [CrossRef]
14. Sheppard, D.A.; Paskevicius, M.; Humphries, T.D.; Felderhoff, M.; Capurso, G.; Bellosta von Colbe, J.; Dornheim, M.; Klassen, T.; Ward, P.A.; Teprovich, J.A., Jr.; et al. Metal hydrides for concentrating solar-thermal power energy storage. *Appl. Phys. A* **2016**, *122*, 1–15. [CrossRef]

15. Javadian, P.; Sheppard, D.; Jensen, T.R.; Buckley, C. Destabilization of lithium hydride and the thermodynamic assessment of the Li–Al–H system for solar thermal energy storage. *RSC Adv.* **2016**, *6*, 94927–94933. [CrossRef]

16. Puszkiel, J.; Garroni, S.; Milanese, C.; Gennari, F.; Klassen, T.; Dornheim, M.; Pistidda, C. Tetra- hydroborates: Development and potential as hydrogen storage medium. *Inorganics* **2017**, *5*, 74. [CrossRef]

17. Carrillo-Bucio, J.; Tena-García, J.; Suárez-Alcántara, K. Dehydrogenation of surface-oxidized mixtures of 2LiBH$_4$ + Al/Additives (TiF$_3$ or CeO$_2$). *Inorganics* **2017**, *5*, 82. [CrossRef]

18. Nakagawa, Y.; Isobe, S.; Ohki, T.; Hashimoto, N. Unique hydrogen desorption properties of LiAlH$_4$/*h*-BN composites. *Inorganics* **2017**, *5*, 71. [CrossRef]

19. Ouyang, L.; Zhong, H.; Li, H.-W.; Zhu, M. A recycling hydrogen supply system of NaBH$_4$ based on a facile regeneration process: A review. *Inorganics* **2018**, *6*, 10. [CrossRef]

20. Heere, M.; GharibDoust, S.; Brighi, M.; Frommen, C.; Sørby, M.; Černý, R.; Jensen, T.; Hauback, B. Hydrogen sorption in erbium borohydride composite mixtures with LiBH$_4$ and/or LiH. *Inorganics* **2017**, *5*, 31. [CrossRef]

21. Li, G.; Matsuo, M.; Takagi, S.; Chaudhary, A.-L.; Sato, T.; Dornheim, M.; Orimo, S. Thermodynamic properties and reversible hydrogenation of LiBH$_4$–Mg$_2$FeH$_6$ composite materials. *Inorganics* **2017**, *5*, 81. [CrossRef]

22. Huen, P.; Paskevicius, M.; Richter, B.; Ravnsbæk, D.; Jensen, T.R. Hydrogen storage stability of nanoconfined MgH$_2$ upon cycling. *Inorganics* **2017**, *5*, 57. [CrossRef]

23. Patelli, N.; Calizzi, M.; Pasquini, L. Interface enthalpy-entropy competition in nanoscale metal hydrides. *Inorganics* **2018**, *6*, 13. [CrossRef]

24. Wang, L.; Aguey-Zinsou, K.-F. Synthesis of LiAlH$_4$ nanoparticles leading to a single hydrogen release step upon Ti coating. *Inorganics* **2017**, *5*, 38. [CrossRef]

25. Sleiman, S.; Huot, J. Microstructure and hydrogen storage properties of Ti1V$_{0.9}$Cr$_{1.1}$ alloy with addition of *x* wt % Zr (*x* = 0, 2, 4, 8, and 12). *Inorganics* **2017**, *5*, 86. [CrossRef]

26. Yang, S.; Wang, H.; Ouyang, L.; Liu, J.; Zhu, M. Improvement in the electrochemical lithium storage performance of MgH$_2$. *Inorganics* **2018**, *6*, 2. [CrossRef]

27. Weeks, J.A.; Tinkey, S.C.; Ward, P.A.; Lascola, R.; Zidan, R.; Teprovich, J.A. Investigation of the reversible lithiation of an oxide free aluminum anode by a LiBH$_4$ solid state electrolyte. *Inorganics* **2017**, *5*, 83. [CrossRef]

28. Chong, M.; Autrey, T.; Jensen, C. Lewis base complexes of magnesium borohydride: Enhanced kinetics and product selectivity upon hydrogen release. *Inorganics* **2017**, *5*, 89. [CrossRef]

29. Hansen, B.; Ravnsbæk, D.; Reed, D.; Book, D.; Gundlach, C.; Skibsted, J.; Jensen, T. Hydrogen storage capacity loss in a LiBH$_4$–Al composite. *J. Phys. Chem. C* **2013**, *117*, 7423–7432. [CrossRef]

30. Frommen, C.; Sørby, M.; Heere, M.; Humphries, T.; Olsen, J.; Hauback, B. Rare earth borohydrides-crystal structures and thermal properties. *Energies* **2017**, *10*, 2115. [CrossRef]

31. Mansell, S.; Liddle, S. Rare earth and actinide complexes. *Inorganics* **2016**, *4*, 31. [CrossRef]

32. Crivello, J.-C.; Denys, R.V.; Dornheim, M.; Felderhoff, M.; Grant, D.M.; Huot, J.; Jensen, T.R.; Jongh, P.; Latroche, M.; Walker, G.S.; et al. Mg-based compounds for hydrogen and energy storage. *Appl. Phys. A* **2016**, *122*, 85. [CrossRef]

33. Crivello, J.-C.; Dam, B.; Denys, R.V.; Dornheim, M.; Grant, D.M.; Huot, J.; Jensen, T.R.; Jongh, P.; Latroche, M.; Milanese, C.; et al. Review of magnesium hydride based materials: Development and optimisation. *Appl. Phys. A* **2016**, *122*, 97. [CrossRef]

34. Nielsen, T.; Besenbacher, F.; Jensen, T. Nanoconfined hydrides for energy storage. *Nanoscale* **2011**, *3*, 2086–2098. [CrossRef] [PubMed]

35. Huot, J.; Ravnsbæk, D.B.; Zhang, J.; Cuevas, F.; Latroche, M.; Jensen, T.R. Mechanochemical synthesis of hydrogen storage materials. *Prog. Mater. Sci.* **2013**, *58*, 30–75. [CrossRef]

inorganics

MDPI

Review

Tetrahydroborates: Development and Potential as Hydrogen Storage Medium

Julián Puszkiel [1,2], Sebastiano Garroni [3], Chiara Milanese [4], Fabiana Gennari [2], Thomas Klassen [1,5], Martin Dornheim [1] and Claudio Pistidda [1,*]

[1] Institute of Materials Research, Helmholtz-Zentrum Geesthacht, Max-Planck-Straße 1, D-21502 Geesthacht, Germany; julian.puszkiel@hzg.de (J.P.); klassen@hsu-hh.de (T.K.); martin.dornheim@hzg.de (M.D.)
[2] National Council of Scientific and Technological Research (CONICET), Bariloche Atomic Center (National Commission of Atomic Energy) and Balseiro Institute (University of Cuyo) Av. Bustillo 9500, San Carlos de Bariloche, 8400 Río Negro, Argentina; gennari.fabiana36@gmail.com
[3] International Research Centre in Critical Raw Materials-ICCRAM, University of Burgos, 09001 Burgos, Spain; sgarroni@ubu.es
[4] Pavia Hydrogen Lab, C.S.G.I. & Chemistry Department, Physical Chemistry Section, University of Pavia, Viale Taramelli, 1627100 Pavia, Italy; chiara.milanese@unipv.it
[5] Department of Mechanical Engineering, Helmut Schmidt University, Holstenhofweg 85, D-22043 Hamburg, Germany
* Correspondence: claudio.pistidda@hzg.de

Received: 15 September 2017; Accepted: 22 October 2017; Published: 31 October 2017

Abstract: The use of fossil fuels as an energy supply becomes increasingly problematic from the point of view of both environmental emissions and energy sustainability. As an alternative, hydrogen is widely regarded as a key element for a potential energy solution. However, differently from fossil fuels such as oil, gas, and coal, the production of hydrogen requires energy. Alternative and intermittent renewable energy sources such as solar power, wind power, etc., present multiple advantages for the production of hydrogen. On the one hand, the renewable sources contribute to a remarkable reduction of pollutants released to the air and on the other hand, they significantly enhance the sustainability of energy supply. In addition, the storage of energy in form of hydrogen has a huge potential to balance an effective and synergetic utilization of renewable energy sources. In this regard, hydrogen storage technology is a key technology towards the practical application of hydrogen as "energy carrier". Among the methods available to store hydrogen, solid-state storage is the most attractive alternative from both the safety and the volumetric energy density points of view. Because of their appealing hydrogen content, complex hydrides and complex hydride-based systems have attracted considerable attention as potential energy vectors for mobile and stationary applications. In this review, the progresses made over the last century on the synthesis and development of tetrahydroborates and tetrahydroborate-based systems for hydrogen storage purposes are summarized.

Keywords: Tetrahydroborates; synthesis; decomposition pathways; solid state hydrogen storage

1. Introduction

Since the second industrial revolution in the late 19th century, humankind has experienced an uninterrupted period of industrial and economic growth. The possibility to produce energy using the so far cheap and abundant fossil fuels has represented, beyond doubt, the driving force of this growth period. All of the current energy systems are based on coal, oil, natural gas, as well as nuclear fission, depending on the conditions in a given country and its access to the cheapest primary energy source. The total final consumption of energy worldwide grew more than twice from the 1970s to the 2000s. Unless an alternative to the use of fossil fuels is found, this trend is going to come soon to an end [1]. Moreover, the massive use of fossil fuels has disturbed the equilibrium of the global climate. The transition from a carbon-based economy to a carbon-free economy is perhaps the greatest challenge of the 21st century. Possible candidates to replace the fossil fuels as energy sources are the so called renewable energy sources (e.g., solar-, wind-, geothermal-, wave-, hydroelectric-energy, etc.). However, the main obstacle to the use of these energy sources is the fact that they are intermittent and un-evenly distributed. To fully exploit these energy sources, an efficient energy storage system is required. In this regard, hydrogen is widely considered to be capable of solving both the issues of increasing CO_2 emissions and of future energy sustainability. Hydrogen has a gravimetric energy density of 120 MJ/kg, which is more than twice the energy content of most common fossil fuels (e.g., methane 50 MJ/kg, propane 46 MJ/kg and gasoline 45 MJ/kg). Unfortunately, compared to fossil fuels, hydrogen, being the lightest element of the periodic table, has an extremely low volumetric energy density.

Hydrogen is usually stored in three main forms: as highly pressurized gas under 350 and 700 bar, in liquefied form at −253 °C, and in metal hydrides, chemically bonded to metals. Although, the volumetric energy density improves when hydrogen is compressed and/or liquefied, it still remains low as compared to that of fossil fuels. A significant improvement of the volumetric energy density is achieved when hydrogen is bonded to another element in solid state form [2–5]. It is well known that several metals, and in particular transitions metals, have a high affinity for hydrogen. This high affinity leads to the reaction between hydrogen (H_2) and the metal (M) to form metal hydrides (MH_x). This process can be described as follows:

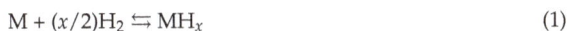

$$M + (x/2)H_2 \leftrightarrows MH_x \tag{1}$$

Metal hydrides such as those of Mg [6–13] and Pd [14–18] have been exhaustively studied for decades. Mg has a relative high gravimetric hydrogen density of 7.6 wt %, it is abundant, and it is cheap [12,13], however, the Mg/MgH_2 system presents a high thermodynamic stability ($\Delta H_{dehydrogenation}$ = −74 kJ/mol H_2) [6], leading to operative temperatures above 300 °C. Thus, the thermal management for large scale use of MgH_2 becomes an important issue, constraining the practical applications of MgH_2 to the stationary ones. In the case of Pd, the absorption of hydrogen already occurs at room temperature and presents excellent catalytic properties for the dissociation of the hydrogen molecule. However, its gravimetric hydrogen density is extremely low (~0.6 wt %) and it is quite costly. Therefore, Pd is not suitable for any large scale hydrogen storage application.

Other examples of hydrides with high gravimetric hydrogen capacity but improper enthalpy of formation are aluminum hydride (AlH_3, gravimetric hydrogen capacity of ~10.0 wt %), and lithium hydride (LiH, gravimetric hydrogen capacity of ~12.6 wt %). The low formation enthalpy of AlH_3 (5–8 kJ/mol H_2) requires extreme pressure conditions for the hydrogenation of Al [19]. On the contrary, the high formation enthalpy of LiH (140 kJ/mol H_2) leads to a harsh dehydrogenation temperature above 700 °C at 1 bar of hydrogen [4]. Therefore, tailoring the metal hydrides reaction enthalpies and/or developing hydrogen containing materials that are different from the conventional metal hydrides are key issues for the design of vessels based on solid-state hydrogen storage materials towards practical applications. Recently, because of their high hydrogen storage capacity, "complex hydrides" attracted considerable attention as potential hydrogen storage materials [20–23]. The name "complex hydrides" originates from the presence of an anionic non metal–hydrogen complex ($[BH_4]^-$, $[NH_2]^-$, $[AlH_4]^-$) or metal–hydrogen complex ($[NiH_4]^{4-}$, $[CoH_5]^{4-}$, $[FeH_6]^{4-}$, $[MnH_6]^{5-}$, and $[ZnH_4]^{2-}$) bonded to a cationic alkali, alkali-earth, or transition metal. Taking into account this formulation, the complex metal hydrides are classified as non-transition metal complex hydrides (such as $LiBH_4$, $LiNH_2$, $LiAlH_4$) and transition metal complex hydrides (such as Mg_2NiH_4, Mg_2CoH_5, Mg_2FeH_6, Mg_3MnH_7, K_2ZnH_4). Although this class of hydrides has been known for a long time since the first report on pure metal amides was published in the 1809 [24,25], they had initially not been considered as potential hydrogen storage materials. In the case of the non-transition metal complex hydrides, this lack of initial interest can be traced to their apparent irreversibility and more in general to the difficulty to produce them in a large scale. The interest on hydrogen as an energy carrier began around the 1960s, and it has grown significantly since the 1990s. In 1997, Bogdanovič and Schwickardi were the first to demonstrate the concrete possibility to reversibly store hydrogen in titanium-based doped $NaAlH_4$ at moderate temperature and pressure conditions [26]. Since then, many efforts have been done to investigate and optimize the hydrogen storage properties of complex hydrides.

There are several review works about light complex hydrides in which different aspects of tetrahydroborates are described [20–37]. In some works the synthesis, structure, stability, kinetics, thermodynamics, as well as tailoring of the tetrahydroborates and their hydride mixtures are covered [20–31]. Others focus on the synthesis of tetrahydroborates, high metal boranes and rare-earth borohydrides [28,32–34]. There is a review that describes the hydrogen storage properties of $NaBH_4$ [35]. In addition, some review works are mainly devoted to the crystal structures of tetrahydroborates [28,36,37].

The aim of this work is to highlight selected aspects of tetrahydroborates, such as (1) a historical overview about their synthesis; (2) the different decompositions mechanisms; and (3) the conceptual background of alternative approaches to tailor their hydrogen storage properties, mainly the concept of the reactive hydride composites (RHC), along with the most relevant achievement in their use as hydrogen storage materials and future prospects.

2. From Boron to Tetrahydroborates

In the last two centuries, boron has gained a significant importance in our daily life. This low-abundance element and its compounds are used in a variety of branches, including agriculture, medicine, electronics, chemical synthesis, catalysis, and energy. Highly impure boron was isolated for the first time in 1808 by H. Davy [38] J. L. Gay Lussac and L. J. Thénard [39]. In 1892 H. Moissan obtained boron with a purity of roughly 95% by the reduction of borax with magnesium, as shown in reaction (2) [40]:

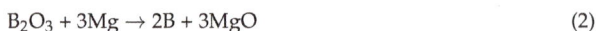

$$B_2O_3 + 3Mg \rightarrow 2B + 3MgO \tag{2}$$

However, high-purity boron (>99.9%) was isolated only in 1922 when a new synthetic pathway that was capable to ensure a kilograms scale production was introduced. In that process volatile boron compounds (i.e., BBr_3 and BCl_3) were reduced by H_2 at temperatures between 800 °C and 1200 °C (reaction (3)). The temperature of the process was selective for the final boron allotrope [41].

$$BX_3 + 3/2H_2 \rightarrow B + 3HX \quad (X = Br, Cl) \tag{3}$$

Together with the availability of good purity boron, the preparation and characterization of boron hydrides started in 1912. Thanks to the work of A. Stock and co-workers in the period between the 1912 and 1936, six boron hydrides, i.e., B_2H_6, B_4H_{10}, B_5H_9, B_5H_{11}, B_6H_{10}, and $B_{10}H_{14}$, were isolated and their chemical features investigated [42–44]. The early syntheses of boron hydrides were based upon the reaction between magnesium boride and hydrochloric acid. This preparation pathway was largely inefficient often giving mixtures of products (B_yH_x, y = 4, 5, 6, 10, x = 9, 10, 14), with yields of only 4–5% [44]. A first improvement in the reaction yield (11%) was achieved replacing the hydrochloric acid with 8*N*-phosphoric acid.

At this time, diborane was obtained in millimolar quantities by the thermal decomposition of B_4H_{10}, and later in the 1931 H. Schlesinger and A. B. Burg reported the synthesis of B_2H_6 from BCl_3 and H_2, using a high voltage discharge [45]. Despite the fact that the amount of B_2H_6 produced by this method was still small, the reaction provided a considerable yield of 75%. It was necessary to wait until 1947 to find an effective and practical method to synthesize B_2H_6 with a purity of 99% [46]. This method was based on a halide-hydride exchange reaction performed in ethyl ether according to reaction (4):

$$3LiAlH_4 + 4BCl_3 \rightarrow 2B_2H_6 + 3LiCl + 3AlCl_3 \tag{4}$$

In the following years, several useful routes to synthesize B_2H_6 via halide-hydride exchange reaction were found, as for example, the reactions between NaH with $B(OCH_3)_3$, BH_4^- with I_2 and BH_4^- with H_2PO_4 [47–51].

Along with the increasing availability of B_2H_6, the synthesis of new compounds was enabled. The first tetrahydroborate, i.e., $Al(BH_4)_3$, was produced in 1940 starting from trimethylaluminium and diborane (reaction (5) [52]):

$$Al_2(CH_3)_6 + 4B_2H_6 \rightarrow 2B(CH_3)_3 + 2Al(BH_4)_3 \tag{5}$$

To synthesize significant quantities of tetrahydroborates and to limit the formation of complex mixtures as side-products, the reaction had to be performed in an excess of diborane. The stoichiometric formula of the aluminum tetrahydroborate was obtained by the characterization of the vapor density and the evaluation of the reaction between the newly synthesized compound with water and hydrogen chloride. The isolated material was a colorless solid below $-64.5\,^\circ$C, its vapor pressure measured at $0\,^\circ$C was 119.5 mm Hg, and its extrapolated boiling point was $44.5\,^\circ$C.

Encouraged by the close chemical similarities between aluminum and beryllium, A. B. Burg and H. I. Schlesinger [53] successfully synthesized beryllium tetrahydroborate, following the same approach as used for the synthesis of aluminum tetrahydroborate. The reaction of dimethylberyllium with diborane proceeds in several steps towards the formation of beryllium tetrahydroborate, i.e., $Be(BH_4)_2$. In the first step, a solid methyl-rich compound with unknown composition is formed. Then, this compound turns into a non-volatile liquid. Upon further reaction with B_2H_6, an easily sublimable solid is produced. This solid has an approximate composition formula CH_3BeBH_4. The addition of further B_2H_6 leads to the formation of beryllium tetrahydroborate coexisting with a mixture of other derivates of diborane compounds in gas form, such as boron trimethyl and methyl plus a non-volatile beryllium hydroborate species. Due to its complexity, the synthesis of $Be(BH_4)_2$ through this route cannot be summarized in a simple chemical equation.

In the same year, H. I. Schlesinger and H. C. Brown [54] investigated the possibility of forming tetrahydroborates of alkali metals. Because of its simple preparation and purification routes, ethyllithium (LiC_2H_5) was chosen as starting material. Once ethyllithium and B_2H_6 were put into contact, they reacted to form several ethyl derivates of diborane plus the solid tetrahydroborates of lithium ($LiBH_4$). This compound is characterized by a rather high stability. In fact, differently from the aluminum and beryllium tetrahydroborate, that ignite spontaneously when exposed to air at room temperature, $LiBH_4$ is quite stable in dry air.

In 1941, H. I. Schlesinger and co-workers [55] were asked to undertake the synthesis of new volatile compounds of uranium. As a consequence of the fact that the tetrahydroborates of aluminum and beryllium are the most volatile compounds of these elements, the synthesis of uranium tetrahydroborate was attempted [53,56]. Although, the tetrahydroborates of aluminum, beryllium, and lithium were prepared by the reaction of diborane with metal alkyls, no alkyl of uranium was known at that time. Therefore, a new synthetic procedure was developed. Uranium tetrahydroborate was obtained by the interaction of uranium (IV) fluoride with aluminum tetrahydroborate, as shown in reaction (6):

$$UF_4 + 2Al(BH_4)_3 \rightarrow U(BH_4)_4 + 2Al(BH_4)F_2 \tag{6}$$

At the time of its first synthesis, uranium tetrahydroborate was the most volatile compound of the known uranium (IV) derivatives.

In 1950, E. Wiberg and R. Bauer were the first to synthesize $Mg(BH_4)_2$ by the production of fine magnesium hydride powder, reaction (7), and then its subsequent reaction with diborane, reaction (8) [57]:

$$3Mg(C_2H_5)_2 + B_2H_6 \rightarrow 3MgH_2 + 2B(C_2H_5)_3 \tag{7}$$

$$3MgH_2 + 3B_2H_6 \rightarrow 3Mg(BH_4)_2 \tag{8}$$

Until that year, the following tetrahydroborates were prepared by the interaction of diborane with metal derivatives: $HZn(BH_4)$, $Ga(BH_4)$, $Ti(BH_4)_3$, $Pu(BH_4)_3$, $Zr(BH_4)_4$, $Hf(BH_4)_4$, $Th(BH_4)_4$, $Np(BH_4)_4$ [57].

In 1952–1953, U.S. scientists were requested to synthesize larger amounts of borohydride by a more efficient method. The complexity of the process utilized for their synthesis involved slow reaction and repeated treatment with large excess of diborane. Thus, a more satisfactory synthetic method was developed for preparing lithium borohydride, and from it, sodium and potassium tetrahydroborate.

In solvent environment (ethyl ether) lithium hydride absorbs large quantities of diborane. After the filtration of the insoluble LiH impurities, the evaporation of the solvent from the filtrate leads to the precipitation of $LiBH_4 \cdot (C_2H_5)_2O$ first, and then to high purity $LiBH_4$. At that time, it was believed that in the absence of solvents, lithium hydride does not react with diborane even at an elevated temperature [55]. However, recently Friedrichs et al. [58] showed that the solvent-free synthesis of $LiBH_4$ from LiH in a diborane atmosphere at 120 °C is possible.

Differently from lithium hydride, sodium hydride does not react with diborane even in the presence of solvents. In order to circumvent this inconvenience the use of trimethoxyborohydride as hydride ions source was proposed [55,59]. The reaction between trimethoxyborohydride and diborane proceeds rapidly and quantitatively forming sodium tetrahydroborate and trimethylborate, according to reaction (9):

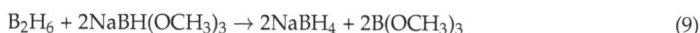

$$B_2H_6 + 2NaBH(OCH_3)_3 \rightarrow 2NaBH_4 + 2B(OCH_3)_3 \tag{9}$$

It was also reported that dimethoxiborine reacts with sodium trimethoxyborohydride to form sodium tetrahydroborate and trimethylborate as shown in reaction (10):

$$3(CH_3O)_2BH + NaBH(OCH_3)_3 \rightarrow NaBH_4 + 3B(OCH_3)_3 \tag{10}$$

Similarly, diborane reacts with sodium tetramethoxyborohydride to form sodium tetrahydroborate plus trimethylborate (reaction (11)):

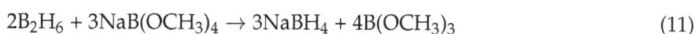

$$2B_2H_6 + 3NaB(OCH_3)_4 \rightarrow 3NaBH_4 + 4B(OCH_3)_3 \tag{11}$$

Following the same approach applied for the previously discussed tetrahydroborates, potassium tetrahydroborate was also synthesized for the first time by the reaction of diborane with potassium tetramethoxyborohydride (reaction (13)) [59]. The latter was synthesized from potassium methoxyde and methyl borate as shown in reaction (12):

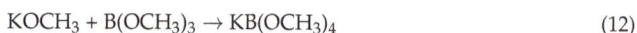

$$KOCH_3 + B(OCH_3)_3 \rightarrow KB(OCH_3)_4 \tag{12}$$

$$3KB(OCH_3)_4 + B_2H_6 \rightarrow 3KBH_4 + 4B(OCH_3)_3 \tag{13}$$

Although, these methods allowed Schlesinger and co-workers to synthesize sodium tetrahydroborates in laboratory scale, they were not appropriate for an industrial scale production. In particular, the risks connected to the use of diborane represented a serious limitation to the scaling up the above mentioned processes. The first attempt to synthesize sodium tetrahydroborate without the use of diborane was carried out heating sodium trimethoxyborohydride to about 230 °C [59]. At this temperature, the disproportion of the starting material to $NaBH_4$ and $NaB(OCH_3)_4$ takes place via reaction (14):

$$4NaBH(OCH_3)_3 \rightarrow NaBH_4 + 3NaB(OCH_3)_4 \tag{14}$$

However, reaction (14) does not proceed to completeness at about 230 °C. Then, the attempts to improve the yield of this reaction were stopped when in the same laboratory it was observed that sodium hydride and methyl borate quickly reacted at high temperature, i.e., 225–275 °C, to form $NaBH_4$ and $NaOCH_3$ (reaction (15)):

$$4NaH + B(OCH_3)_3 \rightarrow NaBH_4 + 3NaOCH_3 \tag{15}$$

As a consequence of the extremely good yield of this reaction, i.e., 94%, and the high purity of the obtained $NaBH_4$ up to 96%, this method is still the most used synthesis process to obtain sodium tetrahydroborate. Similarly, high purity $LiBH_4$, i.e., 95%, was synthesized by the reaction between lithium hydride and methyl borate with an overall reaction yield of 70% (reaction (16)):

$$4LiH + B(OCH_3)_3 \rightarrow LiBH_4 + 3LiOCH_3 \tag{16}$$

The possibility to produce $NaBH_4$ and $LiBH_4$ in large quantities led to the development of a new method to successfully synthesize metal tetrahydroborates. The method consists in the metathesis between metal halides and alkali tetrahydroborates in solvents, reaction (17) [60–62]:

$$yMX_n + yM_a(BH_4)_n \rightarrow yM(BH_4)_n + yM_aX_n \quad (M, M_a = \text{metals}, X = \text{halide}) \tag{17}$$

The first example of this reaction reported in literature was the metathesis between lithium tetrahydroborate and aluminum chloride as shown in reaction (18):

$$AlCl_3 + 3LiBH_4 \rightarrow Al(BH_4)_3 + 3LiCl \tag{18}$$

The interaction between tetrahydroborates and metal halides leads also to the formation of multi-cation tetrahydroborates. E. Wiber and W. Henle [61,62] reported on the synthesis of $Li[ZnCl(BH_4)_2]$ and $Li[CdCl(BH_4)_2]$ as intermediates of the syntheses of $Zn(BH_4)_2$ and $Cd(BH_4)_2$ via metathesis between lithium tetrahydroborate and $ZnCl_2$ and $CdCl_2$, respectively, as seen in reactions (19) and (20):

$$ZnCl_2 + 2LiBH_4 \rightarrow Li[ZnCl(BH_4)_2] + LiCl \rightarrow Zn(BH_4)_2 + 2LiCl \tag{19}$$

$$CdCl_2 + 2LiBH_4 \rightarrow Li[CdCl(BH_4)_2] + LiCl \rightarrow Cd(BH_4)_2 + 2LiCl \tag{20}$$

In 1955, E. Wiberg and R. Hartwimmer [63] reported for the first time on the synthesis of calcium, strontium and barium tetrahydroborates, i.e., $Ca(BH_4)_2$, $Sr(BH_4)_2$, $Ba(BH_4)_2$, respectively. The synthesis was performed by the reaction between the hydrides of the mentioned elements and diborane (reaction (21)).

$$MH_2 + 2B_2H_6 \rightarrow M(BH_4)_2 \ (M = \text{metal}) \tag{21}$$

In 1961, H. Nöth [64] synthesized several new multi-cation tetrahydroborates via metathesis between lithium, sodium and potassium tetrahydroborates and metal halides or by direct reaction between tetrahydroborates, i.e., $Li[Zn(BH_4)_3]$, $Li_2[Cd(BH_4)_4]$, $Li[Ti(BH_4)_4]$, $Li[Fe(BH_4)_3]$, $Na[Zn(BH_4)_3]$, $K_2[Zn_3(BH_4)_8]$, plus a long series of halogen derivatives as for example $Li_2[ZnI_2(BH_4)_2]$, $Li_3[Mn(BH_4)_3I_2]$, etc. The syntheses of some of these multi-cation tetrahydroborates are described in reactions (22)–(28):

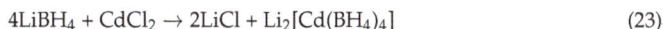

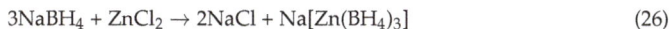

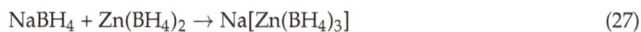

$$3LiBH_4 + ZnCl_2 \rightarrow 2LiCl + Li[Zn(BH_4)_3] \tag{22}$$

$$4LiBH_4 + CdCl_2 \rightarrow 2LiCl + Li_2[Cd(BH_4)_4] \tag{23}$$

$$LiBH_4 + Ti(BH_4)_3 \rightarrow Li[Ti(BH_4)_4] \tag{24}$$

$$3LiBH_4 + FeCl_2 \rightarrow 2LiCl + Li[Fe(BH_4)_3] \tag{25}$$

$$3NaBH_4 + ZnCl_2 \rightarrow 2NaCl + Na[Zn(BH_4)_3] \tag{26}$$

$$NaBH_4 + Zn(BH_4)_2 \rightarrow Na[Zn(BH_4)_3] \tag{27}$$

$$2KBH_4 + 3Zn(BH_4)_2 \rightarrow K_2[Zn_3(BH_4)_8] \tag{28}$$

In the following years, it was possible to observe an increasing attention towards this class of materials both as laboratory reagents and as energy storage materials. As a consequence, several new tetrahydroborates were synthesized. Nowadays, the mono-cation tetrahydroborates of 32 chemical elements, plus an undefined number of multi-cation tetrahydroborates are known [23,38,42–45,52–57,59–78].

The development of the borane chemistry in general, and of the tetrahydroborates in particular, was from the early beginning connected with the research in the field of energy storage. In fact, the research in this field was stirred by the need of finding new propellants for military applications by the U.S. and Soviet Union [79]. Although these war research plans were dismissed around the 1950s, new pacific applications for this class of materials came to light in the field of energy storage.

3. Decomposition Reactions of Li, Na, K, Mg, Ca and U Tetrahydroborates

Homoleptic tetrahydroborates have a high gravimetric hydrogen capacity. However, those compounds, stable at room temperature, can release hydrogen under thermal input. Predictions by first-principle calculations suggested that the charge transfer from M^+ to $[BH_4]_n{}^-$ is a crucial factor for the stability of $M(BH_4)_n$ (M = metal). Moreover, a linear relationship between the enthalpy of $M(BH_4)_n$ and the Pauling electronegativity (X_p) of M was underlined [80–82]. Experimental results also demonstrated that the desorption temperatures (T_d) of the tetrahydroborates are generally correlated with the X_p value of the metal. Gas chromatography experiments showed a decrease of the desorption temperature with the increase of X_p value of the metal, e.g., T_d: $(NaBH_4) \approx 550\ °C > (LiBH_4) \approx 470\ °C > (Sc(BH_4)_3) \approx 280\ °C > (Zr(BH_4)_4) \approx 170\ °C > (Zn(BH_4)_2) \approx 125\ °C$ with X_p: Na = 0.9 < Li = 1.0 < Sc = 1.3 < Zr = 1.4 < Zn = 1.6 [82,83]. The Pauling electronegativity of the metal is also correlated with the decomposition products of tetrahydroborates. Nakamori et al. [84] reported that stable-covalent metal tetrahydroborates (X_p < 1.5) during decomposition release hydrogen, whereas less stable-covalent tetrahydroborates (X_p > 1.5) release a mixture of diborane and hydrogen. An important general feature of the dehydrogenation reaction of $M(BH_4)_n$ (M: K, Na, Li, Mg, Ca, etc.) is the thermodynamically favorable formation of the $M(B_{12}H_{12})_n$ compounds (reaction (29)). On the one hand, they are considered as intermediate compounds [85–103]. However, on the other hand, their high stability and highly negative enthalpy of formation suggest that they are final decomposition products of $M(BH_4)_n$ [92,102–106].

$$M(BH_4)_n \rightarrow 1/6M(B_{12}H_{12})_n + 5/6MH_n + 13n/12H_2 \tag{29}$$

Here below, the decomposition behaviors of KBH_4, $NaBH_4$, $LiBH_4$, $Mg(BH_4)_2$, $Ca(BH_4)_2$, and $U(BH_4)_4$ are summarized. It is worth mentioning that the decomposition behavior of $U(BH_4)_4$ is here shown because of the scarce information available about this particular borohydride, but it cannot be considered as a potential material for hydrogen storage applications.

3.1. LiBH4

When $LiBH_4$ is heated, a phase transition from the ordered low-temperature orthorhombic phase to the disordered high-temperature hexagonal polymorph takes place at about 115 °C. Increasing the temperature, $LiBH_4$ releases a small amount of hydrogen (about 0.3 wt %) between 100 °C and 200 °C, and melts at around 270 °C [91,107,108]. The two main hydrogen releases occur at 320 °C and 400 °C. When heating up to 600 °C, the final reaction products are solid lithium hydride, solid boron, and gaseous hydrogen, as shown in reaction (30).

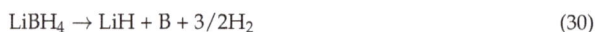

$$LiBH_4 \rightarrow LiH + B + 3/2H_2 \tag{30}$$

The total weight loss associated with the hydrogen release of reaction (30) is 13.8 wt % and its experimental reaction enthalpy is 75 kJ/mol H_2 [109]. However, reaction (30) proceeds through several different intermediate steps, as predicted via theoretical calculations and verified via experimental evidences [85,86,89,91–93]. Computational and experimental works have reported that the decomposition reaction of $LiBH_4$ proceeds through the formation of the intermediate $Li_2B_{12}H_{12}$ as indicated in reaction (31), and finally $Li_2B_{12}H_{12}$ decomposes to solid lithium hydride, solid boron and gaseous hydrogen according to reaction (32). This happens because the enthalpy value for reaction (31) is lower than that for reaction (30), namely 56 kJ/mol H_2 and 75 kJ/mol H_2, respectively [85,86].

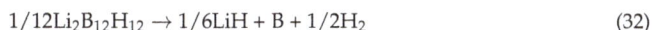

$$LiBH_4 \rightarrow 1/12Li_2B_{12}H_{12} + 10/12LiH + 13/12H_2 \tag{31}$$

$$1/12Li_2B_{12}H_{12} \rightarrow 1/6LiH + B + 1/2H_2 \tag{32}$$

Caputo et al. [92] studied the decomposition reactions of $LiBH_4$ by first-principle approach. They argued that the calculated enthalpy of formation of $Li_2B_{12}H_{12}$ is highly negative. Hence, it is a product that hinders the reversible dehydrogenation/re-hydrogenation process of $LiBH_4$.

Friedrichs et al. [93] reported that together with hydrogen, small quantities of B_2H_6 are released. Once formed, B_2H_6 decomposes to elemental B and hydrogen or it might react with the still present $LiBH_4$ to form $Li_2B_{12}H_{12}$ and possibly $Li_2B_{10}H_{10}$ as in reactions (33) and (34), respectively.

$$2LiBH_4 + 5B_2H_6 \rightarrow Li_2B_{12}H_{12} + 13H_2 \tag{33}$$

$$LiBH_4 + 2B_2H_6 \rightarrow 1/2Li_2B_{10}H_{10} + 11/2H_2 \tag{34}$$

The decomposition pathway either via the direct formation of $Li_2B_{12}H_{12}$ (reactions (31) and (32)), or the interaction of $LiBH_4$ and B_2H_6 depends mainly on the applied gas overpressure. Under vacuum conditions the decomposition reaction can go through the formation of B_2H_6, as shown in reactions (33) and (34) [93]. However, under hydrogen or even argon overpressure the decomposition of $LiBH_4$ may not proceed via the intermediate formation of gaseous diborane (B_2H_6) since the gas overpressure kinetically suppresses its release [110,111]. Thus, $LiBH_4$ decomposes through the intermediate formation of $Li_2B_{12}H_{12}$, as indicated in reactions (31) and (32).

3.2. NaBH4

Upon heating, $NaBH_4$ starts to release hydrogen at 470 °C when it is still in solid form, and continues the dehydrogenation after melting at about 515 °C through a multistep process. From dehydrogenation pressure-composition isotherms (PCI) of $NaBH_4$ measured at 600 °C, 650 °C and 700 °C under dynamic hydrogen flow, just one plateau was observed. Thus, the decomposition of $NaBH_4$ over 600 °C proceeds in one step and the final products are liquid sodium, solid boron, and gaseous hydrogen, with a total gravimetric H_2 capacity of 10.6 wt % according to reaction (35). The experimental estimated decomposition enthalpy of reaction (35) amounts to 108 ± 3 kJ·mol^{-1} H_2, which is in agreement with the observed high stability of $NaBH_4$ [112].

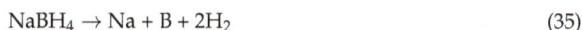

$$NaBH_4 \rightarrow Na + B + 2H_2 \tag{35}$$

Alternatively, $Na_2B_{12}H_{12}$ phase has been reported experimentally as an intermediate of the decomposition of $NaBH_4$ [94–96]. However, the high thermodynamic stability of $Na_2B_{12}H_{12}$ calculated by first-principle approach suggests that it should rather be considered as a product that prevents the subsequent re-hydrogenation because of its poor reactivity towards hydrogen [105]. The mechanism of formation of $Na_2B_{12}H_{12}$ has not been totally understood yet, but theoretical approaches suggest that $Na_2B_{12}H_{12}$ is a product of the reaction between boranes and unreacted $NaBH_4$ as well as in the case of $LiBH_4$ [92,106].

3.3. KBH₄

KBH$_4$ decomposes after melting at about 585 °C by forming potassium in liquid form, solid boron, and gaseous hydrogen, according to reaction (36). The gravimetric H$_2$ capacity is 7.48 wt % [21,92]. The intermediate formation of K$_2$B$_{12}$H$_{12}$ has been also predicted via first-principle calculations [97].

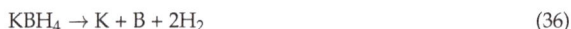

$$KBH_4 \rightarrow K + B + 2H_2 \tag{36}$$

3.4. Mg(BH₄)₂

Magnesium tetrahydroborate, i.e., Mg(BH$_4$)$_2$, has a gravimetric hydrogen storage capacity of 14.9 wt % and an enthalpy of decomposition of about 40 kJ/mol H$_2$ [76,113–115]. Upon heating Mg(BH$_4$)$_2$, the phase transition from the hexagonal to the orthorhombic polymorph takes place at 185 °C. However, at room temperature, different polymorphs of Mg(BH$_4$)$_2$ can be found, depending on the synthesis method. Experimental evidences suggest that the decomposition occurs in several steps between 250 °C and 500 °C, finally forming MgB$_2$ and hydrogen [29,113,115–118]. The mechanism of decomposition is strongly debated and several different reaction paths can be found in the literature, as described in reactions (37)–(40) [72,98–101]. In this case, as for the LiBH$_4$, the final decomposition products and the presence of reaction intermediates/side products depend strongly on the applied experimental conditions.

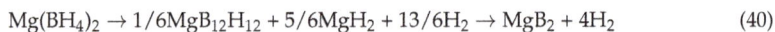

$$Mg(BH_4)_2 \rightarrow Mg + 2B + 4H_2 \tag{37}$$

$$Mg(BH_4)_2 \rightarrow MgH_2 + 2B + 3H_2 \tag{38}$$

$$Mg(BH_4)_2 \rightarrow MgB_2 + 4H_2 \tag{39}$$

$$Mg(BH_4)_2 \rightarrow 1/6MgB_{12}H_{12} + 5/6MgH_2 + 13/6H_2 \rightarrow MgB_2 + 4H_2 \tag{40}$$

Experimentally, it was observed that when the decomposition of Mg(BH$_4$)$_2$ is carried out at temperatures above 450 °C and at a hydrogen pressure below the equilibrium pressure of MgH$_2$, the final products are MgB$_2$ and H$_2$ [98,101,108,113]. The formation of MgB$_2$ upon decomposition is the key for the reversibility of Mg(BH$_4$)$_2$.

3.5. Ca(BH₄)₂

Calcium tetrahydroborate, i.e., Ca(BH$_4$)$_2$, has a gravimetric hydrogen storage capacity of 11.6 wt %. Upon heating, it undergoes a polymorphic transition from the orthorhombic α-phase to the tetragonal β-phase at around 170 °C. However, the polymorphs present in the samples, as well as the structural phase transitions between them, depend on the sample synthesis: for instance ball milling leads to the formation of the metastable orthorhombic γ-phase at room temperature [99,119–123]. Like Mg(BH$_4$)$_2$, the decomposition path of Ca(BH$_4$)$_2$ is debated and several different reaction mechanisms are reported in the literature as shown in reactions (41)–(45) [102–104]:

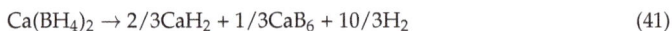

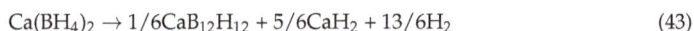

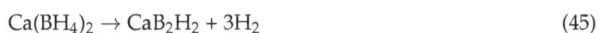

$$Ca(BH_4)_2 \rightarrow 2/3CaH_2 + 1/3CaB_6 + 10/3H_2 \tag{41}$$

$$Ca(BH_4)_2 \rightarrow CaH_2 + 2B + 3H_2 \tag{42}$$

$$Ca(BH_4)_2 \rightarrow 1/6CaB_{12}H_{12} + 5/6CaH_2 + 13/6H_2 \tag{43}$$

$$Ca(BH_4)_2 \rightarrow CaB_2H_6 + H_2 \tag{44}$$

$$Ca(BH_4)_2 \rightarrow CaB_2H_2 + 3H_2 \tag{45}$$

Due to extremely similar reaction enthalpies of some of the reported paths, i.e., reaction (41) = 37.04 kJ/mol H$_2$, reaction (43) = 31.34 − 39.2 kJ/mol H$_2$ and reaction (44) = 31.09 kJ/mol H$_2$, the possibility that the decomposition of the Ca(BH$_4$)$_2$ takes place following more than a single reaction path cannot be excluded [102–104].

3.6. U(BH₄)₄

At temperatures below 70 °C, $U(BH_4)_4$ is stable for long periods. However, at about 100 °C it decomposes to form uranium (III) tetrahydroborate, diborane and hydrogen as shown in reaction (46) [124]:

$$2U(BH_4)_4 \rightarrow 2U(BH_4)_3 + B_2H_6 + H_2 \tag{46}$$

At temperatures above 150 °C, uranium (IV) tetrahydroborate rapidly decomposes to form the uranium boride, i.e., UB_4, or elemental boron and uranium plus hydrogen, as described in reaction (47) [124]:

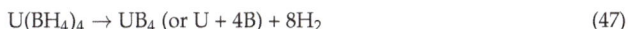

$$U(BH_4)_4 \rightarrow UB_4 \text{ (or U + 4B)} + 8H_2 \tag{47}$$

4. Tailoring the Hydrogen Storage Properties of Tetrahydroborates

Homoleptic tetrahydroborates are characterized by high decomposition temperature and extremely harsh re-hydrogenation conditions, i.e., temperatures above 400 °C and hydrogen pressures higher than 100 bar. Until the end of the 1990s, tetrahydroborates have been mostly used as laboratory reagents and have been investigated just sporadically as possible energy storage materials. However, these compounds have caught special attention owing to their promising hydrogen storage properties (light weight and high hydrogen content). Therefore, intensive research has been carried out to improve the hydrogen storage properties of tetrahydroborates, mainly $LiBH_4$, $NaBH_4$, $Mg(BH_4)_2$, and $Ca(BH_4)_2$. Several strategies have been applied to tailor the thermodynamic and kinetic features of the above mentioned borohydrides, aiming to reach the targets set by U.S. Department of Energy DoE [125]. The most relevant strategies are: (1) addition of catalytic additives [126–130]; (2) thermodynamic tuning via hydrides mixtures [131–135]; and, (3) nanoconfinement of tetrahydroborates and their hydride mixtures [95,132–138].

The first strategy refers to the addition of metals or metal compounds, in general transition metals, which leads to an improvement of the hydrogenation and dehydrogenation kinetic behavior by reducing the activation energy barriers and consequently accelerating the intrinsic rate limiting steps. In general, the addition of some metals and compounds also promotes a thermodynamic modification by changing the reaction pathway. This concept lays on the reduction of the thermodynamic stability and the enhancement of the kinetic behavior of tetrahydroborates (MBH_4) by the partial substitution of M (cation substitution) [126,130], or the partial substitution of hydrogen atoms inside the $[BH_4]^-$ anion (in general by a non-metal such as fluorine; anion substitution) [127–129]. During these processes, it is also possible to attain the formation of a new compound as a product of a favorable thermodynamic interaction between the additive and the tetrahydroborate that provides new catalytic phases. As depicted in the general reactions (48)–(50), the right selection of metal or non-metal can reduce the stability of the tetrahydroborate ($\Delta H_2 < \Delta H_1$ and $\Delta H_3 < \Delta H_1$). On the one hand, the partial cation substitution can reduce the thermodynamic stability and consequently the dehydrogenation temperature by increasing the Pauling electronegative (X_p) (see Section 3) [130]. On the other hand, the partial anion substitution lowers the thermodynamic stability by weakening the B–H bonds. The following general reactions depict these concepts:

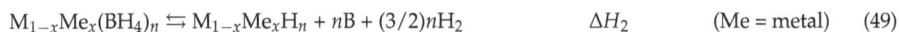

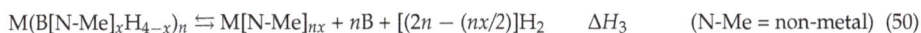

$$M(BH_4)_n \leftrightharpoons MH_n + nB + (3/2)nH_2 \qquad\qquad \Delta H_1 \tag{48}$$

$$M_{1-x}Me_x(BH_4)_n \leftrightharpoons M_{1-x}Me_xH_n + nB + (3/2)nH_2 \qquad \Delta H_2 \qquad \text{(Me = metal)} \tag{49}$$

$$M(B[N\text{-}Me]_xH_{4-x})_n \leftrightharpoons M[N\text{-}Me]_{nx} + nB + [(2n - (nx/2)]H_2 \quad \Delta H_3 \qquad \text{(N-Me = non-metal)} \tag{50}$$

This strategy usually presents limited reversibility since quite stable compounds are formed.

The second strategy is the thermodynamic tuning via the use of hydrides mixtures. This concept was first explored via the formation of alloys in 1958. That year, Libowitz et al. observed that at a given temperature $ZrNiH_3$ has a desorption pressure much higher than the one measured for

ZrH_2. Thus, $ZrNiH_3$ has an enthalpy of decomposition lower than that of ZrH_2 [131]. Similarly, in 1967, Reilly et al. [132] reported on the possibility to change the reaction enthalpy of a hydride by mixing it with compounds with a proper stoichiometric ratio, which reversibly react with the hydride during desorption to form a stable compound. In particular, he showed that the reaction enthalpy of $3MgH_2 + MgCu_2$ is lower than the one of pure MgH_2. Unfortunately, the gained reduction of the reaction enthalpy is at the expense of the system's hydrogen capacity. In the last 15 years, the approach of Libowitz et al. [131] and Reilly et al. [132] has been improved by Chen et al. [133], Vajo et al. [134] and Barkhordarian et al. [135]. These researchers designed light-weight metal hydrides-based system with reduced reaction enthalpy using an appropriate combination of complex metal hydrides and metal hydrides. Such hydride mixtures are the so-called Reactive Hydride Composites (RHC). This strategy allows for the reduction of the thermodynamic stability ($\Delta H_4 < \Delta H_1$) upon dehydrogenation because of the exothermic formation of a new and reversible phase, as shown in reaction (51). It offers the advantage of improved reversibility and high gravimetric storage capacity.

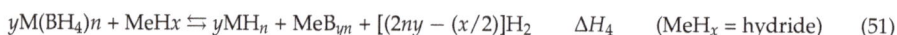

$$yM(BH_4)n + MeHx \leftrightarrows yMH_n + MeB_{yn} + [(2ny - (x/2)]H_2 \qquad \Delta H_4 \qquad (MeH_x = hydride) \qquad (51)$$

Despite the fact that the RHC strategy has shown good results as for example with the mixture $2LiBH_4 + MgH_2$ [134,135], experimental results have proved that the predicted thermodynamic improvement is not reached due to kinetic constraints that lead to competing multi-step reactions [95,136,137].

Nanoconfinement is the third presented strategy that has been recently applied to the pure tetrahydroborates, doped tetrahydroborates and the undoped and doped RHC [138–141]. It is well known that the physical and chemical properties of a material can change markedly when reducing its particle size from the micrometer range to the nanometer range. In fact, on the one hand, the enthalpy of decomposition/formation can be reduced due to the larger impact of the surface energy on the stability of the material [138] and on the other hand, the reduction of the diffusion distances and the larger specific area for the hydrogen interaction account for the kinetic enhancement [139,142].

In the following subsections some relevant results regarding the three above described strategies applied to the $LiBH_4$, $NaBH_4$, $Mg(BH_4)_2$, and $Ca(BH_4)_2$ tetrahydroborates are summarized.

4.1. LiBH₄

The first attempts to influence the dehydrogenation and reversibility properties of $LiBH_4$ were made by A. Züttel et al. [142] by mixing the tetrahydroborate with SiO_2. They lowered the hydrogen desorption temperature by more than 70 °C. However, the formation of stable silicates upon decomposition prevented the material from being re-hydrogenated [143,144]. This effort was followed by several other researchers, who attempted to improve the hydrogen sorption properties and reversibility of $LiBH_4$ by mixing it with metal halides [126,145–150], pure metals [151–154], and carbon based additives [155,156]. Undoubtedly, the addition of these materials led to a general decrease of the $LiBH_4$ desorption temperature, but the reversibility achieved under elevated temperature and hydrogen pressure conditions (e.g., 600 °C and 100 bar H_2 pressure for $LiBH_4$ doped with Ni) was just partial.

The most prominent RHC system based on tetrahydroborates is $2LiBH_4 + MgH_2$ [134,135,157] According to the reaction (52), it has the reversible hydrogen storage capacity of 11.5 wt % and a theoretical reaction enthalpy of 46 kJ/mol H_2, which implies a hydrogen equilibrium pressure of 1 bar at 170 °C [158].

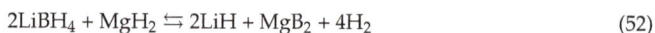

$$2LiBH_4 + MgH_2 \leftrightarrows 2LiH + MgB_2 + 4H_2 \qquad (52)$$

In the last decade this system has been thoroughly investigated with respect to reaction pathway, temperature and hydrogen pressure boundaries, microstructure, effect of additives on the reaction kinetics, and nanoconfinement [159–173].

Inorganics **2017**, *5*, 74

For the $2LiBH_4 + MgH_2$ system, the value of hydrogen pressure applied during the desorption process is fundamental to ensure the reversibility. In fact, when $2LiBH_4 + MgH_2$ is heated under a hydrogen pressure lower than 1 bar, $LiBH_4$ and MgH_2 decompose individually, forming a series of compounds, which under moderate temperature and hydrogen pressures hardly react together to form again $LiBH_4$ (i.e., $Li_2B_{12}H_{12}$, $Li_2B_2H_6$, LiH, elemental B, and Mg). However, the application of a hydrogen back pressure higher than 1 bar but lower than 5 bar ensures the possibility to reversibly release under moderate temperature the hydrogen contained in the system. Under dynamic temperature conditions, the decomposition reaction takes place in two distinguished steps. At first, the decomposition of MgH_2 into Mg and H_2 occurs, and then, after a long incubation period, Mg reacts together with $LiBH_4$ to form MgB_2, LiH and H_2 as shown in reaction (53). Differently from the desorption reaction, the hydrogenation process takes place in a single step [136].

$$2LiBH_4 + Mg \leftrightarrows 2LiH + MgB_2 + 3H_2 \qquad (53)$$

Despite the favorable calculated enthalpy of reaction and the high gravimetric hydrogen capacity, the sluggish kinetics is the primary issue of this system. In the last decade, several studies on the effect of transition metal based additives were carried out. The nucleation of MgB_2 is the key issue for a fast dehydrogenation kinetic behavior and it is enhanced by the presence of transition metal borides that act as nucleation center for the MgB_2 nuclei [159]. Although the reaction times are much different with and without additives, analyses of the reaction kinetics indicate that similar processes limit the sorption reactions for all the composites, independently of the additive and the initial preparation state. The absorption and desorption reactions in the RHC are characterized by significant mass transport through the composite, and therefore, the length scales of the phase separation play a significant role in the reaction kinetics.

Another RHC system that presents interesting hydrogen storage properties is the combination of $LiBH_4$ and Mg_2NiH_4, according to reaction (54). This hydride system exhibits a low reaction enthalpy of 15.4 kJ/mol H_2 and an entropy value of 62.2 J/K mol H_2. Moreover, it starts to release hydrogen at 250 °C [174].

$$4LiBH_4 + 5Mg_2NiH_4 \leftrightarrows 2MgNi_{2.5}B_2 + 4LiH + 8MgH_2 + 8H_2 \qquad (54)$$

Despite the fact that this system does not display a high hydrogen capacity, i.e., 2.5 wt %, it represents another example of reversible RHC systems.

4.2. NaBH₄

The effect of the doping on the reversibility of $NaBH_4$ was studied mostly by Mao et al. [175] and Humphries et al. [176]. They observed a sensible enhancement of the hydrogen desorption properties of $NaBH_4$ when Ti-based (i.e., Ti, TiH_2, and TiF_3) and Ni-based additives (i.e., nano-Ni, Ni_3B, $NiCl_2$, and NiF_2) were used. However, partial reversibility was achieved only in the case of Ti-based additives. In fact, when using TiF_3, a hydrogen content of 4.0 wt % was stored back in the decomposed $NaBH_4$ after hydrogenation at 55 bar and 500 °C.

In order to improve the hydrogen storage properties of $NaBH_4$, nanocomposites of $NaBH_4$-C with a matrix of 2 to 3 nm pore size diameter were synthesized. This allowed the lowering of the dehydrogenation temperature of $NaBH_4$ down to 250 °C. However, the re-hydrogenation under 60 bar and 325 °C was not completed due to the loss of Na upon dehydrogenation [96].

As in the case of the mixture $2LiBH_4 + MgH_2$, the RHC based on $2NaBH_4 + MgH_2$ is reversible because of the formation of MgB_2 upon desorption. The dehydrogenation reactions of the $2NaBH_4 + MgH_2$ can proceed either as reaction (55) with 7.8 wt % H_2 or reaction (56) with 9.8 wt % H_2, depending on the conditions [177–179].

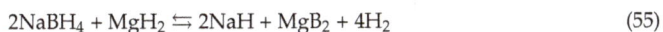

$$2NaBH_4 + MgH_2 \leftrightarrows 2NaH + MgB_2 + 4H_2 \qquad (55)$$

$$2NaBH_4 + MgH_2 \leftrightarrows 2Na + MgB_2 + 5H_2 \tag{56}$$

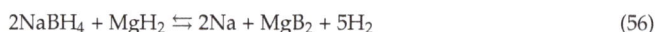

However, the dehydrogenation of this system also goes on in two steps under 1 bar of inert gas, i.e., first the dehydrogenation of MgH_2 and then the subsequent formation of MgB_2 [178,180]. Furthermore, the formation of stable side products such as $Na_2B_{12}H_{12}$ has been also suggested [180]. Under vacuum, the pathway is different and it involves the formation of free Mg and finally free B [181].

As well as the $LiBH_4/Mg_2NiH_4$ mixture, the stoichiometric ratio $4NaBH_4 + 5Mg_2NiH_4$ also leads to the formation of $MgNi_{2.5}B_2$, lowering the enthalpy of the hydride system to 76 kJ/mol H_2 in respect to the one of pure $NaBH_4$ (108 kJ·mol^{-1} H_2), as shown in reaction (57) [112,182]. Consequently, the dehydrogenation temperature is reduced to 360 °C. However, the hydrogen capacity of the system is still low, i.e., 4.5 wt %.

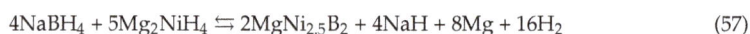

$$4NaBH_4 + 5Mg_2NiH_4 \leftrightarrows 2MgNi_{2.5}B_2 + 4NaH + 8Mg + 16H_2 \tag{57}$$

4.3. Mg(BH$_4$)$_2$

The effect of additives on the reversibility of $Mg(BH_4)_2$ has been an object of intensive investigation [101,183–188]. The result of the doping seems to be a sensible acceleration of hydrogenation and dehydrogenation processes. However, in the literature, just a few hydrogenation-dehydrogenation cycles are reported and the observed enhancement of the reaction kinetics, in particular for the desorption process, is usually connected with the chemical reaction between the hydride and the additive. Recently, Zavarotynska et al. reported on the partial reversibility of γ-Mg(BH$_4$)$_2$ in presence of Co-based additives (Co$_{add}$) [187,188]. They investigated the first three absorption-desorption cycles of γ-Mg(BH$_4$)$_2$ + Co$_{add}$ at 285 °C and 3 bar H$_2$ pressure. The first desorption yielded 4 wt % of H$_2$, and the re-absorption at the same temperature but at the pressure of 120 bar H$_2$ resulted in the uptake of about 2 wt % H$_2$. Upon re-absorption, γ-Mg(BH$_4$)$_2$ was not formed back, but instead crystalline β-Mg(BH$_4$)$_2$ was synthesized. With further cycling, the reversible hydrogen capacity of the samples decreased with an increase in the amount of nano-crystalline MgO and unidentified boron–hydrogen compounds.

Although the addition of selected additives to the tetrahydroborates partially helps to improve the hydrogen absorption and desorption properties, the use of this class of materials as hydrogen storage media is hampered by unfavorable thermal stability.

4.4. Ca(BH$_4$)$_2$

As previously mentioned, for Ca(BH$_4$)$_2$ several decomposition paths (often overlapping) might be observed (Section 3.5), depending on temperature and hydrogen pressure at which the process is carried out [100–102,189]. The reversibility is linked to the final decomposition products. The formation of Ca(BH$_4$)$_2$ was shown to be partially reversible when the pristine material was doped with Ti and Nb-based additives. Kim et al. [190,191] observed that after desorption at 420 °C in static vacuum, Ca(BH$_4$)$_2$ can be formed back with a yield of roughly 50% (at 350 °C and 90 bar H$_2$), if the starting material is milled together with TiCl$_3$ and NbF$_5$. From these works, it is possible to conclude that the key to the reversibility of Ca(BH$_4$)$_2$ is the formation of CaB$_6$. It is known that when using selected additives, Ca(BH$_4$)$_2$ can be synthesized starting from CaH$_2$ and CaB$_6$ [190,192–194]. In fact, Rönnebro et al. successfully synthesized Ca(BH$_4$)$_2$ with the yield of roughly 60% heating a mixture of CaH$_2$ and CaB$_6$ doped with TiCl$_3$ up to 440 °C at a hydrogen pressure of 700 bar [194]. A similar yield of formed Ca(BH$_4$)$_2$ was obtained by Rongeat et al. [193] via reactive ball milling and following heat treatment at 350 °C under 90 bar of hydrogen pressure of a mixture of CaH$_2$ and CaB$_6$ with the addition of TiCl$_3$ or TiF$_3$.

Recently, Bonatto Minella et al. [195], based on the work of Bösenberg et al. [159], proposed an interesting approach to explain the effect of the TM-fluorides on the reversibility of the CaB$_6$/CaH$_2$–Ca(BH$_4$)$_2$ system. The formation of transition-metal boride nanoparticles during milling, and hydrogen desorption of the TM-fluorides doped Ca(BH$_4$)$_2$ are proposed to support the heterogeneous nucleation of CaB_6, maximizing its formation over the formation of irreversible elemental boron and $CaB_{12}H_{12}$. In the case of material doped with NbF_5 and TiF_4, the $\{111\}CaB_6/\{10\bar{1}1\}NbB_2$, $\{111\}CaB_6/\{10\bar{1}0\}NbB_2$, as well as $\{111\}CaB_6/\{10\bar{1}1\}TiB_2$ plane pairs have the potential to be the matching planes because the *d*-value mismatch is well below the *d*-critical mismatch value (6%) [196,197]. This mechanism is similar to that observed for $2LiBH_4 + MgH_2$. Transition metal borides formed by the reaction of the transition metal based additives with the tetrahydroborates act as heterogeneous nucleation sites for MgB_2, thus improving desorption and absorption reaction kinetics [159].

Despite the fact that the TM-boride species assist the formation of CaB_6, the reversibility of the Ca-RHC system, i.e., $Ca(BH_4)_2 + MgH_2$ as shown in reaction (58), is not complete [137,198–200].

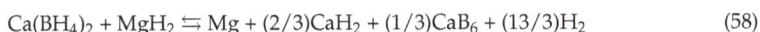

$$Ca(BH_4)_2 + MgH_2 \leftrightarrows Mg + (2/3)CaH_2 + (1/3)CaB_6 + (13/3)H_2 \qquad (58)$$

Instead, every time that the material is cycled, the formation of stable phases such as $CaB_{12}H_{12}$ and elemental B takes place as indicated in reaction (59) and (42), respectively. Consequently, a continuous reduction of the system hydrogen capacity takes place upon cycling.

$$Ca(BH_4)_2 + MgH_2 \leftrightarrows Mg + (5/6)CaH_2 + CaB_6 + (1/6)CaB_{12}H_{12} + (19/6)H_2 \qquad (59)$$

Based on the previous works on $4LiBH_4 + 5Mg_2NiH_4$ [174] and $4NaBH_4 + 5Mg_2NiH_4$ [184], Bergemann et al. [201] proposed as an alternative the combination of $Ca(BH_4)_2$ with Mg_2NiH_4. In this system, upon dehydrogenation, the formation of $MgNi_{2.5}B_2$ as the only B-containing phase is observed according to reaction (60). There was no evidence of the presence of irreversible $CaB_{12}H_{12}$ and B. However, upon re-hydrogenation $Ca(BH_4)_2$ was only partially formed. This limited reversibility was attributed to thermodynamic and kinetic constraints.

$$Ca(BH_4)_2 + 2.5Mg_2NiH_4 \rightarrow CaH_2 + MgNi_{2.5}B_2 + 4Mg + 8H_2 \qquad (60)$$

5. Summary and Future Research Directions

In this review, a detailed account on selected aspect of the most relevant tetrahydroborates and tetrahydroborate-based systems investigated in the last decades as potential hydrogen storage candidates was reported. As a matter of fact, these materials used alone or as a part of RHC systems have some limitations, which make their use for practical applications challenging. Owing to their high gravimetric hydrogen capacity, tetrahydroborates are extremely appealing hydrogen storage materials. However, the contained hydrogen is not easily accessible. Under thermal input, tetrahydroborates such as $LiBH_4$, $NaBH_4$, $K(BH_4)$, $Mg(BH_4)_2$, and $Ca(BH_4)_2$ decompose at temperatures higher than 300 °C, and their decomposition products are partially reversible only under extreme temperature and hydrogen pressure conditions. Additives such as transition metals, transition metal halides or oxides, sensibly influence the decomposition reaction of these materials, often lowering the starting decomposition temperature by several tenths of degrees. However, the reversibility of the material still remains an issue since stable phases are formed upon decomposition. The hitherto most promising approach to effectively store hydrogen in tetrahydroborates is through the use of the RHC strategy. When compared to the pristine materials, this method enables the achievement of full reversibility under milder temperatures and hydrogen pressure conditions, e.g., 350–400 °C and 3 bar–50 bar for $2LiBH_4 + MgH_2$. Despite the fact that these conditions, mainly temperature, are not even near to those required for on board practical application (desired temperature ≈ 90 °C), there has been an attempt to develop a demonstration hydrogen storage tank based on $2LiBH_4 + MgH_2$ [202]. This tank contained 250 g of material. It has been found that during operation, the thermal management becomes an important issue, and the material undergoes a long-range phase separation, which negatively affects the kinetic behavior and the hydrogen capacity of the system. To the best of our knowledge, this was the first tank design based on borohydrides and it has to be considered as a starting point in the development of tetrahydroborates based systems for hydrogen storage. Nowadays, the research is directed to find better systems capable to work near to the targeted temperature condition of about 90 °C and at the same time characterized by a hydrogen storage capacity of about 10 wt %. Besides, the cost of the material is not a minor issue when one wants to employ such materials in large hydrogen storage tanks.

In spite of several efforts to find the proper tetrahydroborate-based systems, the correct additive and suitable material preparation procedure, it has not yet been possible to achieve complete hydrogen desorption/absorption at the theoretically expected temperature and hydrogen pressure conditions. Moreover, the hydrogenation-dehydrogenation cycling stability is still a key issue to be addressed. Although tetrahydroborates are known for more than a century, the major achievements towards their utilization have been obtained in the last 20 years. Based on this perspective, the hydrogen storage properties of tetrahydroborates can be further improved by different strategies that are broadly used nowadays, such as the RHC approach, the use of selected additives, and/or nanoengineering. Thus, in the near future tetrahydroborates will hopefully fulfill the conditions for a practical application.

Acknowledgments: The authors thank CONICET (Consejo Nacional de Invetigaciones Científicas y Técnicas), ANPCyT—(Agencia Nacional de Promoción Científica y Tecnológica) and CNEA (Comisión Nacional de Energía Atómica). The authors also acknowledge Alexander von Humboldt Foundation (J. Puszkiel fellowship holder, No. 1187279 STP).

Author Contributions: Claudio Pistidda and Julián Puszkiel conceieved the design of the structure of the work together with Sebastiano Garroni, Chiara Milanese. Fabiana Gennari contributed also with the writing and correction of the paper. Thomas Klassen and Martin Dorhnheim contributed with the correction of the paper.

Conflicts of Interest: The authors declare no conflict of interest.

References

1. Evers, A.A. *The Hydrogen Society: More Than Just a Vision*; Hydrogeit Verlag: Oberkraemer, Germany, 2010; pp. 22–29.
2. Schlapbach, L.; Züttel, A. Hydrogen-storage materials for mobile applications. *Nature* **2001**, *414*, 353–358. [CrossRef] [PubMed]
3. Züttel, A. Materials for hydrogen storage. *Mater. Today* **2003**, *6*, 24–33. [CrossRef]
4. Dornheim, M.; Eigen, N.; Barkhordarian, G.; Klassen, T.; Bormann, R. Tailoring hydrogen storage materials towards application. *Adv. Eng. Mater.* **2006**, *8*, 377–385. [CrossRef]
5. Dornheim, M.; Doppiu, S.; Barkhordarian, G.; Boesenberg, U.; Klassen, T.; Gutfleisch, O.; Bormann, R. Hydrogen storage in magnesium-based hydrides and hydride composites. *Scr. Mater.* **2007**, *56*, 841–846. [CrossRef]
6. Stampfer, J.F.; Holley, C.E.; Suttle, J.F. The Magnesium-Hydrogen System. *J. Am. Chem. Soc.* **1960**, *82–87*, 3504–3508. [CrossRef]
7. Stander, C.M. Kinetics of formation of magnesium hydride from magnesium and hydrogen. *Z. Phys. Chem.* **1977**, *104*, 229–238. [CrossRef]
8. Stander, C.M. Kinetics of decomposition of magnesium hydride. *J. Inorg. Nucl. Chem.* **1977**, *39*, 221–223. [CrossRef]
9. Vigeholm, B.; Kjoller, J.; Larsen, B.; Pedersen, A.S. Formation and decomposition of magnesium hydride. *J. Less Common Met.* **1983**, *89*, 135–144. [CrossRef]
10. Huot, J.; Liang, G.; Boily, S.; Van Neste, A.; Schulz, R. Structural study and hydrogen sorption kinetics of ball-milled magnesium hydride. *J. Alloys Compd.* **1999**, *293*, 495–500. [CrossRef]
11. Hanada, N.; Ichikawa, T.; Fujii, H. Catalytic Effect of Nanoparticle 3d-Transition Metals on Hydrogen Storage Properties in Magnesium Hydride MgH_2 Prepared by Mechanical Milling. *J. Phys. Chem. B* **2005**, *109*, 7188–7194. [CrossRef] [PubMed]
12. Pistidda, C.; Bergemann, N.; Wurr, J.; Rzeszutek, A.; Møller, K.T.; Hansen, B.R.S.; Garroni, S.; Horstmann, C.; Milanese, C.; Girella, A.; et al. Hydrogen storage systems from waste Mg alloys. *J. Power Sources* **2014**, *270*, 554–563. [CrossRef]
13. Møller, K.T.; Jensen, T.R.; Akiba, E.; Li, H.-W. Hydrogen—A sustainable energy carrier. *Prog. Nat. Sci.* **2017**, *27*, 34–40. [CrossRef]

14. Yamauchi, M.; Ikeda, R.; Kitagawa, H.; Takata, M. Nanosize Effects on Hydrogen Storage in Palladium. *J. Phys. Chem. C* **2008**, *112*, 3294–3299. [CrossRef]

15. Kishore, S.; Nelson, J.A.; Adair, J.H.; Eklund, P.C. Hydrogen storage in spherical and platelet palladium nanoparticles. *J. Alloys Compd.* **2005**, *389*, 234–242. [CrossRef]

16. Pundt, A.; Sachs, C.; Winter, M.; Reetz, M.T.; Fritsch, D.; Kirchheim, R. Hydrogen sorption in elastically soft stabilized Pd-clusters. *J. Alloys Compd.* **1999**, *293*, 480–483. [CrossRef]

17. Wolf, R.J.; Lee, M.W.; Ray, J.R. Pressure-composition isotherms for nanocrystalline palladium hydride. *Phys. Rev. Lett.* **1994**, *73*, 557–560. [CrossRef] [PubMed]

18. Mitsui, T.; Rose, M.K.; Fomin, E.; Ogletree, D.F.; Salmeron, M. Dissociative hydrogen adsorption on palladium requires aggregates of three or more vacancies. *Nature* **2003**, *422*, 705–707. [CrossRef] [PubMed]

19. Konovalov, S.K.; Bulychev, B.N. The P,T-State Diagram and Solid Phase Synthesis of Aluminum Hydride. *Inorg. Chem.* **1995**, *34*, 172–175. [CrossRef]

20. Grochala, W.; Edwards, P.P. Thermal Decomposition of the Non-Interstitial Hydrides for the Storage and Production of Hydrogen. *Chem. Rev.* **2004**, *104*, 1283–1315. [CrossRef] [PubMed]

21. Orimo, S.I.; Nakamori, Y.; Eliseo, J.R.; Züttel, A.; Jensen, C.M. Complex Hydrides for Hydrogen Storage. *Chem. Rev.* **2007**, *107*, 4111–4132. [CrossRef] [PubMed]

22. Chen, P.; Zhu, M. Recent progress in hydrogen storage. *Mater. Today* **2008**, *11*, 36–43. [CrossRef]

23. Ley, M.B.; Jepsen, L.H.; Lee, Y.S.; Cho, Y.W.; von Colbe, J.M.B.; Dornheim, M.; Rokni, M.; Jensen, J.O.; Sloth, M.; Filinchuk, Y.; et al. Complex hydrides for hydrogen storage—New perspectives. *Mater. Today* **2014**, *17*, 122–128. [CrossRef]

24. Gay-Lussac, J.L.; Thenard, L.J. Notiz über das Kali-und das Natron-Metall. *Ann. Phys.* **1809**, *32*, 23–39. [CrossRef]

25. Gay-Lussac, J.L.; Thenard, L.J. *Recherches Physico-Chimiques*; Deterville: Paris, France, 1811; Volume 2.

26. Bogdanovic, B.; Schwickardi, M. Ti-doped alkali metal aluminum hydrides as potential novel reversible hydrogen storage materials. *J. Alloys Compd.* **1997**, *253–254*, 1–9. [CrossRef]

27. Rude, L.H.; Nielsen, T.K.; Ravnsbæk, D.B.; Bösenberg, U.; Ley, M.B.; Richter, B.; Arnbjerg, L.M.; Dorbheim, M.; Filinchuk, Y.; Besenbacher, F.; et al. Tailoring properties of borohydrides for hydrogen storage: A review. *Phys. Status Solidi A* **2011**, *208*, 1754–1773. [CrossRef]

28. Paskevicius, M.; Jepsen, L.H.; Schouwink, P.; Černý, R.; Ravnsbæk, D.B.; Filinchuk, Y.; Dornheim, M.; Besenbacher, F.; Jensen, T.R. Metal borohydrides and derivatives—Synthesis, structure and properties. *Chem. Soc. Rev.* **2017**, *46*, 1565–1634. [CrossRef] [PubMed]

29. George, L.; Saxena, S.K. Structural stability of metal hydrides, alanates and borohydrides of alkali and alkali-earth elements: A review. *Int. J. Hydrogen Energy* **2010**, *35*, 5454–5470. [CrossRef]

30. Li, H.-W.; Yan, Y.; Orimo, S.-I.; Züttel, A.; Jensen, C.M. Recent Progress in Metal Borohydrides for Hydrogen Storage. *Energies* **2011**, *4*, 185–214. [CrossRef]

31. Mohtadi, R.; Remhof, A.; Jena, P. Complex metal borohydrides: Multifunctional materials for energy storage and conversion. *J. Phys. Condens. Matter* **2016**, *28*, 353001. [CrossRef] [PubMed]

32. Hagemann, H.; Černý, R. Synthetic approaches to inorganic borohydrides. *Dalton Trans.* **2010**, *39*, 6006–6012. [CrossRef] [PubMed]

33. Hansen, B.R.S.; Paskevicius, M.; Li, H.-W.; Akiba, E.; Jensen, T.R. Metal boranes: Progress and applications. *Coord. Chem. Rev.* **2016**, *323*, 60–70. [CrossRef]

34. Gharib Doust, S.P.; Ravnsbæk, D.B.; Černý, R.; Jensen, T.R. Synthesis, structure and properties of bimetallic sodium rare-earth (RE) borohydrides, NaRE(BH$_4$)$_4$, RE = Ce, Pr, Er or Gd. *Dalton Trans.* **2017**, *46*, 13421–13431. [CrossRef] [PubMed]

35. Mao, J.; Gregory, D.H. Recent Advances in the Use of Sodium Borohydride as a Solid State Hydrogen Store. *Energies* **2015**, *8*, 430–453. [CrossRef]

36. Filinchuk, Y.; Chernyshov, D.; Dmitriev, V. Light metal borohydrides: Crystal structures and beyond. *Z. Kristallogr.* **2008**, *223*, 649–659. [CrossRef]

37. Cerný, R.; Schouwink, P. The crystal chemistry of inorganic metal borohydrides and their relation to metal oxides. *Acta Crystallogr. B* **2015**, *B71*, 619–640. [CrossRef] [PubMed]

38. Davy, H. The Bakerian Lecture: An account of some new analytical researches on the nature of certain bodies, particularly the alkalies, phosphorus, sulphur, carbonaceous matter, and the acids hitherto undecompounded, with some general observations on chemical theory. *Philos. Trans. R. Soc.* **1809**, *99*, 39–104.

39. Gay Lussac, J.L.; Thenard, L.J. Notice sur la décomposition et la recomposition de l'acide boracique. *Ann. Chim.* **1808**, *68*, 169–174.
40. Moissan, H. Préparation du bore amorphe. *C. R. Acad. Sci.* **1892**, *114*, 392–397.
41. Greenwood, N.N.; Earnshaw, A. *Chemistry of the Elements, Boron*; Butterworth Heinemann: Leeds, UK, 1998.
42. Shore, S.G. Systematic Approaches to the Preparation of Boron Hydrides and Their Derivatives. *Am. Chem. Soc.* **1983**, 1–16. [CrossRef]
43. Stock, A. *Hydrides of Boron and Silicon*; Cornell University Press: Ithaca, NY, USA, 1933.
44. Stock, A.; Nassenz, C. Hydrogen boride. *Berichte* **1912**, *45*, 3529.
45. Schlesinger, H.I.; Burg, A.B. Hydrides of Boron. VIII. The Structure of the Diammoniate of Diborane and its Relation to the Structure of Diborane. *Chem. Rev.* **1942**, *31*, 1–41. [CrossRef]
46. Finholt, A.E.; Bond, A.C.; Schlesinger, H.I. The Preparation and Some Properties of Hydrides of Elements of the Fourth Group of the Periodic System and of their Organic Derivatives. *J. Am. Chem. Soc.* **1947**, *69*, 2692–2696. [CrossRef]
47. Parry, R.M.; Walter, M.K. *Boron Hydrides Preparative Inorganic Reactions*; Jolly, W., Ed.; Interscience: New York, NY, USA, 1968; Volume 5, pp. 45–102.
48. Adams, R.M. Preparation of Diborane. *Adv. Chem. Ser.* **1961**, *32*, 60.
49. Nainan, K.C.; Ryschkewitsch, G.E. A new synthesis of $B_3H_8^-$ ion. *Inorg. Nucl. Chem. Lett.* **1970**, *6*, 765–766. [CrossRef]
50. Duke, B.J.; Gilbert, J.R.; Read, I.A. Preparation and purification of diborane. *J. Chem. Soc.* **1964**, 540–541.
51. Freeguard, G.F.; Long, L.H. Improved preparation of diborane. *Chem. Ind.* **1965**, *11*, 471.
52. Schlesinger, H.I.; Thomas Sanderson, R.; Burg, A.B. Metallo Borohydrides. I. Aluminum Borohydride. *J. Am. Chem. Soc.* **1940**, *62*, 3421–3425. [CrossRef]
53. Burg, A.B.; Schlesinger, H.I. Metallo Borohydrides. II. Beryllium Borohydride. *J. Am. Chem. Soc.* **1940**, *62*, 3425–3429. [CrossRef]
54. Schlesinger, H.I.; Brown, H.C. Metallo Borohydrides. III. Lithium Borohydride. *J. Am. Chem. Soc.* **1940**, *62*, 3429–3435. [CrossRef]
55. Schlesinger, H.I.; Brown, H.C.; Abraham, B.; Bond, A.C.; Davidson, N.; Finholt, A.E.; Gilbreath, J.R.; Hoekstra, H.; Horvitz, L.; Hyde, E.K.; et al. New Developments in the Chemistry of Diborane and the Borohydrides. I. General Summary. *J. Am. Chem. Soc.* **1953**, *75*, 186–190. [CrossRef]
56. Schlesinger, H.I.; Sanderson, R.T.; Burg, A.B. A volatile compound of aluminum, boron and hydrogen. *J. Am. Chem. Soc.* **1939**, *61*, 536. [CrossRef]
57. Wiberg, E.; Bauer, R. Zur Kenntnis eines Magnesium-bor-wasserstoffs $Mg(BH_4)_2$. *Z. Naturforsch. B* **1950**, *5*, 397. [CrossRef]
58. Friedrichs, O.; Borgschulte, A.; Kato, S.; Buchter, F.; Gremaud, R.; Remhof, A.; Züttel, A. Low-Temperature Synthesis of $LiBH_4$ by Gas–Solid Reaction. *Chem. Eur. J.* **2009**, *15*, 5531–5534. [CrossRef] [PubMed]
59. Schlesinger, H.I.; Brown, H.C.; Hoekstra, H.R.; Rapp, L.R. Reactions of Diborane with Alkali Metal Hydrides and Their Addition Compounds. New Syntheses of Borohydrides. Sodium and Potassium Borohydrides. *J. Am. Chem. Soc.* **1953**, *75*, 199–204. [CrossRef]
60. Schlesinger, H.I.; Brown, H.C.; Hyde, E.K. The Preparation of Other Borohydrides by Metathetical Reactions Utilizing the Alkali Metal Borohydrides1. *J. Am. Chem. Soc.* **1953**, *75*, 209–213. [CrossRef]
61. Wiberg, E.; Henle, W. Zur Kenntnis eines Cadmium-bor-wasserstoffs $Cd(BH_4)_2$. *Z. Naturforsch. B* **1952**, *7*, 582. [CrossRef]
62. Wiberg, E.; Henle, W. Zur Kenntnis eines ätherlöslichen Zink-bor-wasser-stoffs $Zn(BH_4)_2$. *Z. Naturforsch. B* **1952**, *7*, 579–580. [CrossRef]
63. Wiberg, E.; Hartwimmer, R. Zur Kenntnis von Erdalkaliboranaten $Me[BH_4]_2$ III. Synthese aus Erdalkalihydriden und Diboran. *Z. Naturforsch. B* **1955**, *10*, 295–296. [CrossRef]
64. Nöth, H. Anorganische Reaktionen der Alkaliboranate. *Angew. Chem.* **1961**, *73*, 371–383. [CrossRef]
65. Černý, R.; Chul Kim, K.; Penin, N.; D'Anna, V.; Hagemann, H.; Sholl, D.S. $AZn_2(BH_4)_5$ (A = Li, Na) and $NaZn(BH_4)_3$: Structural Studies. *J. Phys. Chem. C* **2010**, *114*, 19127–19133. [CrossRef]
66. Černý, R.; Penin, N.; D'Anna, V.; Hagemann, H.; Durand, E.; Růžička, J. $Mg_xMn_{(1-x)}(BH_4)_2$ (x = 0–0.8), a cation solid solution in a bimetallic borohydride. *Acta Mater.* **2011**, *59*, 5171–5180. [CrossRef]
67. Černý, R.; Penin, N.; Hagemann, H.; Filinchuk, Y. The First Crystallographic and Spectroscopic Characterization of a 3d-Metal Borohydride: $Mn(BH_4)_2$. *J. Phys. Chem. C* **2009**, *113*, 9003–9007. [CrossRef]

68. Černý, R.; Ravnsbæk, D.B.; Schouwink, P.; Filinchuk, Y.; Penin, N.; Teyssier, J.; Smrčok, L.; Jensen, T.R. Potassium Zinc Borohydrides Containing Triangular $[Zn(BH_4)_3]^-$ and Tetrahedral $[Zn(BH_4)_xCl_{4-x}]^{2-}$ Anions. *J. Phys. Chem. C* **2012**, *116*, 1563–1571. [CrossRef]

69. Černý, R.; Ravnsbæk, D.B.; Severa, G.; Filinchuk, Y.; D'Anna, V.; Hagemann, H.; Haase, D.; Skibsted, J.; Jensen, C.M.; Jensen, T.R. Structure and Characterization of $KSc(BH_4)_4$. *J. Phys. Chem. C* **2010**, *114*, 19540–19549. [CrossRef]

70. Černý, R.; Schouwink, P.; Sadikin, Y.; Stare, K.; Smrčok, L.; Richter, B.; Smrčok, L.; Richter, B.; Jensen, T.R. Trimetallic Borohydride $Li_3MZn_5(BH_4)_{15}$ (M = Mg, Mn) Containing Two Weakly Interconnected Frameworks. *Inorg. Chem.* **2013**, *52*, 9941–9947. [CrossRef] [PubMed]

71. Černý, R.; Severa, G.; Ravnsbæk, D.B.; Filinchuk, Y.; D'Anna, V.; Hagemann, H.; Haase, D.; Jensen, C.M.; Jensen, T.R. $NaSc(BH_4)_4$: A Novel Scandium-Based Borohydride. *J. Phys. Chem. C* **2010**, *114*, 1357–1364. [CrossRef]

72. Her, J.H.; Stephens, P.W.; Gao, Y.; Soloveichik, G.L.; Rijssenbeek, J.; Andrus, M.; Zhao, J.-C. Structure of unsolvated magnesium borohydride $Mg(BH_4)_2$. *Acta Crystallogr. B* **2007**, *63*, 561–568. [CrossRef] [PubMed]

73. Ravnsbæk, D.B.; Filinchuk, Y.; Černý, R.; Ley, M.B.; Haase, D.; Jakobsen, H.J.; Skibsted, J.; Jensen, T.R. Thermal Polymorphism and Decomposition of $Y(BH_4)_3$. *Inorg. Chem.* **2010**, *49*, 3801–3809. [CrossRef] [PubMed]

74. Ravnsbæk, D.B.; Nickels, E.A.; Černý, R.; Olesen, C.H.; David, W.I.F.; Edwards, P.P.; Filinchuk, Y.; Jensen, T.R. Novel Alkali Earth Borohydride $Sr(BH_4)_2$ and Borohydride-Chloride $Sr(BH_4)Cl$. *Inorg. Chem.* **2013**, *52*, 10877–10885. [CrossRef] [PubMed]

75. Ravnsbœk, D.; Filinchuk, Y.; Cerenius, Y.; Jakobsen, H.J.; Besenbacher, F.; Skibsted, J.; Jensen, T.R. A Series of Mixed-Metal Borohydrides. *Angew. Chem.* **2009**, *48*, 6659–6663. [CrossRef] [PubMed]

76. Sarner, S.F. *Propellant Chemistry*, 1st ed.; Reinhold Publishing Corporation: New York, NY, USA, 1966.

77. Schlesinger, H.I.; Brown, H.C.; Finholt, A.E.; Gilbreath, J.R.; Hoekstra, H.R.; Hyde, E.K. Sodium Borohydride, Its Hydrolysis and Its Use as a Reducing Agent and in the Generation of Hydrogen. *J. Am. Chem. Soc.* **1953**, *75*, 215–219. [CrossRef]

78. Makhaev, V.D.; Antsyshkina, A.S.; Petrova, L.A.; Sadikov, G.G. Interaction of Zirconium, Yttrium, and Zinc Tetrahydroborate Complexes $NaMn(BH_4)_n + 1(DME)_m$ (M = Zr, Y, Zn) with Triethylcarbinol: Crystal and Molecular Structure of $B[OC(C_2H_5)_3]_3$. *Russ. J. Inorg. Chem.* **2004**, *49*, 1154–1157.

79. Clark, J.D.; New Brunswick, N.J. *Ignition: An Informal History of Liquid Rocket Propellants*; Rutgers University Press: New Brunswick, NJ, USA, 1972.

80. Miwa, K.; Ohba, N.; Towata, S.; Nakamori, Y.; Orimo, S. First-principles study on lithium borohydride $LiBH_4$. *Phys. Rev. B* **2004**, *69*, 245120. [CrossRef]

81. Miwa, K.; Ohba, N.; Towata, S.; Nakamori, Y.; Orimo, S. First-principles study on copper-substituted lithium borohydride, $(Li_{1-x}Cu_x)BH_4$. *J. Alloys Compd.* **2005**, *404–406*, 140–143. [CrossRef]

82. Schrauzer, G.N. Über ein Periodensystem der Metallboranate. *Naturwissenschaften* **1995**, *42*, 438. [CrossRef]

83. Nakamori, Y.; Li, H.-W.; Miwa, K.; Towata, S.; Orimo, S. Syntheses and Hydrogen Desorption Properties of Metal-Borohydrides $M(BH_4)_n$ (M = Mg, Sc, Zr, Ti, and Zn; n = 2–4) as Advanced Hydrogen Storage Materials. *Mater. Trans.* **2006**, *47*, 1898–1901. [CrossRef]

84. Nakamori, Y.; Miwa, K.; Ninomiya, A.; Li, H.; Ohba, N.; Towata, S.I.; Züttel, A.; Orimo, S. Correlation between thermodynamical stabilities of metal borohydrides and cation electronegativites: First-principles calculations and experiments. *Phys. Rev. B Condens. Matter Mater. Phys.* **2006**, *74*, 045126. [CrossRef]

85. Orimo, S.; Nakamori, Y.; Ohba, N.; Miwa, K.; Aoki, M.; Towata, S.; Züttel, A. Experimental studies on intermediate compound of $LiBH_4$. *Appl. Phys. Lett.* **2006**, *89*, 021920. [CrossRef]

86. Ohba, N.; Miwa, K.; Aoki, M.; Noritake, T.; Towata, S.I.; Nakamori, Y.; Orimo, S.; Züttel, A. First-principles study on the stability of intermediate compounds of $LiBH_4$. *Phys. Rev. B Condens. Matter Mater. Phys.* **2006**, *74*, 075110. [CrossRef]

87. Her, J.H.; Yousufuddin, M.; Zhou, W.; Jalisatgi, S.S.; Kulleck, J.G.; Zan, J.A.; Hwang, S.-J.; Bowman, R.C.; Udovic, T.J. Crystal Structure of $Li_2B_{12}H_{12}$: A Possible Intermediate Species in the Decomposition of $LiBH_4$. *Inorg. Chem.* **2008**, *47*, 9757–9759. [CrossRef] [PubMed]

88. Li, H.-W.; Kikuchi, K.; Nakamori, Y.; Ohba, N.; Miwa, K.; Towata, S.; Orimo, S. Dehydriding and rehydriding processes of well-crystallized $Mg(BH_4)_2$ accompanying with formation of intermediate compounds. *Acta Mater.* **2008**, *56*, 1342–1347. [CrossRef]

89. Hwang, S.J.; Bowman, R.C.; Reiter, J.W.; Rijssenbeek, J.; Soloveichik, G.L.; Zhao, J.-C.; Kabbour, H.; Ahn, C.C. NMR Confirmation for Formation of $[B_{12}H_{12}]^{2-}$ Complexes during Hydrogen Desorption from Metal Borohydrides. *J. Phys. Chem. C* **2008**, *112*, 3164–3169. [CrossRef]

90. Li, H.-W.; Miwa, K.; Ohba, N.; Fujita, T.; Sato, T.; Yan, Y.; Towata, S.; Chen, M.W.; Orimo, S. Formation of an intermediate compound with a $B_{12}H_{12}$ cluster: Experimental and theoretical studies on magnesium borohydride $Mg(BH_4)_2$. *Nanotechnology* **2009**, *20*, 204013. [CrossRef] [PubMed]

91. Züttel, A.; Wenger, P.; Rentsch, S.; Sudan, P.; Mauron, P.; Emmenegger, C. LiBH$_4$ a new hydrogen storage material. *J. Power Sources* **2003**, *118*, 1–7. [CrossRef]

92. Caputo, R.; Züttel, A. First-principles study of the paths of the decomposition reaction of LiBH$_4$. *Mol. Phys.* **2010**, *108*, 1263–1276. [CrossRef]

93. Friedrichs, O.; Remhof, A.; Hwang, S.J.; Züttel, A. Role of $Li_2B_{12}H_{12}$ for the formation and decomposition of LiBH$_4$. *Chem. Mater.* **2010**, *22*, 3265–3268. [CrossRef]

94. Mao, J.; Guo, Z.; Yu, X.; Liu, H. Improved Hydrogen Storage Properties of NaBH$_4$ Destabilized by CaH$_2$ and $Ca(BH_4)_2$. *J. Phys. Chem. C* **2011**, *115*, 9283–9290. [CrossRef]

95. Garroni, S.; Milanese, C.; Pottmaier, D.; Mulas, G.; Nolis, P.; Girella, A.; Caputo, R.; Olid, D.; Teixdor, F.; Baricco, M.; et al. Experimental Evidence of $Na_2[B_{12}H_{12}]$ and Na Formation in the Desorption Pathway of the $2NaBH_4 + MgH_2$ System. *J. Phys. Chem. C* **2011**, *115*, 16664–16671. [CrossRef]

96. Ngene, P.; van den Berg, R.; Verkuijlen, M.H.W.; de Jong, K.P.; de Jongh, P.E. Reversibility of the hydrogen desorption from NaBH$_4$ by confinement in nanoporous carbon. *Energy Environ. Sci.* **2011**, *4*, 4108–4115. [CrossRef]

97. Kim, C.K.; Scholl, D.S. Crystal Structures and Thermodynamic Investigations of $LiK(BH_4)_2$, KBH$_4$, and NaBH$_4$ from First-Principles Calculations. *J. Phys. Chem. C* **2010**, *114*, 678–686. [CrossRef]

98. Hanada, N.; Chlopek, K.; Frommen, C.; Lohstroh, W.; Fichtner, M. Thermal decomposition of $Mg(BH_4)_2$ under He flow and H$_2$ pressure. *J. Mater. Chem.* **2008**, *18*, 2611–2614. [CrossRef]

99. Riktor, M.D.; Sorby, M.H.; Chlopek, K.; Fichtner, M.; Buchter, F.; Zuttel, A.; Hauback, B.C. In situ synchrotron diffraction studies of phase transitions and thermal decomposition of $Mg(BH_4)_2$ and $Ca(BH_4)_2$. *J. Mater. Chem.* **2007**, *17*, 4939–4942. [CrossRef]

100. Li, H.W.; Kikuchi, K.; Nakamori, Y.; Miwa, K.; Towata, S.; Orimo, S. Effects of ball milling and additives on dehydriding behaviors of well-crystallized $Mg(BH_4)_2$. *Scr. Mater.* **2007**, *57*, 679–682. [CrossRef]

101. Pistidda, C.; Garroni, S.; Dolci, F.; Bardají, E.G.; Khandelwal, A.; Nolis, P.; Dornheim, M.; Gosalawit, R.; Jensen, T.; Cerenius, Y.; et al. Synthesis of amorphous $Mg(BH_4)_2$ from MgB$_2$ and H$_2$ at room temperature. *J. Alloys Compd.* **2010**, *508*, 212–215. [CrossRef]

102. Zhang, Y.; Majzou, E.; Ozoliņš, V.; Wolverton, C. Theoretical prediction of different decomposition paths for $Ca(BH_4)_2$ and $Mg(BH_4)_2$. *Phys. Rev. B* **2010**, *82*, 174107. [CrossRef]

103. Kim, Y.; Reed, D.; Lee, Y.-S.; Lee, J.Y.; Shim, J.-H.; Book, D.; Cho, Y.W. Identification of the Dehydrogenated Product of $Ca(BH_4)_2$. *J. Phys. Chem. C* **2009**, *113*, 5865–5871. [CrossRef]

104. Ozolins, V.; Majzoub, E.H.; Wolverton, C. First-Principles Prediction of Thermodynamically Reversible Hydrogen Storage Reactions in the Li-Mg-Ca-B-H System. *J. Am. Chem. Soc.* **2009**, *131*, 230–237. [CrossRef] [PubMed]

105. Caputo, R.; Garroni, S.; Olid, D.; Teixidor, F.; Suriñach, S.; Dolors Baró, M. Can $Na_2[B_{12}H_{12}]$ be a decomposition product of NaBH$_4$? *Phys. Chem. Chem. Phys.* **2010**, *12*, 15093–15100. [CrossRef] [PubMed]

106. Cakır, D.; Wijs, G.A.D.; Brocks, G. Native Defects and the Dehydrogenation of NaBH$_4$. *J. Phys. Chem. C* **2011**, *115*, 24429–24434. [CrossRef]

107. Fedneva, E.M.; Alpatova, V.L.; Mikheeva, V.I. Thermal stability of lithium borohy-dride. *Transl. Zh. Neorg. Khim. Russ. J. Inorg. Chem.* **1964**, *9*, 826–827.

108. Stasinevich, D.S.; Egorenko, G.A. Thermographic investigation of alkali metal and magnesium tetrahydroborates at pressures up to 10 atm. *Russ. J. Inorg. Chem.* **1968**, *13*, 341–343.

109. Mauron, P.; Buchter, F.; Friedrichs, O.; Remhof, A.; Bielmann, M.; Zwicky, C.N.; Züttel, A. Stability and Reversibility of LiBH$_4$. *J. Phys. Chem. B* **2008**, *112*, 906–910. [CrossRef] [PubMed]

110. Yan, Y.; Li, H.-W.; Maekawa, H.; Miwa, K.; Towata, S.; Orimo, S. Formation of Intermediate Compound $Li_2B_{12}H_{12}$ during the Dehydrogenation Process of the LiBH$_4$–MgH$_2$ System. *J. Phys. Chem. C* **2011**, *115*, 19419. [CrossRef]

111. Kim, K.-B.; Shim, J.-H.; Park, S.-H.; Choi, I.-S.; Oh, K.H.; Cho, Y.W. Dehydrogenation Reaction Pathway of the LiBH$_4$–MgH$_2$ Composite under Various Pressure Conditions. *J. Phys. Chem. C* **2015**, *119*, 9714. [CrossRef]

112. Martelli, P.; Caputo, R.; Remhof, A.; Mauron, P.; Borgschulte, A.; Züttel, A. Stability and Decomposition of NaBH$_4$. *J. Phys. Chem. C* **2010**, *114*, 7173–7177. [CrossRef]

113. Matsunaga, T.; Buchter, F.; Mauron, P.; Bielman, M.; Nakamori, Y.; Orimo, S.; Ohba, N.; Miwa, K.; Towata, S.; Züttel, A. Hydrogen storage properties of Mg[BH$_4$]$_2$. *J. Alloys Compd.* **2008**, *459*, 583–588. [CrossRef]

114. Kuznetsov, V.A.; Dymova, T.N. Evaluation of the standard enthalpies and isobaric potentials of the formation of certain complex hydrides. *Russ. Chem. Bull.* **1971**, *20*, 204–208. [CrossRef]

115. Chlopek, K.; Frommen, C.; Leon, A.; Zabara, O.; Fichtner, M. Synthesis and properties of magnesium tetrahydroborate, Mg(BH$_4$)$_{(2)}$. *J. Mater. Chem.* **2007**, *17*, 3496–3503. [CrossRef]

116. Sartori, S.; Knudsen, K.D.; Zhao-Karger, Z.; Bardaij, E.G.; Fichtner, M.; Hauback, B.C. Small-angle scattering investigations of Mg-borohydride infiltrated in activated carbon. *Nanotechnology* **2009**, *20*, 505702. [CrossRef] [PubMed]

117. Severa, G.; Rönnebro, E.; Jensen, C.M. Direct hydrogenation of magnesium boride to magnesium borohydride: Demonstration of >11 weight percent reversible hydrogen storage. *Chem. Commun.* **2010**, *46*, 421–423. [CrossRef] [PubMed]

118. Soloveichik, G.L.; Gao, Y.; Rijssenbeek, J.; Andrus, M.; Kniajanski, S.; Bowman, R.C., Jr.; Hwang, S.-J.; Zhao, J.-C. Magnesium borohydride as a hydrogen storage material: Properties and dehydrogenation pathway of unsolvated Mg(BH$_4$)$_2$. *Int. J. Hydrogen Energy* **2009**, *34*, 916–928. [CrossRef]

119. Barkhordarian, G.; Jensen, T.R.; Doppiu, S.; Bösenberg, U.; Borgschulte, A.; Gremaud, R.; Cerenius, Y.; Dornheim, M.; Klassen, T.; Bormann, R. Formation of Ca(BH$_4$)$_2$ from Hydrogenation of CaH$_2$ + MgB$_2$ Composite. *J. Phys. Chem. C* **2008**, *112*, 2743–2749. [CrossRef]

120. Buchter, F.; Lodziana, Z.; Remhof, A.; Friedrichs, O.; Borgschulte, A.; Mauron, P.; Züttel, A.; Sheptyakov, D.; Barkhordarian, G.; Bormann, R.; et al. Structure of Ca(BD$_4$)$_2$ beta-phase from combined neutron and synchrotron X-ray powder diffraction data and density functional calculations. *J. Phys. Chem. B* **2008**, *112*, 8042–8048. [CrossRef] [PubMed]

121. Buchter, F.; Łodziana, Z.; Remhof, A.; Friedrichs, O.; Borgschulte, A.; Mauron, P.; Züttel, A. Structure of the Orthorhombic γ-Phase and Phase Transitions of Ca(BD$_4$)$_2$. *J. Phys. Chem. C* **2009**, *113*, 17223–17230. [CrossRef]

122. Filinchuk, Y.; Ronnebro, E.; Chandra, D. Crystal structures and phase transformations in Ca(BH$_4$)$_{(2)}$. *Acta Mater.* **2009**, *57*, 732–738. [CrossRef]

123. Nickels, E.A.; Jones, M.O.; David, W.I.F.; Johnson, S.R.; Lowton, R.L.; Sommariva, M.; Edwards, P.P. Tuning the Decomposition Temperature in Complex Hydrides: Synthesis of a Mixed Alkali Metal Borohydride. *Angew. Chem.* **2008**, *47*, 2817–2819. [CrossRef] [PubMed]

124. Schlesinger, H.I.; Brown, H.C. Sodium borohydride, its hydrolysis and its use as a reducing agent and in the generation of hydrogen. *J. Am. Chem. Soc.* **1953**, *75*, 219–221. [CrossRef]

125. Energy.com. Available online: https://energy.gov/eere/fuelcells/downloads/hydrogen-storage-materials-requirements-meet-2017-board-hydrogen-storage (accessd on 25 July 2017).

126. Au, M.; Jurgensen, A.; Zeigler, K. Modified Lithium Borohydrides for Reversible Hydrogen Storage. *J. Phys. Chem. B* **2006**, *110*, 26482–26487. [CrossRef] [PubMed]

127. Eigen, N.; Bösenberg, U.; Bellosta von Colbe, J.M.; Jensen, T.R.; Cerenius, Y.; Dornheim, M.; Klassen, T.; Bormann, R. Reversible hydrogen storage in NaF–Al composites. *J. Alloys Compd.* **2009**, *477*, 76–80. [CrossRef]

128. Brinks, H.W.; Fossdal, A.; Hauback, B.C. Adjustment of the Stability of Complex Hydrides by Anion Substitution. *J. Phys. Chem. C* **2008**, *112*, 5658–5661. [CrossRef]

129. Corno, M.; Pinatel, E.; Ugliengo, P.; Baricco, M. A computational study on the effect of fluorine substitution in LiBH$_4$. *J. Alloys Compd.* **2011**, *509*, s679–s683. [CrossRef]

130. Li, H.-W.; Orimo, S.; Nakamori, Y.; Miwa, K.; Ohba, N.; Towata, S.; Züttel, A. Materials designing of metal borohydrides: Viewpoints from thermodynamical stabilities. *J. Alloys Compd.* **2007**, *446–447*, 315–318. [CrossRef]

131. Libowitz, G.G.; Hayes, H.F.; Gibb, T.R.P. The System Zirconium-Nickel and Hydrogen. *J. Phys. Chem.* **1958**, *62*, 76–79. [CrossRef]

132. Reilly, J.J.; Wiswall, R.H. The reaction of hydrogen with alloys of magnesium and nikel and formation of Mg$_2$NiH$_4$. *Inorg. Chem.* **1968**, *7*, 2254–2256. [CrossRef]

133. Chen, P.; Xiong, Z.; Luo, J.; Lin, J.; Tan, K.L. Interaction of hydrogen with metal nitrides and imides. *Nature* **2002**, *420*, 302–304. [CrossRef] [PubMed]

134. Vajo, J.J.; Mertens, F.; Ahn, C.C.; Bowman, R.C.; Fultz, B. Altering Hydrogen Storage Properties by Hydride Destabilization through Alloy Formation: LiH and MgH$_2$ Destabilized with Si. *J. Phys. Chem. B* **2004**, *108*, 13977–13983. [CrossRef]

135. Barkhordarian, G.; Klassen, T.; Dornheim, M.; Bormann, R. Unexpected kinetic effect of MgB$_2$ in reactive hydride composites containing complex borohydrides. *J. Alloys Compd.* **2007**, *440*, L18–L21. [CrossRef]

136. Bösenberg, U.; Ravnsbæk, D.B.; Hagemann, H.; D'Anna, V.; Minella, C.B.; Pistidda, C.; Beek, W.; Jensen, T.R.; Bormann, R.; Dornheim, M. Pressure and temperature influence on the desorption pathway of the LiBH$_4$–MgH$_2$ composite system. *J. Phys. Chem. C* **2010**, *114*, 15212–15217. [CrossRef]

137. Bonatto Minella, C.; Garroni, S.; Olid, D.; Teixidor, F.; Pistidda, C.; Lindemann, I.; Gutfleisch, O.; Baró, M.D.; Bormann, R.; Klassen, T.; et al. Experimental evidence of Ca[B$_{12}$H$_{12}$] formation during decomposition of a Ca(BH$_4$)$_2$ + MgH$_2$ based reactive hydride composite. *J. Phys. Chem. C* **2011**, *115*, 18010–18014. [CrossRef]

138. Fichtner, M. Properties of nanoscale metal hydrides. *Nanotechnology* **2009**, *20*, 204009. [CrossRef] [PubMed]

139. Kim, K.C.; Dai, B.; Johnson, J.K.; Sholl, D.S. Assessing nanoparticle size effects on metal hydride thermodynamics using the Wulff construction. *Nanotechnology* **2009**, *20*, 204001. [CrossRef] [PubMed]

140. Vajo, J.J. Influence of nano-confinement on the thermodynamics and dehydrogenation kinetics of metal hydrides. *Curr. Opin. Solid State Mater. Sci.* **2011**, *15*, 52–61. [CrossRef]

141. Nielsen, K.N.; Besenbacher, F.; Jensen, T.R. Nanoconfined hydrides for energy storage. *Nanoscale* **2011**, *3*, 2086–2098. [CrossRef] [PubMed]

142. Züttel, A.; Rentsch, S.; Fischer, P.; Wenger, P.; Sudan, P.; Mauron, P.; Emmenegger, C. Hydrogen storage properties of LiBH$_4$. *J. Alloy Compd.* **2003**, *356–357*, 515–520. [CrossRef]

143. Zhang, Y.; Zhang, W.-S.; Fan, M.-Q.; Liu, S.-S.; Chu, H.-L.; Zhang, Y.-H.; Gao, X.-Y.; Sun, L.-X. Enhanced Hydrogen Storage Performance of LiBH$_4$–SiO$_2$–TiF$_3$ Composite. *J. Phys. Chem. C* **2008**, *112*, 4005–4010. [CrossRef]

144. Mosegaard, L.; Møller, B.; Jørgensen, J.-E.; Filinchuk, Y.; Cerenius, Y.; Hanson, J.C.; Dimasi, E.; Besenbacher, F.; Jensen, T.R. Reactivity of LiBH$_4$: In Situ Synchrotron Radiation Powder X-ray Diffraction Study. *J. Phys. Chem. C* **2008**, *112*, 1299–1303. [CrossRef]

145. Au, M.; Jurgensen, A.R.; Spencer, W.A.; Anton, D.L.; Pinkerton, F.E.; Hwang, S.-J.; Kim, C.; Bowman, R.C. Stability and Reversibility of Lithium Borohydrides Doped by Metal Halides and Hydrides. *J. Phys. Chem. C* **2008**, *112*, 18661–18671. [CrossRef]

146. Vajo, J.J.; Skeith, S.L.; Mertens, F. Reversible Storage of Hydrogen in Destabilized LiBH$_4$. *J. Phys. Chem. B* **2005**, *109*, 3719–3722. [CrossRef] [PubMed]

147. Zavorotynska, O.; Corno, M.; Pinatel, E.; Rude, L.H.; Ugliengo, P.; Jensen, T.R.; Baricco, M. Theoretical and Experimental Study of LiBH$_4$–LiCl Solid Solution. *Crystals* **2012**, *2*, 144–158. [CrossRef]

148. Rude, L.H.; Zavorotynska, O.; Arnbjerg, L.M.; Ravnsbæk, D.B.; Malmkjær, R.A.; Grove, H.; Hauback, B.C.; Baricco, M.; Filinchuk, Y.; Besenbacher, F.; et al. Bromide substitution in lithium borohydride, LiBH$_4$–LiBr. *Int. J. Hydrogen Energy* **2011**, *36*, 15664–15672. [CrossRef]

149. Rude, L.H.; Groppo, E.; Arnbjerg, L.M.; Ravnsbæk, D.B.; Malmkjær, R.A.; Filinchuk, Y.; Baricco, M.; Besenbacher, F.; Jensen, T.R. Iodide substitution in lithium borohydride, LiBH$_4$–LiI. *J. Alloys Compd.* **2011**, *509*, 8299–8305. [CrossRef]

150. Gennari, F.C.; Albanesi, L.F.; Puszkiel, J.A.; Larochette, P.A. Reversible hydrogen storage from 6LiBH$_4$–MCl$_3$ (M = Ce, Gd) composites by in-situ formation of MH$_2$. *Int. J. Hydrogen Energy* **2011**, *36*, 563–570. [CrossRef]

151. Xia, G.L.; Guo, Y.H.; Wu, Z.; Yu, X.B. Enhanced hydrogen storage performance of LiBH$_4$–Ni composite. *J. Alloys Compd.* **2009**, *479*, 545–548. [CrossRef]

152. Xu, J.; Yu, X.; Zou, Z.; Li, Z.; Wu, Z.; Akins, D.L.; Yang, H. Enhanced dehydrogenation of LiBH$_4$ catalyzed by carbon-supported Pt nanoparticles. *Chem. Commun.* **2008**, *44*, 5740–5742. [CrossRef] [PubMed]

153. Kang, X.-D.; Wang, P.; Ma, L.-P.; Cheng, H.-M. Reversible hydrogen storage in LiBH$_4$ destabilized by milling with Al. *Appl. Phys. A* **2007**, *89*, 963–966. [CrossRef]

154. Ngene, P.; van Zwienen, M.; de Jongh, P.E. Reversibility of the hydrogen desorption from LiBH$_4$: A synergetic effect of nanoconfinement and Ni addition. *Chem. Commun.* **2010**, *46*, 8201–8203. [CrossRef] [PubMed]

155. Fang, Z.-Z.; Kang, X.-D.; Wang, P.; Cheng, H.-M. Improved Reversible Dehydrogenation of Lithium Borohydride by Milling with As-Prepared Single-Walled Carbon Nanotubes. *J. Phys. Chem. C* **2008**, *112*, 17023–17029. [CrossRef]

156. Shao, J.; Xiao, X.; Fan, X.; Zhang, L.; Li, S.; Ge, H.; Wang, Q.; Chen, L. Low-Temperature Reversible Hydrogen Storage Properties of LiBH$_4$: A Synergetic Effect of Nanoconfinement and Nanocatalysis. *J. Phys. Chem. C* **2014**, *118*, 11252–11260. [CrossRef]

157. Vajo, J.J.; Olson, G.L. Hydrogen storage in destabilized chemical systems. *Scr. Mater.* **2007**, *56*, 829–834. [CrossRef]

158. Alapati, S.V.; Johnson, J.K.; Sholl, D.S. Identification of destabilized metal hydrides for hydrogen storage using first principles calculations. *J. Phys. Chem. B* **2006**, *110*, 8769–8776. [CrossRef] [PubMed]

159. Bösenberg, U.; Kim, J.W.; Gosslar, D.; Eigen, N.; Jensen, T.R.; Bellosta von Colbe, J.M.; Zhou, Y.; Dahms, M.; Kim, D.H.; Günther, R.; et al. Role of additives in LiBH$_4$–MgH$_2$ reactive hydride composites for sorption kinetics. *Acta Mater.* **2010**, *58*, 3381–3389. [CrossRef]

160. Deprez, E.; Justo, A.; Rojas, T.C.; López-Cartés, C.; Bonatto Minella, C.; Bösenberg, U.; Dornheim, M.; Bormann, R.; Fernández, A. Microstructural study of the LiBH$_4$–MgH$_2$ reactive hydride composite with and without Ti-isopropoxide additive. *Acta Mater.* **2010**, *58*, 5683–5694. [CrossRef]

161. Deprez, E.; Munoz-Márquez, M.A.; Rolán, M.A.; Prestipino, C.; Palomares, F.J.; Minella, C.B.; Bösenberg, U.; Dornheim, M.; Bormann, R.; Fernández, A. Oxidation state and local structure of Ti-based additives in the reactive hydride composite 2LiBH$_4$ + MgH$_2$. *J. Phys. Chem. C* **2010**, *114*, 3309–3317. [CrossRef]

162. Busch, N.; Jepsen, J.; Pistidda, C.; Puszkiel, J.A.; Karimi, F.; Milanese, C.; Tolkiehn, M.; Chaudhary, A.-L.; Klassen, T.; Dornheim, M. Influence of milling parameters on the sorption properties of the LiH + MgB$_2$ system doped with TiCl$_3$. *J. Alloys Compd.* **2015**, *645*, S299–S303. [CrossRef]

163. Gosalawit-Utke, R.; Milanese, C.; Javadian, P.; Jepsen, J.; Laipple, D.; Karmi, F.; Puszkiel, J.; Jensen, T.R.; Marini, A.; Klassen, T.; et al. Nanoconfined 2LiBH$_4$–MgH$_2$–TiCl$_3$ in carbon aerogel scaffold for reversible hydrogen storage. *Int. J. Hydrogen Energy* **2013**, *38*, 3275–3282. [CrossRef]

164. Gosalawit-Utke, R.; Nielsen, T.K.; Saldan, I.; Laipple, D.; Cerenius, Y.; Jensen, T.R.; Klassen, T.; Dornheim, M. Nanoconfined 2LiBH$_4$–MgH$_2$ Prepared by Direct Melt Infiltration into Nanoporous Materials. *J. Phys. Chem. C* **2011**, *115*, 10903–10910. [CrossRef]

165. Jepsen, J.; Bellosta von Colbe, J.M.; Klassen, T.; Dornheim, M. Economic potential of complex hydrides compared to conventional hydrogen storage systems. *Int. J. Hydrogen Energy* **2012**, *37*, 4204–4214. [CrossRef]

166. Jepsen, J.; Milanese, C.; Girella, A.; Lozano, G.A.; Pistidda, C.; Bellosta Von Colbe, J.M.; Marini, A.; Klassen, T.; Dornheim, M. Compaction pressure influence on material properties and sorption behaviour of LiBH$_4$–MgH$_2$ composite. *Int. J. Hydrogen Energy* **2013**, *38*, 8357–8366. [CrossRef]

167. Shim, J.H.; Lim, J.H.; Rather, S.; Lee, Y.S.; Reed, D.; Kim, Y.; Book, D.; Cho, Y.W. Effect of hydrogen back pressure on dehydrogenation behavior of LiBH$_4$-based reactive hydride composites. *J. Phys. Chem. Lett.* **2009**, *1*, 59–63. [CrossRef]

168. Puszkiel, J.A.; Gennari, F.C.; Larochette, P.A.; Ramallo-López, J.M.; Vainio, U.; Karimi, F.; Pranzas, P.K.; Troiani, H.; Pistidda, C.; Jepsen, J.; et al. Effect of Fe additive on the hydrogenation-dehydrogenation properties of 2LiH + MgB$_2$/2LiBH$_4$ + MgH$_2$ system. *J. Power Sources* **2015**, *284*, 606–616. [CrossRef]

169. Cova, F.; Rönnebro, E.C.E.; Choi, Y.J.; Gennari, F.C.; Arneodo Larochette, P. New Insights into the Thermodynamic Behavior of 2LiBH$_4$–MgH$_2$ Composite for Hydrogen Storage. *J. Phys. Chem. C* **2015**, *119*, 15816–15822. [CrossRef]

170. Zhong, Y.; Wan, X.; Ding, Z.; Shaw, L.L. New dehydrogenation pathway of LiBH$_4$ + MgH$_2$ mixtures enabled by nanoscale LiBH$_4$. *Int. J. Hydrogen Energy* **2016**, *41*, 22104. [CrossRef]

171. Xia, G.; Tan, Y.; Wu, F.; Fang, F.; Sun, D.; Guo, Z.; Liu, H.; Yu, X. Mixed-metal (Li, Al) amidoborane: Synthesis and enhanced hydrogen storage properties. *Nano Energy* **2016**, *26*, 488. [CrossRef]

172. Puszkiel, J.; Castro Riglos, M.V.; Karimi, F.; Santoru, A.; Pistidda, C.; Klassen, T.; Bellosta von Colbe, J.M.; Dornheim, M. Changing the dehydrogenation pathway of LiBH$_4$–MgH$_2$ via nanosized lithiated TiO$_2$. *Phys. Chem. Chem. Phys.* **2017**, *19*, 7455–7460. [CrossRef] [PubMed]

173. Puszkiel, J.; Castro Riglos, M.V.; Ramallo-López, J.M.; Mizrahi, M.; Karimi, F.; Santoru, A.; Hoell, A.; Gennari, F.C.; Arneodo Larochette, P.; Pistidda, C.; et al. A novel catalytic route for hydrogenation–dehydrogenation of 2LiH + MgB$_2$ via in situ formed core–shell Li$_x$TiO$_2$ nanoparticles. *J. Mater. Chem. A* **2017**, *5*, 12922. [CrossRef]

174. Vajo, J.J.; Li, W.; Liu, P. Thermodynamic and kinetic destabilization in LiBH$_4$/Mg$_2$NiH$_4$: Promise for borohydride-based hydrogen storage. *Chem. Commun.* **2010**, *46*, 6687–6689. [CrossRef] [PubMed]

175. Mao, J.; Guo, Z.; Nevirkovets, I.P.; Liu, H.K.; Dou, S.X. Hydrogen De-/Absorption Improvement of NaBH$_4$ Catalyzed by Titanium-Based Additives. *J. Phys. Chem. C* **2012**, *116*, 1596–1604. [CrossRef]

176. Humphries, T.D.; Kalantzopoulos, G.N.; Llamas-Jansa, I.; Olsen, J.E.; Hauback, B.C. Reversible Hydrogenation Studies of NaBH$_4$ Milled with Ni-Containing Additives. *J. Phys. Chem. C* **2013**, *117*, 6060–6065. [CrossRef]

177. Mao, J.F.; Yu, X.B.; Guo, Z.P.; Liu, H.K.; Wu, Z.; Ni, J. Enhanced hydrogen storage performances of NaBH$_4$-MgH$_2$ system. *J. Alloys Compd.* **2009**, *479*, 619–623. [CrossRef]

178. Garroni, S.; Pistidda, C.; Brunelli, M.; Vaughan, G.B.M.; Suriñach, S.; Baró, M.D. Hydrogen desorption mechanism of 2NaBH$_4$ + MgH$_2$ composite prepared by high-energy ball milling. *Scr. Mater.* **2009**, *60*, 1129–1132. [CrossRef]

179. Dornheim, M. *Handbook of Hydrogen Storage*; Hirscher, M., Ed.; Wiley-VCH: New York, NY, USA, 2010; pp. 187–214.

180. Garroni, S.; Milanese, C.; Girella, A.; Marini, A.; Mulas, G.; Menéndez, E.; Pistidda, C.; Dornheim, M.; Suriñach, S.; Baró, M.D. Sorption properties of NaBH$_4$/MH$_2$ (M = Mg, Ti) powder systems. *Int. J. Hydrogen Energy* **2010**, *35*, 5434–5441. [CrossRef]

181. Li, G.Q.; Matsuo, M.; Deledda, S.; Hauback, B.C.; Orimo, S. Dehydriding Property of NaBH$_4$ Combined with Mg$_2$FeH$_6$. *Mater. Trans.* **2014**, *55*, 1141–1143. [CrossRef]

182. Afonso, G.; Bonakdarpour, A.; Wilkinson, D.P. Hydrogen Storage Properties of the Destabilized 4NaBH$_4$/5Mg$_2$NiH$_4$ Composite System. *J. Phys. Chem. C* **2013**, *117*, 21105–21111. [CrossRef]

183. Al-Kukhun, A.; Hwang, H.T.; Varma, A. NbF$_5$ additive improves hydrogen release from magnesium borohydride. *Int. J. Hydrogen Energy* **2012**, *37*, 17671–17677. [CrossRef]

184. Bardaji, E.G.; Hanada, N.; Zabara, O.; Fichtner, M. Effect of several metal chlorides on the thermal decomposition behaviour of α-Mg(BH$_4$)$_2$. *Int. J. Hydrogen Energy* **2011**, *36*, 12313–12318. [CrossRef]

185. Saldan, I.; Frommen, C.; Llamas-Jansa, I.; Kalantzopoulos, G.N.; Hino, S.; Arstad, B.; Heyn, R.H.; Zavorotynska, O.; Deledda, S.; Sørby, M.H.; et al. Hydrogen storage properties of γ-Mg(BH$_4$)$_2$ modified by MoO$_3$ and TiO$_2$. *Int. J. Hydrogen Energy* **2015**, *40*, 12286–12293. [CrossRef]

186. Saldan, I.; Hino, S.; Humphries, T.D.; Zavorotynska, O.; Chong, M.; Jensen, C.M.; Deledda, D.; Hauback, B.C. Structural changes observed during the reversible hydrogenation of Mg(BH$_4$)$_2$ with Ni-based additives. *J. Phys. Chem. C* **2014**, *118*, 23376–23384. [CrossRef]

187. Zavorotynska, O.; Deledda, S.; Vitillo, J.; Saldan, I.; Guzik, M.; Baricco, M.; Walmsley, J.C.; Muller, J.; Hauback, B.C. Combined X-ray and Raman Studies on the Effect of Cobalt Additives on the Decomposition of Magnesium Borohydride. *Energies* **2015**, *8*, 9173–9190. [CrossRef]

188. Zavorotynska, O.; Saldan, I.; Hino, S.; Humphries, T.D.; Deledda, S.; Hauback, B.C. Hydrogen cycling in [gamma]-Mg(BH$_4$)$_2$ with cobalt-based additives. *J. Mater. Chem. A* **2015**, *3*, 6592–6602. [CrossRef]

189. Yan, Y.; Remhof, A.; Rentsch, D.; Züttel, A.; Giri, S.; Jena, P. A novel strategy for reversible hydrogen storage in Ca(BH$_4$)$_2$. *Chem. Commun.* **2015**, *51*, 11008–11011. [CrossRef] [PubMed]

190. Kim, J.-H.; Jin, S.-A.; Shim, J.-H.; Cho, Y.W. Reversible hydrogen storage in calcium borohydride Ca(BH$_4$)$_2$. *Scr. Mater.* **2008**, *58*, 481–483.

191. Kim, J.-H.; Shim, J.-H.; Cho, Y.W. On the reversibility of hydrogen storage in Ti- and Nb-catalyzed Ca(BH$_4$)$_2$. *J. Power Sources* **2008**, *181*, 140–143. [CrossRef]

192. Bonatto Minella, C.; Garroni, S.; Pistidda, C.; Gosalawit-Utke, R.; Barkhordarian, G.; Rongeat, C.; Lindemann, I.; Gutfleisch, O.; Jensen, T.R.; Cerenius, Y.; et al. Effect of Transition Metal Fluorides on the Sorption Properties and Reversible Formation of Ca(BH$_4$)$_2$. *J. Phys. Chem. C* **2011**, *115*, 2497–2504. [CrossRef]

193. Rongeat, C.; D'Anna, V.; Hagemann, H.; Borgschulte, A.; Züttel, A.; Schultz, L.; Gutfleisch, O. Effect of additives on the synthesis and reversibility of Ca(BH$_4$)$_2$. *J. Alloys Compd.* **2010**, *493*, 281–287. [CrossRef]

194. Ronnebro, E.; Majzoub, E.H. Calcium borohydride for hydrogen storage: Catalysis and reversibility. *J. Phys. Chem. B* **2007**, *111*, 12045–12047. [CrossRef] [PubMed]

195. Bonatto Minella, C.; Pellicer, E.; Rossinyol, E.; Karimi, F.; Pistidda, C.; Garroni, S.; Milanese, C.; Nolis, P.; Dolors Baro, M.; Gutfleisch, O.; et al. Chemical state, distribution, and role of Ti- and Nb-based additives on the Ca(BH$_4$)$_2$ system. *J. Phys. Chem. C* **2013**, *117*, 4394–4403. [CrossRef]

196. Kelly, P.M.; Zhang, M.X. Edge-to-edge matching—The fundamentals. *Metall. Mater. Trans.* **2006**, *37*, 833–839. [CrossRef]

197. Zhang, M.X.; Kelly, P.M. Edge-to-edge matching model for predicting orientation relationships and habit planes-the improvements. *Scr. Mater.* **2005**, *52*, 963–968. [CrossRef]

198. Karimi, F.; Klaus Pranzas, P.; Pistidda, C.; Puszkiel, J.A.; Milanese, C.; Vainio, U.; Paskevicius, M.; Emmler, T.; Santoru, A.; Utke, R.; et al. Structural and kinetic investigation of the hydride composite Ca(BH$_4$)$_2$ + MgH$_2$ system doped with NbF$_5$ for solid-state hydrogen storage. *Phys. Chem. Chem. Phys.* **2015**, *17*, 27328–27342. [CrossRef] [PubMed]

199. Karimi, F.; Pranzas, P.K.; Hoell, A.; Vainio, U.; Welter, E.; Raghuwanshi, V.S.; Pistidda, C.; Dornheim, M.; Klassen, T.; Schreyer, A. Structural analysis of calcium reactive hydride composite for solid state hydrogen storage. *J. Appl. Crystallogr.* **2014**, *47*, 67–75. [CrossRef]

200. Bonatto Minella, C.; Pistidda, C.; Garroni, S.; Nolis, P.; Baró, M.D.; Gutfleisch, O.; Klassen, T.; Bormann, R.; Dornheim, M. Ca(BH$_4$)$_2$ + MgH$_2$: Desorption Reaction and Role of Mg on Its Reversibility. *J. Phys. Chem. C* **2013**, *117*, 3846–3852. [CrossRef]

201. Bergemann, N.; Pistidda, C.; Milanese, C.; Emmler, T.; Karimi, F.; Chaudhary, A.-L.; Chierotti, M.R.; Klassen, T.; Dornheim, M. Ca(BH$_4$)$_2$–Mg$_2$NiH$_4$: On the pathway to a Ca(BH$_4$)$_2$ system with a reversible hydrogen cycle. *Chem. Commun.* **2016**, *52*, 4836–4839. [CrossRef] [PubMed]

202. Jepsen, J. Technical and Economic Evaluation of Hydrogen Storage Systems based on Light Metal Hydrides. Ph.D. Thesis, Helmut Schmidt University, Holstenhofweg, Hamburg, Germany, 17 December 2013.

inorganics

MDPI

Article

Dehydrogenation of Surface-Oxidized Mixtures of 2LiBH$_4$ + Al/Additives (TiF$_3$ or CeO$_2$)

Juan Luis Carrillo-Bucio, Juan Rogelio Tena-García and Karina Suárez-Alcántara *

Morelia Unit of the Materials Research Institute of the National Autonomous University of Mexico, Antigua Carretera a Pátzcuaro 8701, Col. Ex Hacienda de San José de la Huerta, Morelia 58190, Michoacán, Mexico; jlcarrillob@iim.unam.mx (J.L.C.-B.); juanrogelio_tenagarcia@iim.unam.mx (J.R.T.-G.)
* Correspondence: karina_suarez@iim.unam.mx; Tel.: +52-5623-7300 (ext. 37889)

Received: 28 September 2017; Accepted: 16 November 2017; Published: 21 November 2017

Abstract: Research for suitable hydrogen storage materials is an important ongoing subject. LiBH$_4$–Al mixtures could be attractive; however, several issues must be solved. Here, the dehydrogenation reactions of surface-oxidized 2LiBH$_4$ + Al mixtures plus an additive (TiF$_3$ or CeO$_2$) at two different pressures are presented. The mixtures were produced by mechanical milling and handled under welding-grade argon. The dehydrogenation reactions were studied by means of temperature programmed desorption (TPD) at 400 °C and at 3 or 5 bar initial hydrogen pressure. The milled and dehydrogenated materials were characterized by scanning electron microscopy (SEM), X-ray diffraction (XRD), and Fourier transformed infrared spectroscopy (FT-IR) The additives and the surface oxidation, promoted by the impurities in the welding-grade argon, induced a reduction in the dehydrogenation temperature and an increase in the reaction kinetics, as compared to pure (reported) LiBH$_4$. The dehydrogenation reactions were observed to take place in two main steps, with onsets at 100 °C and 200–300 °C. The maximum released hydrogen was 9.3 wt % in the 2LiBH$_4$ + Al/TiF$_3$ material, and 7.9 wt % in the 2LiBH$_4$ + Al/CeO$_2$ material. Formation of CeB$_6$ after dehydrogenation of 2LiBH$_4$ + Al/CeO$_2$ was confirmed.

Keywords: hydrogen storage; borohydrides; reactive mixtures

1. Introduction

LiBH$_4$ is an outstanding material regarding its hydrogen content (18.4 wt %) [1]. However, its dehydrogenation temperature is too high for any practical application in hydrogen storage. Pure LiBH$_4$ presents two dehydrogenation steps; one minor at approximately 320 °C and the main one at 500 °C [1]. To reduce the dehydrogenation temperature, improve reaction kinetics or reversibility, LiBH$_4$ has been mixed with several compounds in different proportions. The list includes but is not limited to other borohydrides such as: Ca(BH$_4$)$_2$ [2], NaBH$_4$ [3], or Mg(BH$_4$)$_2$ [4], other complex hydrides such as LiNH$_2$ [5], alanates of Li or Na [6,7], binary hydrides such as MgH$_2$ [8–10], CaH$_2$ [11–13], TiH$_2$ [12], halide salts such as LiCl [14], oxides [15], scaffolds [16], and metals such as Mg, Ti, V, Cr, Sc, or Al [12,17]. The main characteristics of these mixtures are collated in Table 1.

The last system, the 2LiBH$_4$ + Al, is of potential interest. Siegel et al. [18] anticipated, based on first-principles calculations, that the reaction:

$$2LiBH_4 + Al \rightarrow AlB_2 + 2LiH + 3H_2 \tag{1}$$

would release 8.6 wt % hydrogen at 277 °C and p(H$_2$) = 1 bar. Experimentally, Zhang et al. performed the dehydrogenation of 2LiBH$_4$ + Al at a pressure of 0.001 bar H$_2$ and from roughly 325–525 °C [19]. Zhang et al. demonstrated a hydrogen release of approximately 4 wt % by means of isothermal dehydrogenation at 375 °C, meanwhile complete dehydrogenation was obtained up to 577 °C by means of thermogravimetric measurement [19]. Kang et al. achieved the release of 7.2 wt % hydrogen when the mixture was catalyzed with TiF_3, at a pressure of 0.001 bar, 450 °C, and 3 h [20]. Hansen et al. [17] demonstrated that the dehydrogenation reaction of $2LiBH_4 + 3Al$ occurs at $p(H_2) = 10^{-2}$ bar when the material is heated up to 500 °C; and that the re-hydrogenation demonstrated only partial reversibility. Other examples include the experiments of Ravnsbaek et al., where the dehydrogenation reaction of $LiBH_4 + Al$ (1:0.5) was characterized by in-situ synchrotron radiation powder X-ray diffraction from room temperature to 500 °C and dynamic vacuum [21]. All these experiments have in common a very low or vacuum dehydrogenation pressure (summarized in Table 1). The vacuum dehydrogenation pressure is inadequate for any practical application. Additionally, it is well-known that the hydrogen pressure can affect the dehydrogenation products of $LiBH_4$ mixtures. For example with the system $2LiBH_4 + MgH_2 \rightarrow 2LiH + MgB_2 + 4H_2$, a correlation was demonstrated between the dehydrogenation pressure and re-hydrogenation [22]. With the $2LiBH_4$–Al system, Yang et al. proposed that a desorption backpressure of 3 bar could contribute to the formation of AlB_2 [12]; which would improve, in principle, the reversibility of hydrogenation/dehydrogenation reactions. A study of the dehydrogenation backpressure with $2LiBH_4$–Al systems has been poorly explored.

Among the additives for improving reaction kinetics or reversibility; TiF_3 is the material most commonly used, and it is almost mandatory to test TiF_3 in all new mixtures. In their part, oxides such as TiO_2, ZrO_2, Nb_2O_5 or MoO_3 have resulted in successful accelerators for hydrogen desorption reactions [23,24]. CeO_2 is not a commonly used additive for hydrogenation/dehydrogenation reactions. However, interactions of hydrogen with CeO_2 have been reported [25]. The possible effects of CeO_2 as an additive for hydrogen storage materials deserve research.

Conventionally, hydrogen storage materials are produced and handled in a high or ultra-high purity inert atmosphere, i.e., high-purity argon. This is done to avoid the deactivation of materials caused by the formation of thick oxide films that impede hydrogen diffusion from/to the bulk of the storage material. However, some reports have expressed that the use of high-purity inert atmosphere can be relaxed [26] and that allowing some surface oxidation can be helpful to the dehydrogenation kinetics [27].

Here, the dehydrogenation behavior is presented of $2LiBH_4 + Al$ added with 5 wt % of TiF_3 or CeO_2, and surface-modified (oxidized) by the effects of impurities in welding-grade argon. The dehydrogenation reactions were performed at 3 bar and 5 bar of hydrogen. The effects of the borohydride–Al mixture, additives, and surface-oxide effects are discussed.

Table 1. Reported hydrogen desorption conditions for several $LiBH_4$ mixtures.

Material and/or Proposed Reaction, and Reported ΔH^0 (If Available) (kJ/mol H_2)	Desorption Conditions p (bar) and T (°C)	Comments
$LiBH_4 \rightarrow LiH + B + 3/2H_2$ [1]	p: not specified T: 320 °C and 500 °C	Multi-step dehydrogenation reaction
$LiBH_4 \rightarrow Li + B + 2H_2$ 95.1 kJ/mol H_2 [28]	p: 1 bar T: 25 °C	From standard formation enthalpy of $LiBH_4$
$xLiBH_4 + (1 - x) Ca(BH_4)_2$ [2]	p: not specified T: 370 °C for $x = 0.4$	$x = 0 - 1$, eutectic melting at 200 °C
$0.62LiBH_4$-$0.38NaBH_4$ [3]	p: not specified T: onset at 287 °C, peaks at 488 °C and 540 °C	Multi-step dehydrogenation reaction
$xLiBH_4 + (1 - x) Mg(BH_4)_2$ [4]	p: 5 bar T: 170 °C and 215 °C	$x = 0 - 1$, eutectic melting at 180 °C Multi-step dehydrogenation reaction
$LiBH_4 + 2LiNH_2 \rightarrow Li_3BN_2$ + $4H_2$ 23 kJ/mol H_2 [5]	p: 100–0.01 bar T: 430 °C	From pressure composition isotherm.
$LiBH_4 + LiAlH_4$ [6]	p: 0.2 bar T: 118 °C and 210 °C	2:1 mixture, two-step dehydrogenation. Dehydrogenation temperature reduced if TiF_3 addition.
$LiBH_4 + NaAlH_4$ [7]	p: 1 bar He T: from room temperature up to 210 °C for the doped systems and 110–250 °C for the undoped systems.	Molar ratios 1:1, 2:3 and 1:3; with and without $TiCl_3$ additive. Multi-step dehydrogenation reaction.

Table 1. *Cont.*

Material and/or Proposed Reaction, and Reported ΔH^0 (If Available) (kJ/mol H$_2$)	Desorption Conditions p (bar) and T (°C)	Comments
$2LiBH_4 + MgH_2 \rightarrow 2LiH + MgB_2$ $+ 4H_2$ 50.4 kJ/mol H$_2$ [10]	p: 3 bar H$_2$ T: 350–400 °C	Multi-step dehydrogenation reaction
$6LiBH_4 + CaH_2 \leftrightarrow 6LiH + CaB_6$ $+ 10H_2$ 59 kJ/mol H$_2$ [11]	p: 1.3 bar flowing He T: onset at 150 °C, maximum at 350 °C, finished at 450 °C	-
$LiBH_4 + TiH_2$ [12]	p: not specified (argon) T: ~410 °C	-
$LiBH_4 + LiCl$ (1:1) to give $Li(BH_4)_{1-x}Cl_x$ ($x \approx 0.23$) [14]	p: not specified (argon) T: 300–550 °C	Cl$^-$ to BH$_4^-$ substitution at LiBH$_4$
$2LiBH_4 + Al \rightarrow 2LiH + AlB_2$ $+ 3H_2$ 57.9 kJ/mol H$_2$ [18]	277 °C [18]	Theoretical desorption temperature
	Dehydrogenation: p: 0.001 bar H$_2$ T: 325 °C to 525 °C [19]	H$_2$ release of about 4 wt % Multi-step dehydrogenation reaction
	p: 0.001 bar T: 450 °C [20]	Catalyzed with TiF$_3$
	p: dynamic vacuum T: up to 500 °C [21]	Formation of Li$_x$Al$_{1-x}$B$_2$

2. Results

2.1. Characterization of As-Milled Materials

Scanning Electron Microscopy. Some of the descriptions below contain remarks about SEM images that are shown in the Supplementary Materials. Also for comparison purposes, the SEM images of LiBH$_4$ and Al without milling are presented in the Supplementary Materials. Non-milled LiBH$_4$ is composed of large crystals embedded in an amorphous phase. Non-milled Al is composed of particles of approximately 50 μm, heavily agglomerated and forming flakes of 1 mm (Supplementary Materials). Figure 1 presents the most representative SEM images of the as-milled materials. The as-milled 2LiBH$_4$ + Al material, Figure 1a, formed a three-dimensional, porous structure. Interestingly the surface of 2LiBH$_4$ + Al is covered with crystals of approximately 2–3 μm. The material 2LiBH$_4$ + Al/TiF$_3$, Figure 1b, also presented a three-dimensional structure. The material consisted of elongated crystals of 10 μm length and 2 μm width. Elemental analysis by energy-dispersive X-ray spectroscopy (EDS) of those crystals revealed a B-rich phase, i.e., the LiBH$_4$. The addition of CeO$_2$, Figure 1c, to the base material, produced an agglomerated, spherical, and compacted material of about 50 μm in diameter, with some surface formations approximating flake-shape. Here, spots of CeO$_2$ are clearly distinguishable from the base material. In principle, the morphology characteristics of hydrogen storage materials must be reduced particle size, low agglomeration, homogeneity of component materials, and the formation of porous structures allowing the inflow/outflow of hydrogen while maintaining good thermal conductivity. SEM images showed interesting morphologies of the LiBH$_4$ + Al/additives, depending on the additive material.

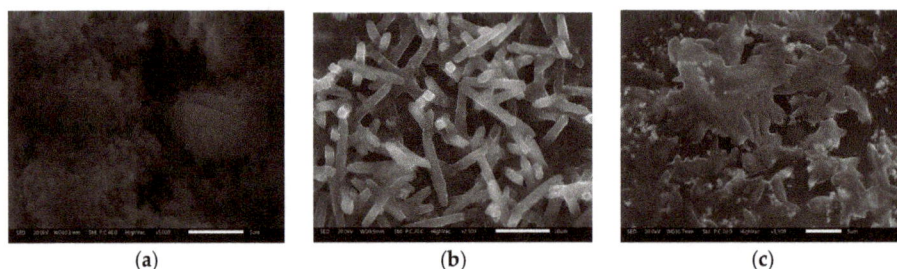

(a)   (b)   (c)

Figure 1. Scanning electron microscopy (SEM) images of the as-milled materials: (a) 2LiBH$_4$ + Al; (b) 2LiBH$_4$ + Al/TiF$_3$; (c) 2LiBH$_4$ + Al/CeO$_2$.

Powder X-ray diffraction. Figure 2 shows the X-ray diffraction patterns of as-milled materials. In the three studied materials, the presence of Al is evidenced by the strong diffraction peaks. By contrast, the $LiBH_4$ presented attenuated peaks as as-milled materials. $LiBH_4$ diffraction peaks in Figure 2 are a mixture of orthorhombic (main) and hexagonal (minor) phases to a different degree in the different mixtures. The material $2LiBH_4 + Al/CeO_2$ presented strong peaks of CeO_2, in agreement with the CeO_2 particles found in the SEM images. The material added with TiF_3 did not present the characteristic peaks of TiF_3. Some researchers have proposed the in-situ formation of $Ti(BH_4)_3$ by the reaction of $LiBH_4$ and TiF_3, followed by a rapid decomposition of $Ti(BH_4)_3$ [29]. Reactions to form other stable fluorine salts such as TiF_2 or AlF_3 cannot be discarded. These side reactions during ball-milling can be the reason for the absence of TiF_3 diffraction peaks. The materials also presented a small amount of aluminum oxide. Refinement of the Al and CeO_2 phases (not including $LiBH_4$ phases, $R_{wp} \approx 7$) produced the following Al-crystal sizes: Al in $2LiBH_4 + Al$: 309.6 ± 16.7 nm; Al in $2LiBH_4 + Al/TiF_3$: 129.0 ± 5.0 nm; Al in $2LiBH_4 + Al/CeO_2$: 142.8 ± 9.2 nm. The CeO_2 crystal size was determined as 120.3 ± 4.2 nm. All these facts point to the formation of fine mixtures of $2LiBH_4 + Al$/additive and an influence of the additives on the crystalline characteristics of the samples.

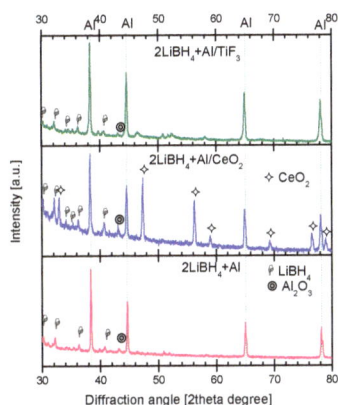

Figure 2. X-ray diffraction patterns of the as-milled materials: (**a**) $2LiBH_4 + Al$; (**b**) $2LiBH_4 + Al/TiF_3$; (**c**) $2LiBH_4 + Al/CeO_2$.

Fourier Transformed Infrared Spectroscopy. An effect of the ball milling is the increase of the local pressure and temperature during ball-ball or ball-vial collisions that induce the decomposition of sensitive materials. In the case of the $2LiBH_4 + Al$ mixtures, it is intended to decrease the decomposition temperature during dehydrogenation, but not to decompose during ball-milling. Thus FT-IR was performed to check the "survival" of $LiBH_4$ after ball-milling. Borohydrides present two regions of interest B–H bending (1000–1600 cm^{-1}) and H–B–H stretching (2000–2500 cm^{-1}) [29,30]. Both IR active modes are presented in all the as-milled samples (Supplementary Materials), indicating sufficient thermal stability during ball milling.

X-ray photoelectron spectroscopy. XPS results are presented in Figure 3. Frame 3a presents the Li 1s XPS spectra. The Li 1s XPS spectrum of $LiBH_4$ with a clean surface was reported to present a peak at 57.1 eV; meanwhile, the oxygen-exposed samples presented a contribution (shoulder) of Li_2O at 55.5 eV [27]. The studied materials present the main peak in a range between 57.2 and 57.5 eV. Because of that, the main contribution of Li binding is not the Li–O interactions. Frame 3(b) presents the B 1s XPS spectra. The reported peak for pure $LiBH_4$ is located at 188.3–188.4 eV [31]. After exposing $LiBH_4$ to oxygen or moisture, a new peak located at 191.5–191.4 eV emerged and it was attributed to $LiBO_2$ [31]. The studied materials presented both peaks, the $LiBH_4$ and the $LiBO_2$ at 188.3–188.0 eV and 191.9–191.3 eV, respectively. The peak intensity ratio of the $LiBH_4$ and $LiBO_2$ is not homogeneous throughout the set of studied materials. The mixture of 2LiBH$_4$ + Al presented the lowest intensity of the LiBO$_2$ peak. Frame c of Figure 3 presents the O 1s XPS spectra; as a reference, the O 1s spectra at the Al raw-material was included. In that spectrum, the peak at 533.6 eV can be related to an Al$_2$O$_3$ layer [32]. The XPS spectra of LiBH$_4$, 2LiBH$_4$ + Al/TiF$_3$, and 2LiBH$_4$ + Al/CeO$_2$ presented two main peaks; one at 533.9 eV and other at 532.6 eV. This last peak can be attributed to the LiBO$_2$ [32]. The LiBO$_2$ peak is very much attenuated in the 2LiBH$_4$ + Al, in agreement with the result at the B 1s spectra. The Al 2p spectrum of the as-received Al and as-milled 2LiBH$_4$ + Al/additive materials is shown in the Supplementary Materials. The (as-received) Al curve presents the characteristic metallic and oxide peaks of Al [33]. Meanwhile, no signal above noise was detected for the mixtures, indicating Al segregation to sub-surface layers.

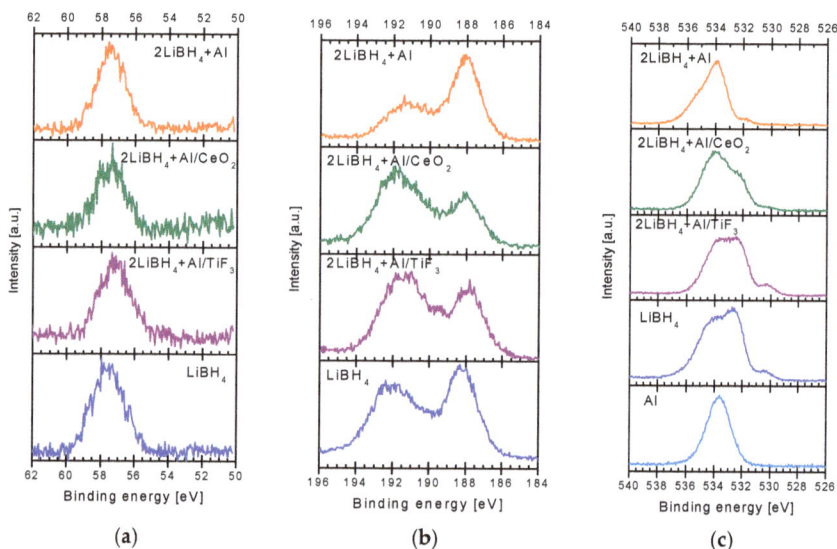

Figure 3. X-ray photoelectron spectroscopy (XPS) spectra of as-milled 2LiBH$_4$ + Al, 2LiBH4 + Al/TiF$_3$, and 2LiBH$_4$/CeO$_2$. (a) Li 1s edge; (b) B 1s edge; (c) O 1s edge.

2.2. Dehydrogenation Reactions

Figure 4 presents the dehydriding reactions traced by thermal-programed control. For two frames, the sample temperature also was plotted. The dehydrogenations of 2LiBH$_4$ + Al and 2LiBH$_4$ + Al/TiF$_3$ at 3 bar and 5 bar initial pressure are presented in Figure 4a. The dehydrogenations were performed up to 400 °C. The dehydrogenations occurred as multiple-step reactions. The first step with both materials was observed to start at 110 °C and produced a small hydrogen release. In the 2LiBH$_4$ + Al material, the second step was observed at 210 °C, and a slight change in reaction rate at roughly 325 °C. The dehydrogenation reactions in the 2LiBH$_4$ + Al material were essentially finished at 400 °C and reached a hydrogen release of −5.8 wt % for the 3 bar initial pressure reaction and −6.0 wt % for the 5 bar initial pressure reaction. The dehydrogenations of 2LiBH$_4$ + Al/TiF$_3$ at 3 bar and 5 bar initial pressure also are multiple-step reactions. At both initial pressures, the dehydrogenation first step onset is situated at about 110 °C. At the 3 bar dehydrogenation, the second reaction step is located at 240 °C. The last step initiated at 340 °C, after a clear-defined plateau. At 3 bar initial pressure the release of hydrogen accounted for −9.3 wt %. The situation is very different for the same material at 5 bar initial pressure, here the onset of the second dehydrogenation step was observed at 200 °C. A slowdown of the dehydrogenation rate was observed at 300 °C. Finally, the dehydrogenation finished after reaching 400 °C, releasing −6.0 wt %. This last quantity is greater than the theoretical value of −8.6 wt %, thus a release of B-compounds is probable. It means that higher initial pressure in combination with lower final temperature needs to be explored for this material.

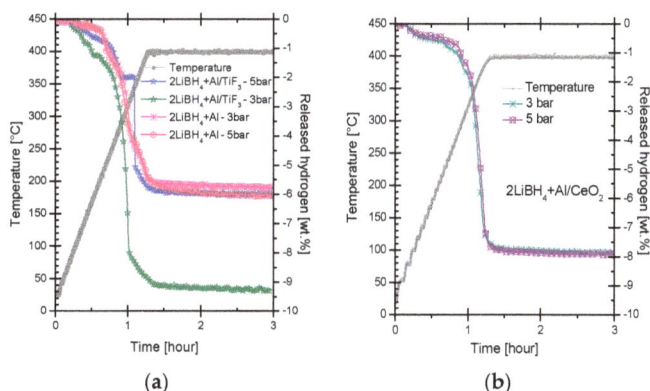

Figure 4. Temperature programed hydrogen desorption of (a) $2LiBH_4 + Al$, and $2LiBH_4 + Al/TiF_3$; (b) $2LiBH_4 + Al/CeO_2$ at 3 and 5 bar initial pressure of hydrogen.

Figure 4b presents the dehydrogenation reactions of $2LiBH_4 + Al/CeO_2$ at 3 bar and 5 bar initial hydrogen pressure up to 400 °C. The dehydrogenations reactions are similar at both pressures and they are also multistep reactions. The onset temperature of the first dehydrogenation step is below 100 °C. The dehydrogenation reactions presented the second step starting at 220 °C. The material $2LiBH_4 + Al/CeO_2$ released −7.9 wt % of hydrogen in both cases. In the material $2LiBH_4 + Al/CeO_2$, dehydrogenation at 350 °C was achieved and can be observed in the Supplementary Materials. The dehydrogenation is a multi-step reaction with a hydrogen release of −5.6 wt % at 5 bar and −6.5 wt % at 3 bar after 3 h of reaction.

2.3. Characterization of Dehydrogenated Materials

Figure 5 presents the most representative SEM images of the $2LiBH_4 + Al$, $2LiBH_4 + Al/TiF_3$, and $2LiBH_4 + Al/CeO_2$ materials dehydrogenated at 5 bar, other SEM images are available in the Supplementary Materials. In general, dehydrogenated materials presented important agglomerations of about 200 µm size. Thus an important increase in particle size after dehydrogenation was observed. The agglomerated materials were composed of an amorphous matrix with crystalline zones of cubic or hexagonal morphologies. Elemental analysis of those crystals revealed the presence of B and Al as major components; and Ti, F, Ce, and O as minor components, accordingly with the additive. An important observation is that the atomic composition of Al and B in the crystals of all the tested materials was not consistent. The elemental analysis was performed in small areas on the crystals, and the results range from roughly the expected 1:2 atomic ratio of AlB_2 to 1:12 atomic ratio of AlB_{12}. The loss of available B for the re-hydrogenation reaction is a common drawback for all borohydrides and their mixtures. B can be lost as borane compounds (BH_3, B_2H_6, etc.), form B-clusters or other boranes that are unreactive at moderate pressures and temperature for further re-hydrogenation reaction. Elemental analysis during SEM data collection indicates good B retention with the dehydrogenated samples.

Due to the detection limit of elemental analysis with the SEM microscope, Li-compounds were not located. The SEM images of the materials with CeO_2 deserve some attention; the as-milled materials presented clear CeO_2 bright spots (Figure 1c); meanwhile, the dehydrogenated materials (Supplementary Materials) presented a more homogeneous distribution of Ce along the sample image. Another interesting point was a tendency to formation of crystals with higher dehydrogenation pressure. A greater quantity of crystals was observed at samples dehydrogenated at 5 bar initial pressure than at 3 bar initial pressure.

Figure 5. SEM images of the dehydrogenated (DHH) materials: (**a**) 2LiBH$_4$ + Al-DHH 5 bar, 400 °C; (**b**) 2LiBH$_4$ + Al/TiF$_3$-DHH 5 bar, 400 °C; (**c**) 2LiBH$_4$ + Al/CeO$_2$-DHH 5 bar, 400 °C.

Powder X-ray diffraction patterns of the dehydrogenated (labeled as DHH) materials are presented in Figure 6. The diffractograms of the dehydrogenated materials at 3, and 5 bar are presented in Frame (a,b) respectively. In general, the exact determination of the reaction extent based purely on the X-ray diffraction (XRD) results is difficult: this is due to Al and LiH sharing the same crystal symmetry and close crystal cell dimensions. Thus the peaks of the reactant and the reaction product overlap, particularly at low diffraction angles. Additionally, the (101) peak of AlB$_2$ (P6MMM) is pretty close to the Al (200) and LiH (200) peak, these peaks also overlap. The rest of the AlB$_2$ peaks are appreciable when zooming the plot with data-processing software; otherwise, these peaks appear rather small. All the dehydrogenated materials presented minor peaks of Li$_2$O and LiOH. The dehydrogenated 2LiBH$_4$ + Al and 2LiBH$_4$ + Al/TiF$_3$ presented Al$_2$O$_3$. The oxides and hydroxide could evolve/crystalize from the initial surface oxides present in the as-milled materials and the hydrogen atmosphere during heating. The dehydrogenated 3 bar- and 5 bar-2LiBH$_4$ + Al/CeO$_2$ materials are very interesting regarding the oxides content; they no longer present peaks of CeO$_2$ after the dehydrogenation reaction. CeO$_2$ reacted to form CeB$_6$ (COD-5910033).

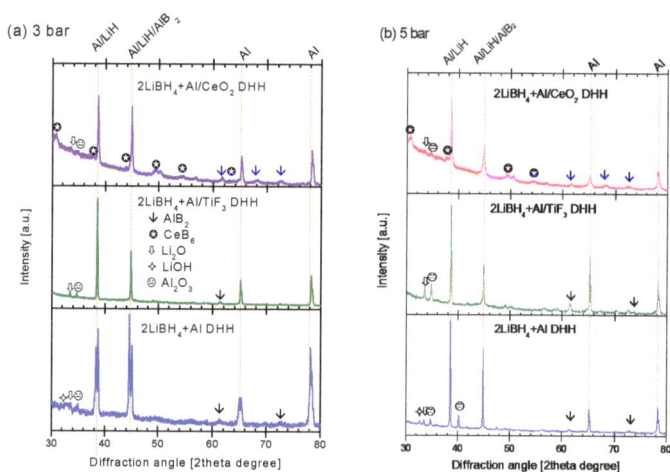

Figure 6. X-ray diffraction patterns of dehydrogenated materials. Frame (**a**) after dehydrogenation at 3 bar initial hydrogen pressure; Frame (**b**) after dehydrogenation at 5 bar initial hydrogen pressure.

FT-IR spectra of all dehydrogenated products are presented in Figure 7. This figure confirms the incomplete dehydrogenation of the materials after reactions. The dehydrogenated materials presented differences in the H–B–H bending region compared to the as-milled materials. First, a shift of the peaks to higher wavenumbers was observed, from 1100–1200 cm^{-1} in as-milled materials to 1400–1560 cm^{-1} in dehydrogenated materials. Secondly, the intensity of these peaks was different with the additive; the 2LiBH$_4$ + Al material presented the highest peak intensity meanwhile the materials added with TiF$_3$ and CeO$_2$ presented reduced signal intensity. The B–H stretching region (2000–2500 cm^{-1}) also presented shifts in wavenumber, however, this shift was not as marked as the H–B–H region. It amounts to approximately 75 cm^{-1}. The materials dehydrogenated at 5 bar presented more intense H–B–H peaks than the materials dehydrogenated at 3 bar. This result points to a different reaction mechanism.

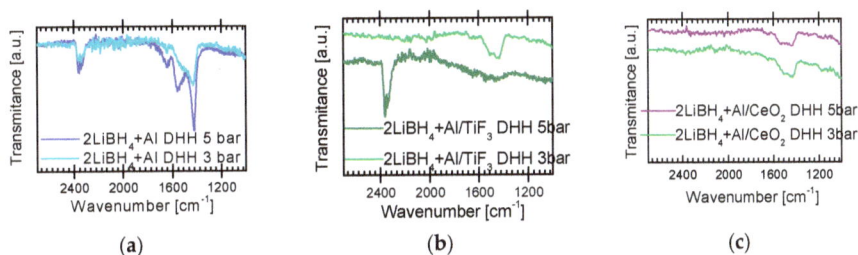

Figure 7. Fourier Transformed Infrared Spectroscopy (FT-IR) of dehydrogenated (DHH) samples: (a) 2LiBH$_4$ + Al-DHH; (b) 2LiBH$_4$ + Al/TiF$_3$; (c) 2LiBH$_4$ + Al/CeO$_2$-DHH.

3. Discussion

Milling effects. A proper integration of LiBH$_4$ and Al in the mixtures was observed after the ball-milling. The milling of Al is not an easy process. Al has the tendency of sintering instead of dispersing unless using proper milling conditions [34]. The milling conditions used here represent a good balance between the optimum milling of Al and reduced degradation of LiBH$_4$.

Protective atmosphere effects. Customarily; the storage, production, processing, and handling of hydrogen storage materials are performed with a high purity argon atmosphere, with oxygen and moisture levels below 0.1 ppm. This would raise significantly the cost of the hydrogen storage materials if commercialization is intended. In our experiments the argon purity was not so high, the supplier-guaranteed oxygen and moisture levels were 10 ppm. This allowed surface oxidation as demonstrated by means of XPS experiments. Kato et al. [27] demonstrated that the surface oxidation of LiBH$_4$ produces Li surface segregation, the formation of Li$_2$O over LiBH$_4$, reduction of diborane desorption, and enhancement of the rate of hydrogen desorption. XPS suggests the formation of LiBO$_2$ rather than Li$_2$O on the surface of our samples and the segregation of Al to sub-surface layers. As a result of the surface oxidation, a reduction of the dehydrogenation temperature was observed here and compared with similar materials carefully protected against surface oxidation [17–22]. A reduction of the dehydrogenation activation barrier was proposed by Kang et al. [35] if LiBH$_4$ or the intermediate LiBH donate one electron (each), to a catalyst on their surfaces. Oxygen is well-known as an electron acceptor. Thus surface oxidation could reduce the activation barrier and help to reach thermodynamic-predicted temperatures for the dehydrogenation reaction.

Dehydrogenation kinetics. The dehydrogenations of the 2LiBH$_4$ + Al, 2LiBH$_4$ + Al/TiF$_3$, and 2LiBH$_4$ + Al/CeO$_2$ materials are multi-step reactions. The multistep nature of the dehydrogenation reaction of the 2LiBH$_4$ + Al is shared with the dehydrogenation of LiBH$_4$ [36] and the RHC LiBH$_4$ + MgH$_2$ [22,37]. Reports of LiBH$_4$-Al dehydrogenation in several molar proportions also described a multistep mechanism for the dehydrogenation reaction [17,19–21,38]. In the materials presented here, the first dehydrogenation step occurred at low temperature, i.e., 100–110 °C. And the main dehydrogenation step occurred between 200–300 °C, finishing at 400 °C. The main dehydrogenation temperature interval is close to the temperature predicted by Siegel et al. for the dehydrogenation reaction of $2LiBH_4 + Al$, 277 °C [18]. A reduction of activation barrier can be responsible for the reduced dehydrogenation temperatures, as pointed out above. Another good point concerning all the studied materials is that the dehydrogenation is rapid, completed within 1 h, mainly during the heating period.

The effect of dehydrogenation pressure. It must be pointed out that this is the first report of dehydrogenation of $2LiBH_4 + Al$/additive mixtures at non-vacuum pressures, i.e., at fuel cell compatible working pressures. Thus the possibility of using these mixtures has been demonstrated for disposable or non-rechargeable hydrogen storage applications as long as further re-hydrogenation can be proved. In general, our materials dehydrogenated at 3 bar presented a better hydrogen release quantity. Further research will be conducted to prove if re-hydrogenation is possible.

The effect of the additives. TiF_3 is a well-known additive in the hydrogen storage area, it is almost mandatory to try TiF_3 as an additive in hydrogen storage systems. Meanwhile, CeO_2 is not commonly used as an additive. CeO_2 has been anticipated to interact with hydrogen [25], and a revision of its effects as additive deserve attention. However, few examples have been published. Ceria could uptake small amounts of hydrogen below 391 °C [39]. Even more, Lin et al. [39] demonstrated an interfacial effect of $CeH_{2.73}/CeO_2$ functioning as "hydrogen pump" and reducing the hydrogen desorption temperature of MgH_2–Mg_2NiH_4–$CeH_{2.73}/CeO_2$. In that work [39], $CeH_{2.73}$ and CeO_2 were formed during successive hydrogenation and oxidation reactions. $LiBH_4$ or $LiBH_4 + 1/2MgH_2$ were mixed with $CeCl_3$, CeF_3, or CeH_2 to improve the hydrogenation/dehydrogenation kinetics, reversibility, or to reduce dehydrogenation temperature [40–44]. In the present work, CeO_2 demonstrated its effectiveness reducing the dehydrogenation temperature. The characterization of dehydrogenated materials by powder X-ray diffraction demonstrated the formation of CeB_6. This boride has been observed after dehydrogenation of $LiBH_4$ or $LiBH_4 + 1/2MgH_2$ destabilized with $CeCl_3$, CeF_3, or CeH_2 [40–44]. On the other hand, X-ray diffraction did not show additional formation of Li or Al-oxides by liberation of the oxygen of CeO_2. The expected formation of oxides is low: for example, a mass balance indicates that if all the oxygen of CeO_2 forms Li_2O, the lithium oxide would be 1.7 wt %. To conclude this part, CeO_2 additive was effective in reducing the dehydrogenation temperature and producing good hydrogen release.

4. Materials and Methods

4.1. Sample Preparation

All reactives were purchased from Sigma-Aldrich and used without further purification. The Al was granular, with a particle size roughly of 1 mm and 99.7% purity. The $LiBH_4$ purity was ≥95%, meanwhile, the purity of TiF_3 and CeO_2 were 99% and 99.995% respectively. The molar ratio of $LiBH_4$ and Al was 2:1. The amount of the additive in each sample was 5 wt %. The mixtures of $2LiBH_4 + Al$ + additives were produced by mechanical milling. The milling was performed in batches of 1 gram of mixture as needed for performing reactions or characterization. The milling was performed in a planetary mill (Across-International) with a rotation of the main plate of 2400 rpm. The milling vials were machinated in stainless steel 316 L with an internal volume of 100 mL, with bolted lids. The milling balls were of yttrium-stabilized zirconium oxide (1 cm diameter). The powder to ball ratio was 1:15. The total milling time was 5 h divided into periods of 1 h milling and 10 min resting. In each cycle of milling-pause, the rotation direction of the planetary mill was inverted. The handling and storage of materials were performed inside a glove box filled with welding grade argon.

4.2. Characterization of the Ball-Milled and Dehydrogenated Materials

Scanning electron microscopy (SEM) images were obtained in a JSM-IT300 microscope (JEOL, Tokio, Japan). Samples were dispersed on carbon tape over a Cu sample holder. The SEM samples were prepared inside the argon glove box and transferred to the microscope by means of a glove bag to avoid oxidation; however, slight oxidation could have been possible. SEM images were obtained by backscattered or secondary electrons and 10 kV or 20 kV of acceleration voltage, accordingly to each sample characteristics.

Powder X-ray diffraction (PXRD) characterization was performed in a BrukerB8 diffractometer (Cu Kα = 1.540598 Å, Bruker AXS, Karlsruhe, Germany). The powders of as-milled and dehydrogenated materials were compacted in a dedicated sample-holder, then they were covered with a Kapton foil for protection against ambient oxygen and moisture. Data processing and phase identification were performed with Diffract Suite Eva (Version 4.2.1.10, Bruker AXS, Karlsruhe, Germany, 2016) or MAUD software (Version 2.55, Trento University, Trento, Italy, 2015). ICSD (Inorganic Crystal Structure Database, Karlsruhe) or COD (Crystallography Open Database) databases were used for phase identification.

Fourier transformed infrared spectroscopy characterization was performed in a Varian 640-IR, FT-IR Spectrometer (Agilent Technologies, Santa Clara, CA, USA). The studied materials were compacted in KBr pellets. The KBr was purchased from Sigma-Aldrich and dried just before pellet preparation. About 2.5 mg of each material was dispersed in 50 mg of dry KBr. FT-IR data was collected in attenuated total reflection (ATR) mode.

X-ray photoelectron spectroscopy analyses were performed only on the as-milled materials. XPS experiments were accomplished in an ultra-high vacuum (UHV) system Scanning XPS microprobe PHI 5000 Versa Probe II (Physical Electronics, Minneapolis, United States of America), with an Al Kα X-ray source (photon energy of 1486.6 eV) monochromatic at 25.4 W, and an multichannel detector. The surface of the samples was etched for 5 min with 1 kV Ar$^+$ at 55.56 nA·mm^{-2}. The XPS spectra were obtained at 45° to the normal surface in the constant pass energy mode, E_0 = 117.40 and 11.75 eV for survey surface and high-resolution narrow scan, respectively. The peak positions were referenced to the background silver 3d$_{5/2}$ photo-peak at 368.2 eV, having an FWHM of 0.56 eV, and C 1s hydrocarbon groups at 285.0 eV, Au 4f$_{7/2}$ in 84.0 eV central peak core level position. XPS characterization was performed on the as-milled materials and raw materials NaBH$_4$ and Al as necessary. Powder samples were compacted in an adequate sample holder and transferred to the equipment by means of a glove bag.

4.3. Dehydrogenation Reaction

Dehydrogenations of 2LiBH$_4$ + Al + additive materials were performed in a Sievert's-type reactor. This reactor was designated and constructed by the research group. It consists basically of twins of a sample holder and a reference holder, a well-known-volume reservoir of H$_2$ for sample and a well-known-volume reservoir of H$_2$ for reference, high precision pressure transducers for sample and reference, and delicate control of the reservoirs, reference-holder and sample-holder temperatures. The registered temperatures and pressures versus time were converted to hydrogen release in wt % with the following formula [45]:

$$\text{wt \%}(H_2) = 100 \times \frac{M(H_2) \times \Delta p \times V_{sample}}{m \times R \times T_{sample} \times Z_{fact}} + 100 \times \frac{M(H_2) \times \Delta p \times V_{reservoir}}{m \times R \times T_{reservoir} \times Z_{fact}}, \tag{2}$$

where $M(H_2)$ is the hydrogen molar mass (2.01588 g·mol^{-1}). $\Delta p = \Delta p_{sample} - \Delta p_{reference}$ is in bar, where Δp_{sample} and $\Delta p_{reference}$ mean the actual pressure minus the initial pressure of sample and reference. At zero-time both sample and reference initial pressures are equal. This performs as a differential pressure transducer and helps reducing small (1 × 10^{-3} bar) variations of the pressures caused by thermal effects. The V and T are the volume and temperature of the reservoir and sample holder, in cm^3 and Kelvin degrees respectively; m is the sample mass in g; R is the gas constant (83.14459 cm^3·bar·K^{-1}·mol^{-1}) and, Z_{fact} is the hydrogen compressibility factor [46,47]. It is necessary to mention that the sample holder and reference holder volume account for less than 1% of the reservoirs volume, meeting the appropriate conditions for hydrogen sorption/desorption and Sieverts law.

Samples were transferred to/from the Sieverts-type reactor without oxygen contact by means of a closing valve at the sample holder. Dehydrogenations were performed by a temperature-controlled process. The dehydrogenation initial pressure was fixed manually at 3 or 5 bar, the reservoir temperature was fixed at 40 °C. The sample temperature was raised from room temperature to 350 °C or 400 °C with a heating rate of 5 or 6 °C/min. The total dehydrogenation time was 3 h. Then, the system was cooled down and the remaining hydrogen was released. The dehydrogenation reaction was marked by a significant and sudden increase of the registered pressure beyond the temperature effects. The gases, hydrogen and argon, used during the dehydrogenation experiments were of chromatographic and high purity grade.

5. Conclusions

The 2LiBH4 + Al/additives mixtures were prepared by ball milling. The milling conditions were optimized for the integration of Al and preservation of $2LiBH_4$. The use of additives TiF_3 and CeO_2 produced different morphologies with as-milled materials. The studied materials presented a significant reduction of the dehydrogenation temperature that can be related to the surface-oxidation. The surface oxidation was the result of the use of welding-grade argon. The dehydrogenation reactions were observed to take place in two main steps, with onsets at 100 °C and 200–300 °C. The maximum released hydrogen was 9.3 wt % in the $2LiBH_4$ + Al/TiF_3 material, and 7.9 wt % in the $2LiBH_4$ + Al/CeO_2 material. Formation of CeB_6 after dehydrogenation of $2LiBH_4$ + Al/CeO_2 was confirmed.

Supplementary Materials: The following are available online at www.mdpi.com/2304-6740/5/4/82/s1, Figure S1: SEM image of $LiBH_4$ (not ball-milled), Figure S2: SEM image of Al (not ball-milled), Figure S3: SEM of as-milled $2LiBH_4$ + Al material, Figure S4: SEM of as-milled $2LiBH_4$ + Al/TiF_3 material, Figure S5: SEM of as-milled $2LiBH_4$ + Al/CeO_2 material, Figure S6: FT-IR of the as-milled $2LiBH_4$ + Al, $2LiBH_4$ + Al/TiF_3 and $2LiBH_4/CeO_2$, Figure S7: Al 1s XPS of the as-milled $2LiBH_4$ + Al, $2LiBH_4$ + Al/TiF_3 and $2LiBH_4/CeO_2$, Figure S8: SEM of dehydrogenated $2LiBH_4$ + Al at 5 bar 400 °C, Figure S9: SEM of dehydrogenated $2LiBH_4$ + Al at 3 bar 400 °C, Figure S10: SEM and EDS of dehydrogenated $2LiBH_4$ + Al/TiF_3 at 5 bar 400 °C, Figure S11: SEM and EDS of dehydrogenated $2LiBH_4$ + Al/TiF_3 at 3 bar 400 °C, Figure S12: SEM of dehydrogenated $2LiBH_4$ + Al/CeO_2 DHH 5 bar 400 °C, Figure S13: SEM of dehydrogenated $2LiBH_4$ + Al/CeO_2 DHH 3 bar 400 °C, Figure S14: Dehydrogenation of $2LiBH_4$ + Al/CeO_2 (3 bar and 5 bar, 350 °C), Figure S15: SEM of dehydrogenated $2LiBH_4$ + Al/CeO_2 DHH 5 bar 350 °C, Figure S16: EDS of dehydrogenated $2LiBH_4$ + Al/CeO_2 DHH 5 bar 350 °C.

Acknowledgments: The present work was supported by SENER-CONACyT Project 215362 "Investigación en mezclas reactivas de hidruro: nanomateriales para almacenamiento de hidrógeno como vector energético". Authors want to thank Omar Solorza-Feria and CINVESTAV-DF for facilitating FT-IR characterization. Authors would also like to thank Orlando Hernández for SEM characterization. XRD was performed at Universidad Autónoma Metropolitana-Iztapalapa, we thank Federico González García. Authors would like to thank Sandra Rodil and Lázaro Huerta Arcos for XPS characterization at IIM-UNAM.

Author Contributions: The findings in this manuscript are part of Juan Luis Carrillo-Bucio master's degree work. Juan Rogelio Tena-García performed the XRD experiments and interpretation. The manuscript was written with contributions of all authors. All authors have given approval to the final version of the manuscript.

Conflicts of Interest: The authors declare no conflict of interest.

References

1. Züttel, A.; Wenger, P.; Rentsch, S.; Sudan, P.; Mauron, P.; Emmenegger, C. LiBH4 a new hydrogen storage material. *J. Power Sources* **2003**, *118*, 1–7. [CrossRef]
2. Lee, J.Y.; Ravnsbæk, D.; Lee, Y.S.; Kim, Y.; Cerenius, Y.; Shim, H.J.; Jensen, T.R.; Hur, N.H.; Cho, Y.W. Decomposition reactions and reversibility of the LiBH4–Ca(BH4)2 composite. *J. Phys. Chem. C* **2009**, *113*, 15080–15086. [CrossRef]
3. Liu, Y.; Reed, D.; Paterakis, C.; Contreras Vasquez, L.; Baricco, M.; Book, D. Study of the decomposition of a 0.62LiBH4–0.38NaBH4 mixture. *Int. J. Hydrog. Energy* **2017**, *42*, 22480–22488. [CrossRef]

4. Gil-Bardají, E.; Zhao-Karger, Z.; Boucharat, N.; Nale, A.; van Setten, M.J.; Lohstroh, W.; Reohm, E.; Catti, M.; Fichtner, M. LiBH₄–Mg(BH₄)₂: A physical mixture of metal borohydrides as hydrogen storage material. *J. Phys. Chem. C* **2011**, *115*, 6095–6101. [CrossRef]

5. Aoki, M.; Miwa, K.; Noritake, T.; Kitahara, G.; Nakamori, Y.; Orimo, S.; Towata, S. Destabilization of LiBH₄ by mixing with LiNH₂. *Appl. Phys. A* **2005**, *80*, 1409–1412. [CrossRef]

6. Mao, J.F.; Guo, Z.P.; Liua, H.K.; Yu, X.B. Reversible hydrogen storage in titanium-catalyzed LiAlH₄–LiBH₄ system. *J. Alloys Compd.* **2009**, *487*, 434–438. [CrossRef]

7. Shi, Q.; Yu, X.; Feidenhans, R.; Vegge, T. Destabilized LiBH₄–NaAlH₄ mixtures doped with titanium based catalysts materials research. *J. Phys. Chem. C* **2008**, *112*, 18244–18248. [CrossRef]

8. Vajo, J.J.; Skeith, S.L.; Mertens, F. Reversible storage of hydrogen in destabilized LiBH₄. *J. Phys. Chem. B* **2005**, *109*, 3719–3722. [CrossRef] [PubMed]

9. Barkhordarian, G.; Klassen, T.; Dornheim, M.; Bormann, R. Unexpected kinetic effect of MgB₂ in reactive hydride composites containing complex borohydrides. *J. Alloys Compd.* **2007**, *440*, L18–L21. [CrossRef]

10. Bösenberg, U.; Doppiu, S.; Mosegaard, L.; Barkhordarian, G.; Eigen, N.; Borgschulte, A.; Jensen, T.R.; Cerenius, Y.; Gutfleisch, O.; Klassen, T.; et al. Hydrogen sorption properties of MgH₂–LiBH₄ composites. *Acta Mater.* **2007**, *55*, 3951–3958. [CrossRef]

11. Pinkerton, F.E.; Meyer, M.S. Reversible hydrogen storage in the lithium borohydride—Calcium hydride coupled system. *J. Alloys Compd.* **2008**, *464*, L1–L4. [CrossRef]

12. Yang, J.; Sudik, A.; Wolverton, C. Destabilizing LiBH₄ with a Metal (M = Mg, Al, Ti, V, Cr, or Sc) or Metal Hydride (MH₂ = MgH₂, TiH₂, or CaH₂). *J. Phys. Chem. C* **2007**, *111*, 19134–19140. [CrossRef]

13. Li, Y.; Li, P.; Qu, X. Investigation on LiBH₄–CaH₂ composite and its potential for thermal energy storage. *Sci. Rep.* **2017**, *7*, 41754–41758. [CrossRef] [PubMed]

14. Arnbjerg, L.M.; Ravnsbæk, D.B.; Filinchuk, Y.; Vang, R.T.; Cerenius, Y.; Besenbacher, F.; Jørgensen, J.E.; Jakobsen, H.J.; Jensen, T.R. Structure and dynamics for LiBH₄–LiCl solid solutions. *Chem. Mater.* **2009**, *21*, 5772–5782. [CrossRef]

15. Yu, X.B.; Grant, D.M.; Walker, G.S. Dehydrogenation of LiBH₄ destabilized with various oxides. *J. Phys. Chem. C* **2009**, *113*, 17945–17949. [CrossRef]

16. Lee, H.S.; Hwang, S.J.; To, M.; Lee, Y.S.; Whan Cho, Y. Discovery of fluidic LiBH₄ on scaffold surfaces and its application for fast co-confinement of LiBH₄–Ca(BH₄)₂ into mesopores. *J. Phys. Chem. C* **2015**, *119*, 9025–9035. [CrossRef]

17. Hansen, B.R.S.; Ravnsbæk, D.B.; Reed, D.; Book, D.; Gundlach, C.; Skibsted, J.; Jensen, T.R. Hydrogen storage capacity loss in a LiBH₄–Al composite. *J. Phys. Chem. C* **2013**, *117*, 7423–7432. [CrossRef]

18. Siegel, D.J.; Wolverton, C.; Ozoliņš, V. Thermodynamic guidelines for the prediction of hydrogen storage reactions and their application to destabilized hydride mixtures. *Phys. Rev. B* **2007**, *76*, 134102–134106. [CrossRef]

19. Zhang, Y.; Tian, Q.; Zhang, J.; Liu, S.S.; Sun, L.X. The dehydrogenation reactions and kinetics of 2LiBH₄–Al composite. *J. Phys. Chem. C* **2009**, *113*, 18424–18430. [CrossRef]

20. Kang, X.D.; Wang, P.; Ma, L.P.; Cheng, H.M. Reversible hydrogen storage in LiBH₄ destabilized by milling with Al. *Appl. Phys. A* **2007**, *89*, 963–966. [CrossRef]

21. Ravnsbæk, D.B.; Jensen, T.R. Mechanism for reversible hydrogen storage in LiBH₄–Al. *J. Appl. Phys.* **2012**, *111*, 112621–112628. [CrossRef]

22. Bosenberg, U.; Ravnsbæk, D.B.; Hagemann, H.; D'Anna, V.; Bonatto-Minella, C.; Pistidda, C.; van Beek, W.; Jensen, T.R.; Bormann, R.; Dornheim, M. Pressure and temperature influence on the desorption pathway of the LiBH₄–MgH₂ composite system. *J. Phys. Chem. C* **2010**, *114*, 15212–15217. [CrossRef]

23. Saldan, I.; Llamas-Jansa, I.; Hino, S.; Frommen, C.; Hauback, B.C. Synthesis and thermal decomposition of Mg(BH₄)₂–TMO (TMO=TiO₂; ZrO₂; Nb₂O₅; MoO₃) composites. *IOP Conf. Ser. Mater. Sci. Eng.* **2015**, *77*, 012041–012047. [CrossRef]

24. Saldan, I.; Frommen, C.; Llamas-Jansa, I.; Kalantzopoulos, G.N.; Hino, S.; Arstad, B.; Heyn, R.; Zavorotynska, O.; Deledda, S.; Sørby, M.H.; et al. Hydrogen storage properties of γ-Mg(BH₄)₂ modified by MoO₃ and TiO₂. *Int. J. Hydrog. Energy* **2015**, *40*, 12286–12293. [CrossRef]

25. Sohlberg, K.; Pantelides, S.T.; Pennycook, S.J. Interactions of hydrogen with CeO₂. *J. Am. Chem. Soc.* **2001**, *123*, 6609–6611. [CrossRef] [PubMed]

26. Suárez-Alcántara, K.; Palacios-Lazcano, A.F.; Funatsu, T.; Cabañas-Moreno, J.G. Hydriding and dehydriding in air-exposed Mg–Fe powder mixtures. *Int. J. Hydrog. Energy* **2016**, *41*, 23380–23387. [CrossRef]

27. Kato, S.; Bielmann, M.; Borgschulte, A.; Zakaznova-Herzog, V.; Remhof, A.; Orimo, S.I.; Zuttel, A. Effect of the surface oxidation of LiBH$_4$ on the hydrogen desorption mechanism. *Phys. Chem. Chem. Phys.* **2010**, *12*, 10950–10955. [CrossRef] [PubMed]

28. JANAF Thermochemical Tables. Available online: http://kinetics.nist.gov/janaf/ (accessed on 15 July 2017).

29. Fang, Z.Z.; Ma, L.P.; Kang, D.; Wang, P.J.; Wang, P.; Cheng, H.M. In situ formation and rapid decomposition of Ti(BH$_4$)$_3$ by mechanical milling LiBH$_4$ with TiF$_3$. *Appl. Phys. Lett.* **2009**, *94*. [CrossRef]

30. D'Anna, V.; Spyratou, A.; Sharma, M.; Hagemann, H. FT-IR spectra of inorganic borohydrides. *Spectrochim. Acta Part A Mol. Biomol. Spectrosc.* **2014**, *128*, 902–906. [CrossRef] [PubMed]

31. Xiong, Z.; Cao, L.; Wang, J.; Mao, J. Hydrolysis behavior of LiBH$_4$ films. *J. Alloys Compd.* **2017**, *698*, 495–500. [CrossRef]

32. Haeberle, J.; Henkel, K.; Gargouri, H.; Naumann, F.; Gruska, B.; Arens, M.; Tallarida, M.; Schmeißer, D. Ellipsometry and XPS comparative studies of thermal and plasma enhanced atomic layer deposited Al$_2$O$_3$-films. *J. Nanotechnol.* **2013**, *4*, 732–742. [CrossRef] [PubMed]

33. Van den Brand, J.; Sloof, W.G.; Terryn, H.; de Wit, J.H.W. Correlation between hydroxyl fraction and O/Al atomic ratio as determined from XPS spectra of aluminum oxide layers. *Surf. Interface Anal.* **2004**, *36*, 81–88. [CrossRef]

34. Ramezani, M.; Neitzert, T. Mechanical milling of aluminum powder using planetary ball milling process. *J. Achiev. Mater. Manuf. Eng.* **2012**, *55*, 790–798.

35. Kang, J.K.; Kim, S.Y.; Han, Y.S.; Muller, R.P.; Goddard, W.A., III. A candidate LiBH$_4$ for hydrogen storage: Crystals structures and reaction mechanisms of intermediate phases. *Appl. Phys. Lett.* **2005**, *87*. [CrossRef]

36. Züttel, A.; Rentsch, S.; Fischer, P.; Wenger, P.; Sudan, P.; Mauron, P.; Emmenegger, C. Hydrogen storage properties of LiBH4. *J. Alloys Compd.* **2003**, *356–357*, 515–520. [CrossRef]

37. Kim, K.B.; Shim, J.H.; Park, S.H.; Choi, I.S.; Oh, K.H.; Cho, Y.W. Dehydrogenation reaction pathway of the LiBH$_4$–MgH$_2$ composite under various pressure conditions. *J. Phys. Chem. C* **2015**, *119*, 9714–9720. [CrossRef]

38. Friedrichs, O.; Kim, J.W.; Remhof, A.; Buchter, F.; Borgschulte, A.; Wallacher, D.; Cho, Y.W.; Fichtner, M.; Oh, K.H.; Zuttel, A. The effect of Al on the hydrogen sorption mechanism of LiBH$_4$. *Phys. Chem. Chem. Phys.* **2009**, *11*, 1515–1520. [CrossRef] [PubMed]

39. Lin, H.J.; Tang, J.J.; Yu, Q.; Wang, H.; Ouyang, L.Z.; Zhao, Y.J.; Liu, J.W.; Wang, W.H.; Zu, M. Symbiotic CeH$_{2.73}$/CeO$_2$ catalyst: A novel hydrogen pump. *Nano Energy* **2014**, *9*, 80–87. [CrossRef]

40. Gennari, F.C.; Fernández Albanesi, L.; Puszkiel, J.A.; Arneodo Larochette, P. Reversible hydrogen storage from 6LiBH$_4$–MCl$_3$ (M = Ce, Gd) composites by in-situ formation of MH$_2$. *Int. J. Hydrog. Energy* **2011**, *36*, 563–570. [CrossRef]

41. Liu, B.H.; Zhang, B.J.; Jiang, Y. Hydrogen storage performance of LiBH$_4$ + 1/2MgH$_2$ composites improved by Ce-based additives. *Int. J. Hydrog. Energy* **2011**, *36*, 5418–5424. [CrossRef]

42. Jin, S.A.; Lee, Y.S.; Shim, J.H.; Cho, Y.W. Reversible hydrogen storage in LiBH$_4$–MH$_2$ (M = Ce, Ca) composites. *J. Phys. Chem. C* **2008**, *112*, 9520–9524. [CrossRef]

43. Gennari, F.C. Destabilization of LiBH$_4$ by MH$_2$ (M = Ce, La) for hydrogen storage: Nanostructural effects on the hydrogen sorption kinetics. *Int. J. Hydrog. Energy* **2011**, *36*, 15231–15238. [CrossRef]

44. Zhang, B.J.; Liu, B.H.; Li, Z.P. Destabilization of LiBH$_4$ by (Ce, La)(Cl, F)$_3$ for hydrogen storage. *J. Alloys Compd.* **2011**, *509*, 751–757. [CrossRef]

45. Peschke, M. Wasserstoffspeicherung in Reaktiven Hydrid-Kompositen Einfluss von fluorbasierten Additiven auf das Sorptionsverhalten des MgB$_2$–LiH-Systems. Diplomarbeit (Undergraduate Thesis), GKSS-Forschungszentrum Geesthacht GmbH Helmut-Schmidt-Universität/Universität der Bundeswehr, Hamburg, Germany, April 2009.

46. Züttel, A.; Borgschulte, A.; Schlapbach, L.; Chorkendorf, I.; Suda, S. Properties if hydrogen. In *Hydrogen as A Future Energy Carrier*; Züttel, A., Borgschulte, A., Schlapbach, L., Eds.; John Wiley & Sons: Wienheim, Germany, 2008; pp. 71–94. ISBN 978-3-527-30817-0.

47. Hemmes, H.; Driessen, A.; Driessen, R. Thermodynamic properties of hydrogen at pressures up to 1 Mbar and Temperatures between 100 and 1000 K. *J. Phys. C Solid State Phys.* **1986**, *19*, 3571–3585. [CrossRef]

inorganics

MDPI

Article

Unique Hydrogen Desorption Properties of LiAlH₄/h-BN Composites

Yuki Nakagawa *, Shigehito Isobe, Takao Ohki and Naoyuki Hashimoto

Graduate School of Engineering, Hokkaido University, N-13, W-8, Sapporo 060-8278, Japan;
isobe@eng.hokudai.ac.jp (S.I.); takao-ohki@eng.hokudai.ac.jp (T.O.); hasimoto@eng.hokudai.ac.jp (N.H.)
* Correspondence: y-nakagawa@eng.hokudai.ac.jp; Tel.: +81-11-706-8175

Received: 30 September 2017; Accepted: 23 October 2017; Published: 25 October 2017

Abstract: Hexagonal boron nitride (h-BN) is known as an effective additive to improve the hydrogen de/absorption properties of hydrogen storage materials consisting of light elements. Herein, we report the unique hydrogen desorption properties of LiAlH₄/h-BN composites, which were prepared by ball-milling. The desorption profiles of the composite indicated the decrease of melting temperature of LiAlH₄, the delay of desorption kinetics in the first step, and the enhancement of the kinetics in the second step, compared with milled LiAlH₄. Li₃AlH₆ was also formed in the composite after desorption in the first step, suggesting h-BN would have a catalytic effect on the desorption kinetics of Li₃AlH₆. Finally, the role of h-BN on the desorption process of LiAlH₄ was discussed by comparison with the desorption properties of LiAlH₄/X (X = graphite, LiCl and LiI) composites, suggesting the enhancement of Li ion mobility in the LiAlH₄/h-BN composite.

Keywords: alanate; h-BN; hydrogen storage; catalyst; Li ion mobility

---

1. Introduction

Hydrogen storage is a key technology for a future hydrogen energy society [1]. However, it is still challenging to develop high performance hydrogen storage materials with high hydrogen density, fast de/absorption kinetics, and high cycle stability under moderate temperature and pressure conditions [2,3]. LiAlH₄ is one of the most promising hydrogen storage materials because of its high hydrogen capacity and relatively low desorption temperature [4]. The hydrogen desorption process of LiAlH₄ is described as follows:

Melting:

LiAlH₄(s) → LiAlH₄(l) Endothermic (150–175 °C)
Decomposition in the first step:

3LiAlH₄(l) → Li₃AlH₆(s) + 2Al + 3H₂ Exothermic (150–200 °C, 5.3 mass % H₂)
Decomposition in the second step:

Li₃AlH₆(s) → 3LiH + Al + 3/2H₂ Endothermic (200–270 °C, 2.6 mass % H₂)

The decomposition in the first step is an exothermic reaction with a ΔH of -10 kJ·mol^{-1} H₂, indicating the reversibility of this step is believed to be thermodynamically difficult [4,5]. In the second step, Li₃AlH₆ decomposes in an endothermic reaction with a ΔH of 25 kJ·mol^{-1} H₂ [4]. Thus, the hydrogenation of LiH/Al to Li₃AlH₆ is thermodynamically possible.

One of the strategies for improving the properties of hydrogen storage materials is the addition of catalysts/dopants [6]. Ti or its compounds are well-known catalysts for the kinetics of alanate [7–9]. Since Bogdanović et al. reported an absence of hysteresis and nearly horizontal pressure plateaus in the TiCl₃-doped NaAlH₄ [7], many researchers have studied complex hydrides including alanate as potential reversible hydrogen storage materials. In the case of $LiAlH_4$, the improved desorption kinetics was reported by the doping of Ti catalyst using mechanically milling [10–12]. Recent study also reported a single step hydrogen release of $LiAlH_4$, which was induced by the synergetic effects of Ti catalytic coating and nanosizing effects [13]. Although the rehydrogenation of the desorbed material was not achieved in the milled sample, the regeneration of $LiAlH_4$ from LiH and Ti-catalyzed Al was possible through the solution synthesis approach using THF and Me_2O [14–16].

Hexagonal boron nitride (h-BN) is known as an effective additive for chemical hydride and complex hydride systems. For instance, NH_3BH_3/h-BN composite released hydrogen at low temperature with minimum induction time and less exothermicity [17]. The remarkable hydrogen de/absorption properties were also achieved in the milled $LiBH_4$/h-BN composites [18,19]. The 30 mol % h-BN doped $LiBH_4$ composite started to release hydrogen from 180 °C, which was 100 °C lower than the onset hydrogen desorption temperature of ball-milled $LiBH_4$ [18]. For the 75 mol % h-BN doped $LiBH_4$ composite, the on-set desorption temperature of $LiBH_4$ was reduced to 175 °C and the peak desorption temperature was reduced by 80 °C compared with milled $LiBH_4$ [19]. Furthermore, under moderate rehydrogenation conditions of 400 °C and 10 MPa H_2 pressure, the dehydrogenation capacity of the composite maintained 3.1 mass % within three cycles, which was very close to its theoretical capacity. It was assumed that the excellent rehydrogenation property of $LiBH_4$ would be related with the enhanced hydrogen and lithium diffusion capability by the nanoscale h-BN, which was synthesized by ball-milling of h-BN at 490 rpm for 20 h [19]. The enhancement of Li^+ and/or H^- diffusion by adding h-BN was firstly reported in the $LiNH_2$/LiH system [20]. Hydrogen was fully desorbed from the $LiNH_2$/LiH/h-BN composite in less than 7 h, whereas the $LiNH_2$/LiH composite desorbed hydrogen in several days. They proposed that h-BN is an efficient catalyst that improves Li^+ diffusion and hence the kinetics of the reaction between $LiNH_2$ and LiH [20]. The mobility of Li^+ ions between LiH and $LiNH_2$ was also enhanced by adding $LiTi_2O_4$ catalyst [21].

Thus, h-BN has attracted much attention as an effective additive to improve the hydrogen de/absorption kinetics of hydrogen storage materials, especially for complex hydrides. However, the addition of h-BN to the alanate system has rarely been reported. In the present study, $LiAlH_4$/h-BN composites were synthesized by planetary ball-milling and their hydrogen desorption properties were analyzed. Also, the desorption process of the composite was investigated by using XRD and FT-IR. Finally, the role of h-BN to the desorption properties of $LiAlH_4$ was discussed by comparison with those of $LiAlH_4$/X (X = graphite, LiCl and LiI) composites.

2. Results

Hydrogen desorption properties of $LiAlH_4$/h-BN composites were analyzed by using TG-DTA-MS. Figure 1 shows the DTA and MS (H_2, $m/z = 2$) profiles of the composites. As shown in Figure 1a, ball-milled $LiAlH_4$ (denoted as 0 mass % in Figure 1) started to melt around 150 °C followed by hydrogen desorption in two steps below 250 °C. In the case of $LiAlH_4$/h-BN composites, desorption profiles were clearly changed compared with $LiAlH_4$. First, the melting temperature (T_m) of $LiAlH_4$ was decreased by adding h-BN. For instance, in the 40 mass % h-BN composite, DTA peak value of T_m was 151 °C, which was 11 °C lower than that of milled $LiAlH_4$. Second, the hydrogen desorption temperature (T_d) in the first step was slightly increased by h-BN addition. As shown in Figure 1b, T_d in the first step became high value with the increasing amount of h-BN. Third, the desorption kinetics in the second step was improved by adding h-BN. As shown in Figure 1b, the desorption peak in the second step became sharp as the amount of h-BN increased up to 14 mass %. However, the peak shape became broad in the 40 mass % h-BN composite, suggesting the addition of too much amount of h-BN could have negative effect on improving the kinetics. Figure 2 shows the TG profiles of $LiAlH_4$/h-BN composites. The total mass loss from TG profile of ball-milled $LiAlH_4$ was 7.7 mass %, which was in good agreement with the theoretical hydrogen desorption amount of $LiAlH_4$ (7.9 mass %) [4]. The hydrogen mass loss of 6.9 mass %, 5.8 mass %, and 3.5 mass % were calculated from the profiles of 4 mass %, 14 mass %, and 40 mass % h-BN composites, respectively. It is noted that theoretical hydrogen desorption capacities of these composites were 7.6 mass %, 6.8 mass %, and 4.7 mass %, respectively, when only considering the hydrogen desorption from LiAlH$_4$. Thus, the experimental values of hydrogen desorption amounts were slightly lower than the theoretical values. This result could originate from the hydrogen desorption during ball-milling or the formation of new H-containing solid compound by the reaction between LiAlH$_4$ and h-BN.

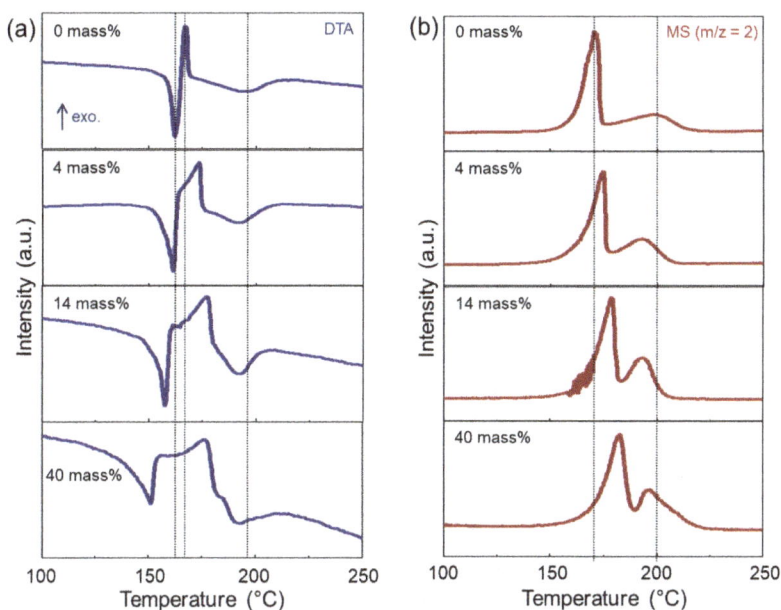

Figure 1. Hydrogen desorption profiles of LiAlH$_4$/x mass % h-BN (x = 0, 4, 14, 40) composites: (**a**) DTA and (**b**) MS (m/z = 2, H$_2$) profiles. Heating rate was 5 °C·min^{-1}.

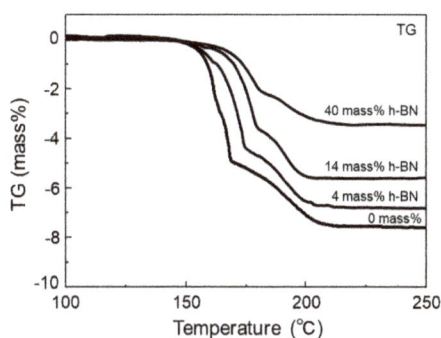

Figure 2. TG profiles of LiAlH$_4$/x mass % h-BN (x = 0, 4, 14, 40) composites. Heating rate was 5 °C·min^{-1}.

To investigate the interaction between LiAlH$_4$ and h-BN, XRD, and FT-IR measurements were performed for the LiAlH$_4$/h-BN composites. Also, the particle size of composite was observed by using SEM and TEM. Figure 3a shows the XRD profiles of the milled LiAlH$_4$ and LiAlH$_4$/h-BN composites. Only LiAlH$_4$ and h-BN phases were observed in the profiles of the composites. Although the diffraction peaks of LiAlH$_4$ were slightly broadened by h-BN addition, the clear relationship between the broadning and the amount of h-BN was not observed. Figure 3b,c shows the XRD profiles of the milled LiAlH$_4$ and 40 mass % h-BN composite after hydrogen desorption. The 40 mass % h-BN composite formed the similar reaction products compared with LiAlH$_4$. In other words, Li$_3$AlH$_6$ and Al were formed after the hydrogen desorption in the first step, and LiH and Al were formed after the second step. The phase of h-BN was also clearly observed after hydrogen desorption. Broad diffraction peaks around 20° and 27° originate from the polyimide film and grease to prevent the sample oxidation. Figure 4 shows SEM and TEM images of 40 mass % h-BN composite and references. As shown in Figure 4a, the milled LiAlH$_4$ contained a lot of large particles with sizes over 10 μm. On the other hand, the 40 mass % h-BN composite showed the average particle size of a few micrometers, indicating the refinement of LiAlH$_4$ particles occurred in the composite. The submicron particles were also observed in the TEM image of the composite, as shown in Figure 4d. The size of as-received h-BN particle was around 1 μm (Figure 4b). Figure 5 shows the FT-IR spectra of 40 mass % h-BN composite and references. The as-milled composite showed the characteristic Al–H vibrations of LiAlH$_4$ [22] around 1795 cm^{-1} and 1644 cm^{-1}. Also, B–N vibrations of h-BN around 1373 cm^{-1} and 818 cm^{-1} were observed. Although the Al–H vibrations of LiAlH$_4$ disappeared after heating, the B–N vibrations of h-BN still remained. These results were consistent with the results of XRD. However, new IR absorption peak was clearly observed around 2300 cm^{-1} after heating up to 183 °C and 300 °C. Also, another new peak appeared around 1100 cm^{-1} after heating up to 300 °C. Although these peaks were not identified in this work, the peak positions were similar to those of LiBH$_4$ [23], suggesting a such kind of Li–B–H phase exist after heating up to 183 °C and/or 300 °C. The unknown peak around 2300 cm^{-1} was also observed for the IR spectra of the composite consisting of BN^{nano}H$_x$ (ball-milled h-BN under 1.0 MPa H$_2$ for 80 h) and LiH [24]. Thus, the formation of new H-containing solid compound could result in the slightly low hydrogen desorption amount in Figure 2. Considering the results of Figure 5, the possible new compound could be covalently functionalized h-BN species. The details were explained in the next discussion part.

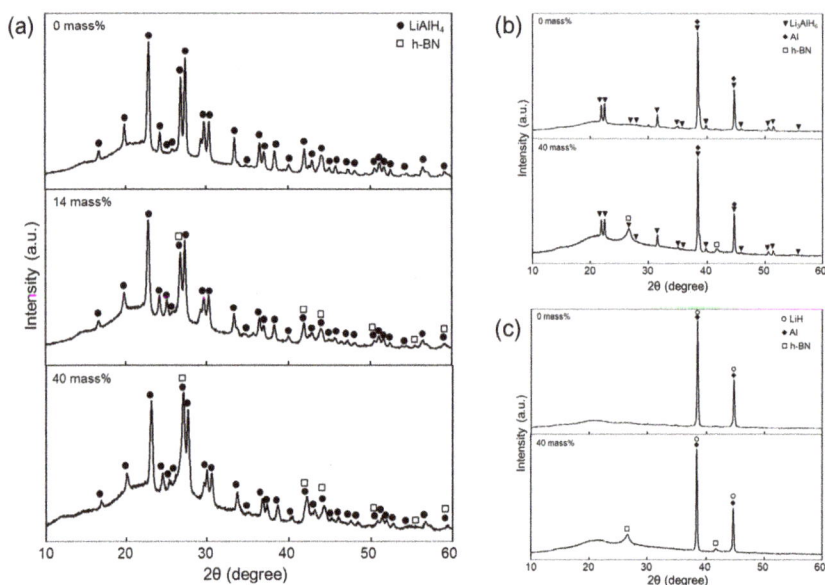

Figure 3. XRD profiles of LiAlH$_4$/x mass % h-BN (x = 0, 14, 40) composites: (**a**) after ball-milling; (**b**) after desorption in the first step; and (**c**) after desorption in the second step. The heating rate was 5 °C·min^{-1}.

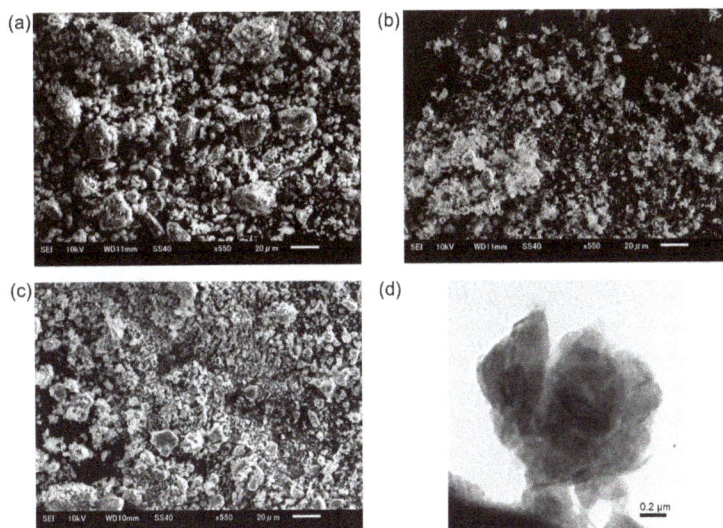

Figure 4. SEM images of (**a**) milled LiAlH$_4$; (**b**) h-BN; (**c**) LiAlH$_4$/40 mass % h-BN composite after milling; and (**d**) TEM image of LiAlH$_4$/40 mass % h-BN composite after milling.

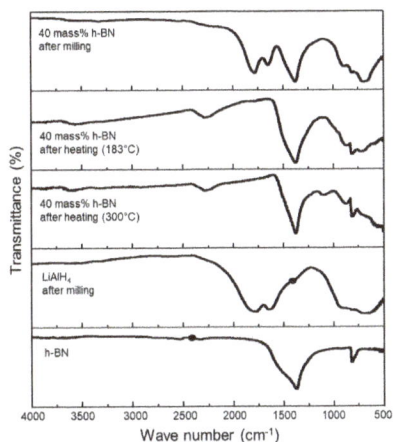

Figure 5. FT-IR spectra of LiAlH$_4$/40 mass % h-BN composite after milling and after hydrogen desorption. LiAlH$_4$ and h-BN spectra are shown as the references. Heating rate was 5 °C·min^{-1}.

As shown in the results of structural characterizaion, LiAlH$_4$/h-BN composites also formed Li$_3$AlH$_6$ as an intermediate product, indicating the similar decomposition pathway with LiAlH$_4$. Thus, h-BN would have the catalytic effect on the hydrogen desorption kinetics of Li$_3$AlH$_6$. The apparent activation energy for hydrogen desorption was calculated by using the Kissinger equation [25],

$$\ln \frac{c}{T_p^2} = -\frac{E_a}{RT_p} + \ln \frac{RA}{E_a}$$

where E_a is the apparent activation energy for hydrogen desorption, c is the heating rate, T_p is the peak temperature, R is gas constant, and A is the frequency factor. Figure 6 shows the Kissinger plots for the hydrogen desorption in the second step of 4 mass % h-BN composite. The obtained apparent activation energy, E_a was 71.5 kJ·mol^{-1}. This value is lower than that reported for Li$_3$AlH$_6$ (92 kJ·mol^{-1}) [26], indicating the desorption kinetics was improved by adding h-BN.

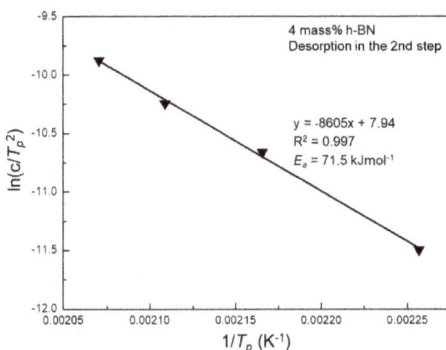

Figure 6. Kissinger plots for the hydrogen desorption in the second step of LiAlH$_4$/4 mass % h-BN composite. Heating rates were 2, 5, 8, and 12 °C·min^{-1}.

In order to understand the role of h-BN, other additives were also ball-milled with LiAlH$_4$ and their desorption properties were analyzed. The detailed results (Figure 7) and the possible role of h-BN on the desorption process were explained in the next discussion part.

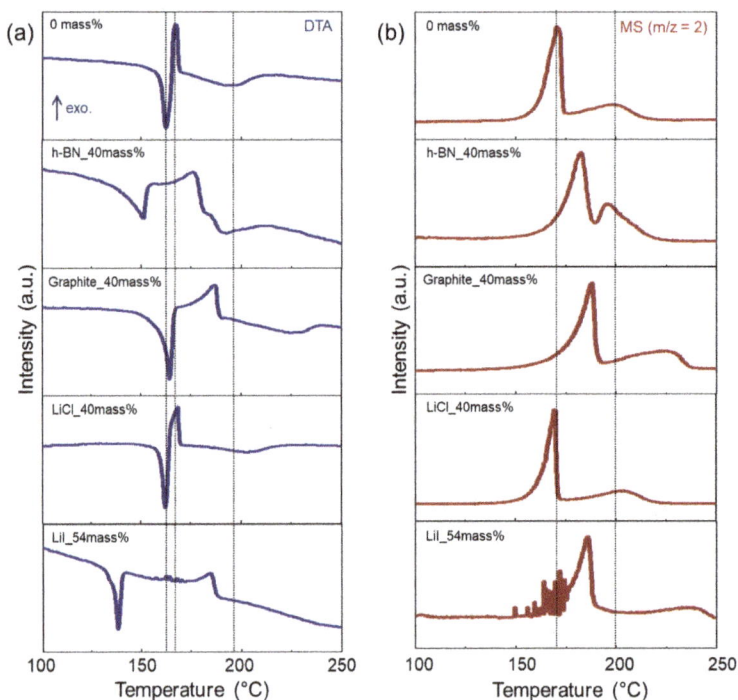

Figure 7. Hydrogen desorption profiles of LiAlH$_4$/X (X = graphite, LiCl, and LiI) composites. The profiles of milled LiAlH$_4$ and LiAlH$_4$/h-BN composites are shown as the references: (**a**) DTA and (**b**) MS (*m/z* = 2, H$_2$) profiles. Heating rate was 5 °C·min^{-1}.

3. Discussion

Figure 7 shows the DTA and MS profiles of LiAlH$_4$/X (X = graphite, LiCl, and LiI) composites. The profiles of LiAlH$_4$ and LiAlH$_4$/h-BN composites are also shown as the references. Graphite was selected as an additive because this compound has a structure similar to h-BN. In spite of its similar structure, hydrogen desorption properties were different from those of h-BN. In the case of LiAlH$_4$/graphite composite, the melting temperature was similar to ball-milled LiAlH$_4$. Also, the desorption kinetics in the second step seemed to be delayed, whereas that of LiAlH$_4$/h-BN was enhanced. Since the both graphite and h-BN are hard materials, the refinement of LiAlH$_4$ particles would occur during the ball-milling process. However, the different desorption profiles were obtained between h-BN and graphite composite, suggesting just the refinement of particles cannot explain this difference. Although the desorption properties of nanoconfined LiAlH$_4$ into graphite with high surface area was reported in the previous study [27], the profiles of graphite composite in this study seemed to be different from those profiles.

Also, LiCl and LiI were selected as additives. Aguey-Zinsou et al. reported that h-BN would enhance the Li ion mobility across the interface of LiNH$_2$ and LiH [20]. Thus, the enhancement of Li ion mobility could be one of the reasons for the unique hydrogen desorption properties of LiAlH$_4$/h-BN composites. Oguchi et al. reported the Li ion conductivity of Li$_3$AlH$_6$/LiI composite at 120 °C was much higher than that of Li$_3$AlH$_6$, but that of Li$_3$AlH$_6$/LiCl at 120 °C was the similar value compared with Li$_3$AlH$_6$ [28]. Their results suggest that LiI additive would be effective for increasing the Li ion mobility of LiAlH$_4$, but LiCl would not be effective near the decomposition temperature range of LiAlH$_4$. As shown in Figure 7, the desorption profiles of LiCl composite was similar to those of milled LiAlH$_4$. On the other hand, those of LiI composite showed the decrease of melting temperature and the delay of the first desorption reaction, which was consistent with the results of h-BN composite. This comparative result suggests that Li ion conductivity would increase in the LiAlH$_4$/h-BN composite. The kinetics in the second step wasdelayed in the LiI composite. The possible origin of the high conductivity of LiAlH$_4$ is the anion substitution from complex anion to I$^-$ [28], which partially took place in the case of LiBH$_4$ [29]. Thus, the high decomposition temperature in the second step could originate from the formation of stable solid solution similar to LiBH$_4$–LiI(LiCl) system [29–33]. For clarifying the total desorption process of these composites, the detailed mechanistic study is needed. The analysis of Li ion conductivity of LiAlH$_4$/h-BN composite is currently in progress.

As shown in Figure 1, all the LiAlH$_4$/h-BN composites showed the melting of LiAlH$_4$. This phenomenon was different from the case of LiAlH$_4$ catalyzed by transition metal (Ti, Fe, Co, Nb, etc. [34–36]), which can release hydrogen below the melting temperature. Thus, h-BN would have little interaction with complex anion of [AlH$_4$]$^-$, whereas transition metal like Ti would destabilize the covalent Al–H bond of complex anion as Sandrock et al. suggested [8]. According to the proposed mechanism by Atakli et al. [37], Li$_3$AlH$_6$ is formed by transferring the alkali cation and a hydrogen anion from the two neighboring alanate molecules to central one. In this context, the diffusion distance of Li$^+$ seems to be very short to form Li$_3$AlH$_6$, suggesting Li$^+$ diffusion would not be the rate-limiting step in the desorption of LiAlH$_4$ in the first step. On the contrary, the desorption properties of h-BN and LiI composite suggested the excess enhancement of Li$^+$ ion mobility might be related to delaying the formation of Li$_3$AlH$_6$. First-principle DFT studies also suggested the formation and migration of [AlH$_4$]$^-$ vacancy would be the rate-limiting step in the decomposition of LiAlH$_4$ [38]. In the second decomposition step, it was proposed that three LiH are formed from Li$_3$AlH$_6$, leaving AlH$_3$, which spontaneously desorbs hydrogen [37]. Thus, the enhanced mobility of Li$^+$ may help the destabilization of complex anions to improve the hydrogen desorption kinetics.

As shown in Figure 5, the presence of new bonds similar to those of LiBH$_4$ or ball-milled h-BN with LiH suggests that covalently functionalized h-BN species could be formed after heating the LiAlH$_4$/h-BN composites. It is known that ball-milling of h-BN with a lot of different materials can generate functionalized h-BN nanosheets [39]. For instance, a one-step method for the preparation and functionalization of few-layer BN was developed based on urea-assisted solid exfoliation of commercially available h-BN [40]. Such kind of functionalized BN nanosheets are attractive for a lot of applications such as polymer matrix composites [41], ion conductors [42], and hydrogen storage [43]. Further investigations of LiAlH$_4$/h-BN composites could pave the way for the covalent functionalization of h-BN nanosheets by interaction with LiAlH$_4$.

4. Materials and Methods

4.1. Synthesis of LiAlH$_4$/X Composites

All samples were handled in an Ar-filled glovebox with O$_2$ and H$_2$O levels below 2 ppm. LiAlH$_4$ (95%, Sigma-Aldrich, Tokyo, Japan), h-BN (98%, Sigma-Aldrich), graphite (99.99%, Kojundo Chemical Lab., Sakado, Japan), LiCl (99.99%, Sigma-Aldrich) and LiI (99.9%, Sigma-Aldrich) were used as starting materials. The LiAlH$_4$/X (X = h-BN, graphite, LiCl and LiI) composites were synthesized by using planetary ball-milling apparatus (Fritsch Pulverisette 7, Yokohama, Japan) with 20 stainless balls (7 mm in diameter) and 300 mg samples (ball: powder mass ratio = 70: 1). The milling pot was equipped with a quick connector for vacuuming and introducing H$_2$ gas. The milling was performed under 0.1 MPa H$_2$ atmosphere with 400 rpm for 2 h with four cycles of 30/15 min operation/interval per each cycle. Also, the milled LiAlH$_4$ was prepared under the same milling conditions for comparison.

4.2. Characterization

Hydrogen desorption properties of the composites were examined by a thermogravimetry and differential thermal analysis equipment (TG-DTA, Bruker, 2000SA, Yokohama, Japan) connected to a mass spectrometer (MS, ULVAC, BGM-102, Chitose, Japan). The desorbed gases were carried from TG-DTA to MS through a capillary by 300 mL·min^{-1} stream of high purity He as a carrier gas. The samples were heated from room temperature to 300 °C with a heating rate of 5 °C·min^{-1}. Structural properties were investigated by powder X-ray diffraction (XRD) measurements (Philips, X'Pert Pro with Cu Kα radiation, Amsterdam, The Netherlands), where all the samples were covered with a polyimide sheet (Kapton, The Nilaco Co., Ltd., Tokyo, Japan) in the glovebox to avoid oxidation during the measurement. Morphology of the composites was observed using scanning electron microscope (SEM, JEOL, JSM-6510LA, Tokyo, Japan) and transmission electron microscope (TEM, JEOL, JEM-2010, Tokyo, Japan). For TEM observations, the samples were dispersed on a molybdenum micro-mesh grid. Fourier transform infrared spectrometer (FT-IR, JASCO, FT/IR 660Plus, Tokyo, Japan) was operated to investigate chemical bonds in the composites. Each sample was put between KBr plates and pressed for measurement.

5. Conclusions

Hydrogen desorption properties of the ball-milled LiAlH$_4$/h-BN composites were investigated. Compared with milled LiAlH$_4$, the composites showed the different desorption profiles, where the decrease of melting temperature (T_m), the delay of desorption kinetics in the first step and the enhancement of the kinetics in the second step were observed. In the 40 mass % h-BN composite, the DTA peak value of T_m was 151 °C, which was 11 °C lower than that of milled LiAlH$_4$. The LiAlH$_4$/h-BN composite formed Li$_3$AlH$_6$ after desorption in the first step similar to LiAlH$_4$. Thus, h-BN would have a catalytic effect on the desorption kinetics of Li$_3$AlH$_6$. The apparent activation energy in the second step desorption was 71.5 kJ·mol^{-1} for the 4 mass % h-BN composite. From SEM and TEM observations, the refinement of LiAlH$_4$ particle was confirmed in the 40 mass % h-BN composite. The particle size of the composite was around a few micrometers. The hydrogen mass loss of the composite was slightly lower than the theoretical value. The new chemical bond similar to Li–B–H species was observed in the FT-IR spectra of the 40 mass % h-BN composite after the desorption. This result suggested covalently functionalized h-BN nanosheets could be formed in the composite. Finally, the desorption properties of LiAlH$_4$/h-BN composite were compared with those of

$LiAlH_4/X$ (X = graphite, LiCl, and LiI) composites, suggesting the enhancement of Li ion mobility in the $LiAlH_4/h$-BN composite compared with $LiAlH_4$. The present work first demonstrates the effect of h-BN addition on the hydrogen desorption properties of alanate.

Acknowledgments: A part of this study was conducted at "Joint-use facilities: Laboratory of Nano-Micro Material Analysis, Laboratory of XPS Analysis, and the Open Facility" at Hokkaido University, supported by the "Material Analysis and Structure Analysis Open Unit (MASAOU)" and "Nanotechnology Platform" program of the Ministry of Education, Culture, Sports, Science and Technology (MEXT), Japan.

Author Contributions: Yuki Nakagawa was involved in all stages of the work, including designing the work, conducting experiments, and analyzing the data; Shigehito Isobe acted as co-supervisor and was involved in the discussion of the results and work planning; Takao Ohki performed the TG-DTA-MS and XRD experiments; Naoyuki Hashimoto acted as the main supervisor and was involved in the discussion of the results; Yuki Nakagawa wrote the paper; and all the authors contributed to the revision of the paper.

Conflicts of Interest: The authors declare no conflict of interest.

References

1. Züttel, A. Materials for hydrogen storage. *Mater. Today* **2003**, *6*, 24–33. [CrossRef]
2. Lai, Q.; Paskevicius, M.; Sheppard, D.A.; Buckley, C.E.; Thornton, A.W.; Hill, M.R.; Gu, Q.; Mao, J.; Huang, Z.; Liu, H.K.; et al. Hydrogen storage materials for mobile and stationary applications: Current state of the art. *ChemSusChem* **2015**, *8*, 2789–2825. [CrossRef] [PubMed]
3. Ley, M.B.; Jensen, L.H.; Lee, Y.S.; Cho, Y.W.; Bellosta von Colbe, J.M.; Dornheim, M.; Rokni, M.; Jensen, J.O.; Sloth, M.; Filinchuk, Y.; et al. Complex hydrides for hydrogen storage—New perspectives. *Mater. Today* **2014**, *17*, 122–128. [CrossRef]
4. Orimo, S.; Nakamori, Y.; Eliseo, J.R.; Züttel, A.; Jensen, C.M. Complex hydrides for hydrogen storage. *Chem. Rev.* **2007**, *107*, 4111–4132. [CrossRef] [PubMed]
5. Jang, J.W.; Shim, J.H.; Cho, Y.W.; Lee, B.J. Thermodynamic calculation of LiH ↔ Li_3AlH_6 ↔ $LiAlH_4$ reactions. *J. Alloys Compd.* **2006**, *420*, 286–290. [CrossRef]
6. Wu, H. Strategies for the improvement of the hydrogen storage properties of metal hydride materials. *ChemPhysChem* **2008**, *9*, 2157–2162. [CrossRef] [PubMed]
7. Bogdanović, B.; Schwickardi, M. Ti-doped alkali metal aluminum hydrides as potential novel reversible hydrogen storage materials. *J. Alloys Compd.* **1997**, *253–254*, 1–9. [CrossRef]
8. Sandrock, G.; Gross, K.; Thomas, G. Effect of Ti-catalyst content on the reversible hydrogen storage properties of the sodium alanates. *J. Alloys Compd.* **2002**, *339*, 299–308. [CrossRef]
9. Gremaud, R.; Borgschulte, A.; Lohstroh, W.; Schreuders, H.; Züttel, A.; Dam, B.; Griessen, R. Ti-catalyzed $Mg(AlH_4)_2$: A reversible hydrogen storage material. *J. Alloys Compd.* **2005**, *404–406*, 775–778. [CrossRef]
10. Easton, D.S.; Schneibel, J.H.; Speakman, S.A. Factors affecting hydrogen release from lithium alanate $(LiAlH_4)$. *J. Alloys Compd.* **2005**, *398*, 245–248. [CrossRef]
11. Amama, P.B.; Grant, J.T.; Shamberger, P.J.; Voevodin, A.A.; Fisher, T.S. Improved dehydrogenation properties of Ti-doped $LiAlH_4$: Role of Ti precursors. *J. Phys. Chem. C* **2012**, *116*, 21886–21894. [CrossRef]
12. Isobe, S.; Ikarashi, Y.; Yao, H.; Hino, S.; Wang, Y.; Hashimoto, N.; Ohnuki, S. Additive effects of $TiCl_3$ on dehydrogenation reaction of $LiAlH_4$. *Mater. Trans.* **2014**, *55*, 1138–1140. [CrossRef]
13. Wang, L.; Aguey-Zinsou, K.F. Synthesis of $LiAlH_4$ nanoparticles leading to a single hydrogen release step upon Ti coating. *Inorganics* **2017**, *5*, 38. [CrossRef]
14. Liu, X.; McGrady, G.S.; Langmi, H.W.; Jensen, C.M. Facile cycling of Ti-doped $LiAlH_4$ for high performance hydrogen storage. *J. Am. Chem. Soc.* **2009**, *131*, 5032–5033. [CrossRef]
15. Graetz, J.; Wegrzyn, J.; Reilly, J.J. Regeneration of lithium aluminum hydride. *J. Am. Chem. Soc.* **2008**, *130*, 17790–17794. [CrossRef] [PubMed]
16. Wang, J.; Ebner, A.D.; Ritter, J.A. Physiochemical pathway for cyclic dehydrogenation and rehydrogenation of $LiAlH_4$. *J. Am. Chem. Soc.* **2006**, *128*, 5949–5954. [CrossRef] [PubMed]
17. Neiner, D.; Karkamkar, A.; Linehan, J.C.; Arey, B.; Autrey, T.; Kauzlarich, S.M. Promotion of hydrogen release from ammonia borane with mechanically activated hexagonal boron nitride. *J. Phys. Chem. C* **2009**, *113*, 1098–1103. [CrossRef]

18. Tu, G.; Xiao, X.; Qin, T.; Jiang, Y.; Li, S.; Ge, H.; Chen, L. Significantly improved de/rehydrogenation properties of lithium borohydride modified with hexagonal boron nitride. *RSC Adv.* **2015**, *5*, 51110–51115. [CrossRef]

19. Zhu, J.; Wang, H.; Cai, W.; Liu, J.; Ouyang, L.; Zhu, M. The milled LiBH$_4$/h-BN composites exhibiting unexpected hydrogen storage kinetics and reversibility. *Int. J. Hydrog. Energy* **2017**, *42*, 15790–15798. [CrossRef]

20. Aguey-Zinsou, K.F.; Yao, J.; Guo, Z.X. Reaction paths between LiNH$_2$ and LiH with effects of nitrides. *J. Phys. Chem. B* **2007**, *111*, 12531–12536. [CrossRef] [PubMed]

21. Zhang, T.; Isobe, S.; Matsuo, M.; Orimo, S.; Wang, Y.; Hashimoto, N.; Ohnuki, S. Effect of lithium ion conduction on hydrogen desorption of LiNH$_2$−LiH composite. *ACS Catal.* **2015**, *5*, 1552–1555. [CrossRef]

22. Ares, J.R.; Aguey-Zinou, K.F.; Porcu, M.; Sykes, J.M.; Dornheim, M.; Klassen, T.; Bormann, R. Thermal and mechanically activated decomposition of LiAlH$_4$. *Mater. Res. Bull.* **2008**, *43*, 1263–1275. [CrossRef]

23. D'Anna, V.; Spyratou, A.; Sharma, M.; Hagemann, H. FT-IR spectra of inorganic borohydrides. Spectrochim. *Acta Part A* **2014**, *128*, 902–906. [CrossRef] [PubMed]

24. Miyaoka, H.; Ichikawa, T.; Fujii, H.; Kojima, Y. Hydrogen desorption reaction between hydrogen-containing functional groups and lithium hydride. *J. Phys. Chem. C* **2010**, *114*, 8668–8674. [CrossRef]

25. Kissinger, H.E. Reaction kinetics in differential thermal analysis. *Anal. Chem.* **1957**, *29*, 1702–1706. [CrossRef]

26. Andreasen, A.; Vegge, T.; Pedersen, A.S. Dehydrogenation kinetics of as-received and ball-milled LiAlH$_4$. *J. Solid State Chem.* **2005**, *178*, 3672–3678. [CrossRef]

27. Wang, L.; Rawal, A.; Quadir, M.Z.; Aguey-Zinsou, K.F. Nanoconfined lithium aluminum hydride (LiAlH$_4$) and hydrogen reversibility. *Int. J. Hydrog. Energy* **2017**, *42*, 14144–14153. [CrossRef]

28. Oguchi, H.; Matsuo, M.; Sato, T.; Takaumura, H.; Maekawa, H.; Kuwano, H.; Orimo, S. Lithium-ion conduction in complex hydrides LiAlH$_4$ and Li$_3$AlH$_6$. *J. Appl. Phys.* **2010**, *107*, 096104. [CrossRef]

29. Maekawa, H.; Matsuo, M.; Takamura, H.; Ando, M.; Noda, Y.; Karahashi, T.; Orimo, S. Halide-stabilized LiBH$_4$, a room-temperature lithium fast-ion conductor. *J. Am. Chem. Soc.* **2009**, *131*, 894–895. [CrossRef] [PubMed]

30. Matsuo, M.; Takamura, H.; Maekawa, H.; Li, H.W.; Orimo, S. Stabilization of lithium superionic conduction phase and enhancement of conductivity of LiBH$_4$ by LiCl addition. *Appl. Phys. Lett.* **2009**, *94*, 084103. [CrossRef]

31. Oguchi, H.; Matsuo, M.; Hummelshøj, J.S.; Vegge, T.; Nørskov, J.K.; Sato, T.; Miura, Y.; Takamura, H.; Maekawa, H.; Orimo, S. Experimental and computational studies on structural transitions in the LiBH$_4$−LiI pseudobinary system. *Appl. Phys. Lett.* **2009**, *94*, 141912. [CrossRef]

32. Mosegaard, L.; Møller, B.; Jørgensen, J.E.; Filinchuk, Y.; Cerenius, Y.; Hanson, J.C.; Dimasi, E.; Besenbacher, F.; Jensen, T.R. Reactivity of LiBH$_4$: In situ synchrotron radiation power X-ray diffraction study. *J. Phys. Chem. C* **2008**, *112*, 1299–1303. [CrossRef]

33. Rude, L.H.; Groppo, E.; Arnbjerg, L.M.; Ravnsbæk, D.B.; Malmkjær, R.A.; Filinchuk, Y.; Baricco, M.; Besenbacher, F.; Jensen, T.R. Iodide substitution in lithium borohydride, LiBH$_4$−LiI. *J. Alloys Compd.* **2011**, *509*, 8299–8305. [CrossRef]

34. Langmi, H.W.; McGrady, G.S.; Liu, X.; Jensen, C.M. Modification of the H$_2$ desorption properties of LiAlH$_4$ through doping with Ti. *J. Phys. Chem. C* **2010**, *114*, 10666–10669. [CrossRef]

35. Li, Z.; Li, P.; Wan, Q.; Zhai, F.; Liu, Z.; Zhao, K.; Wang, L.; Lü, S.; Zou, L.; Qu, X.; et al. Dehydrogenation improvement of LiAlH$_4$ catalyzed by Fe$_2$O$_3$ and Co$_2$O$_3$ nanoparticles. *J. Phys. Chem. C* **2013**, *117*, 18343–18352. [CrossRef]

36. Ismail, M.; Zhao, Y.; Yu, X.B.; Dou, S.X. Effects of NbF$_5$ addition on the hydrogen storage properties of LiAlH$_4$. *Int. J. Hydrog. Energy* **2010**, *35*, 2361–2367. [CrossRef]

37. Atakli, Z.Ö.K.; Callini, E.; Kato, S.; Mauron, P.; Orimo, S.; Züttel, A. The catalyzed hydrogen sorption mechanism in alkali alanates. *Phys. Chem. Chem. Phys.* **2015**, *17*, 20932. [CrossRef] [PubMed]

38. Hoang, K.; Janotti, A; Van de Walle, C.G. Decomposition mechanism and the effects of metal additives on the kinetics of lithium alanate. *Phys. Chem. Chem. Phys.* **2012**, *14*, 2840–2848. [CrossRef] [PubMed]

39. Weng, Q.; Wang, X.; Wang, X.; Bando, Y.; Golberg, D. Functional hexagonal boron nitride nanomaterials: Emerging properties and applications. *Chem. Soc. Rev.* **2016**, *45*, 3989–4012. [CrossRef] [PubMed]

40. Lei, W.; Mochalin, V.N.; Liu, D.; Qin, S.; Gogotsi, Y.; Chen, Y. Boron nitride colloidal solutions, ultralight aerogels and freestanding membranes through one-step exfoliation and functionalization. *Nat. Commun.* **2015**, *6*, 8849. [CrossRef] [PubMed]

41. Liu, D.; He, L.; Lei, W.; Klika, K.D.; Kong, L.; Chen, Y. Multifunctional polymer/porous boron nitride nanosheet membranes for superior trapping emulsified oils and organic molecules. *Adv. Mater. Interfaces* **2015**, *2*, 1500228. [CrossRef]

42. Hu, S.; Lozada-Hidalgo, M.; Wang, F.C.; Mishchenko, A.; Schedin, F.; Nair, R.R.; Hill, E.W.; Boukhvalov, D.W.; Katsnelson, M.I.; Dryfe, R.A.W.; et al. Proton transport through one-atom-thick crystals. *Nature* **2014**, *516*, 227–230. [CrossRef] [PubMed]

43. Lei, W.; Zhang, H.; Wu, Y.; Zhang, B.; Liu, D.; Qin, S.; Liu, Z.; Liu, L.; Ma, Y.; Chen, Y. Oxygen-doped boron nitride nanosheets with excellent performance in hydrogen storage. *Nano Energy* **2014**, *6*, 219–224. [CrossRef]

inorganics

MDPI

Review

A Recycling Hydrogen Supply System of NaBH₄ Based on a Facile Regeneration Process: A Review

Liuzhang Ouyang [1,2,3], Hao Zhong [1,2], Hai-Wen Li [4,*] and Min Zhu [1,2,*]

[1]   School of Materials Science and Engineering, Key Laboratory of Advanced Energy Storage Materials of Guangdong Province, South China University of Technology, Guangzhou 510641, China; meouyang@scut.edu.cn (L.O.); haozhong.gz@gmail.com (H.Z.)

[2]   China-Australia Joint Laboratory for Energy & Environmental Materials, South China University of Technology, Guangzhou 510641, China

[3]   Key Laboratory for Fuel Cell Technology in Guangdong Province, Guangzhou 510641, China

[4]   Kyushu University Platform of Inter/Transdisciplinary Energy Research, International Research Center for Hydrogen Energy and International Institute for Carbon-Neutral Energy, Kyushu University, Fukuoka 819-0395, Japan

*   Correspondence: li.haiwen.305@m.kyushu-u.ac.jp (H.-W.L.); memzhu@scut.edu.cn (M.Z.)

Received: 24 October 2017; Accepted: 5 January 2018; Published: 6 January 2018

Abstract: NaBH₄ hydrolysis can generate pure hydrogen on demand at room temperature, but suffers from the difficult regeneration for practical application. In this work, we overview the state-of-the-art progress on the regeneration of NaBH₄ from anhydrous or hydrated NaBO₂ that is a byproduct of NaBH₄ hydrolysis. The anhydrous NaBO₂ can be regenerated effectively by MgH₂, whereas the production of MgH₂ from Mg requires high temperature to overcome the sluggish hydrogenation kinetics. Compared to that of anhydrous NaBO₂, using the direct hydrolysis byproduct of hydrated NaBO₂ as the starting material for regeneration exhibits significant advantages, i.e., omission of the high-temperature drying process to produce anhydrous NaBO₂ and the water included can react with chemicals like Mg or Mg₂Si to provide hydrogen. It is worth emphasizing that NaBH₄ could be regenerated by an energy efficient method and a large-scale regeneration system may become possible in the near future.

Keywords: sodium borohydride (NaBH₄); hydrolysis; regeneration

1. Introduction

Hydrogen [1–3] has been widely accepted as a clear energy carrier [4–6] due to its high energy density (142 MJ/kg) [7–9] and its environmentally friendly byproduct (water) [10,11]. It can be generated via numerous strategies, such as the electrolysis of water [12–14] and photocatalytic water splitting [15–18]. Supplying hydrogen to end users on demand requires safe and efficient methods of hydrogen storage.

Hydrides, storing hydrogen in a safe and compact way without using high pressure, like 70 MPa, or extremely low temperature, like 20 K (liquid hydrogen), have attracted great interest as promising hydrogen storage materials. Though a great deal of progress has been achieved on the development of solid-state hydrogen storage materials in the previous decades, no material with reasonably good hydrogen absorption and desorption performance at near room temperature has been developed to meet all the requirements for onboard hydrogen storage [19–23]. Hydrolysis of hydrides, such as MgH₂, ammonia borane (AB), and NaBH₄, generating hydrogen with relatively high capacity at room temperature, is attracting increasing interest for hydrogen supply on demand [24–26]. Due to the low cost of Mg and the high capacity of MgH₂ (7.6 wt %) [27–29], much attention has been paid to MgH₂ hydrolysis [30–32]. However, the reaction is interrupted easily by the formation of a magnesium hydroxide layer [33,34]. Compared with MgH_2, AB possesses higher hydrogen capacity (19.6 wt %) [35–37]. AB is stable in water and its solubility is as high as 33.6 g/100 mL [38,39], which provides a simple application of AB aqueous solution. Studies have been focused on the development of catalysts to accelerate and control the reaction [40–42]. However, the high cost of AB [43] and the difficulty of AB regeneration are major blocks for the application of AB hydrolysis [44,45].

$NaBH_4$ hydrolysis is another promising system for hydrogen generation. It has relatively high hydrogen capacity (10.8 wt %) [46–48] and releases hydrogen with high purity at relatively low operational temperature with a controllable process [24,49,50]. Many studies have been focused on the hydrolysis property improvements [51–54]. Unfortunately, a no-go recommendation on $NaBH_4$ hydrolysis for onboard applications was given by the US Department of Energy (DOE) [55]. One of the key reasons are the cost and the regeneration of $NaBH_4$ [56]. As a result, more focus was shifted into the synthesis and regeneration of $NaBH_4$. For commercial $NaBH_4$ production, the Brown-Schlesinger process [57] and the Bayer process [58] are the most popular methods. The Brown-Schlesinger process produces $NaBH_4$ via the reaction between trimethylborate ($B(OCH_3)_3$, TMB) and sodium hydride (NaH), which should be produced by reacting Na and H_2. Different from it, the Bayer process is based on the reaction among borax ($Na_2B_4O_7$), Na, H_2, and silicon oxide (SiO_2) at 700 °C to synthesize $NaBH_4$. Although the above methods are mature technologies and straightforward procedures, the raw materials are too expensive for $NaBH_4$ hydrolysis applications. Thus, the raw materials have been studied to develop suitable $NaBH_4$ synthesis methods. Instead of Na, MgH_2 was used to react with $Na_2B_4O_7$ to synthesize $NaBH_4$ by ball milling at room temperature. Here, Na_2CO_3 addition could increase $NaBH_4$ yield up to 78% [59]. This method provides not only a new reducing agent (MgH_2) for $NaBH_4$ synthesis, but also a new way of ball milling. Enlightened by it, ball milling became popular in $NaBH_4$ synthesis studies, in which Na and MgH_2 reacted with B_2O_3 by ball milling with the $NaBH_4$ yield of only 25% [60]. When Na was substituted by low-cost NaCl, $NaBH_4$ could also be produced [61]. Later, high-pressure ball milling was also tried to synthesize $NaBH_4$, for instance, NaH was reacted with MgB_2 by ball milling under 12 MPa hydrogen pressure with the $NaBH_4$ yield of about 18%.

From the point of cost reduction in synthesis and post-usage of $NaBH_4$, the regeneration of $NaBH_4$ from the byproduct of hydrolysis (see Equation (1) [62]) is in great need for the recycling of the hydrogen supply system of $NaBH_4$:

$$NaBH_4 + xH_2O \rightarrow H_2 + NaBO_2 \cdot xH_2O, (x = 2, 4) \tag{1}$$

According to this, the Brown-Schlesinger process was modified using $NaBO_2$ as source of boric acid to synthesis of $NaBH_4$ [63], the drawback of which the byproduct $NaBO_2 \cdot xH_2O$ of hydrolysis needs to be dried first. As another alternative, the electrochemical route was proposed for recycling $NaBO_2$ to $NaBH_4$. Direct electrolysis of a $NaBO_2$ solution was first proven feasible for regeneration of $NaBH_4$ with using palladium (Pd) or platinum (Pt) as electrodes, where the conversion ratio of $NaBO_2$ was about 17% within 48 h [64]. Later, an Ag electrode was also employed in the recycling of $NaBO_2$; unfortunately, the quantities of reborn $NaBH_4$ were too low to be measured [65]. In contrast to the commercial gas-solid methods, the electrochemical method possesses ultra-low efficiency and complex processes, using precious metal electrodes, although the $NaBO_2$ solution that is the main byproduct of $NaBH_4$ hydrolysis can be used directly without dehydration. Therefore, an efficient and simple route is most urgently needed for the cycling of $NaBO_2$ into $NaBH_4$.

In this paper, we discuss the state-of-the-art progress on the regeneration of $NaBH_4$ from anhydrous $NaBO_2$ or the direct byproduct $NaBO_2 \cdot xH_2O$. In particular, the regeneration steps and the yield of $NaBH_4$ in each process are summarized in Scheme 1 and the facile regeneration process is also proposed. This review can provide important insights for the recycling hydrogen supply system with high efficiency.

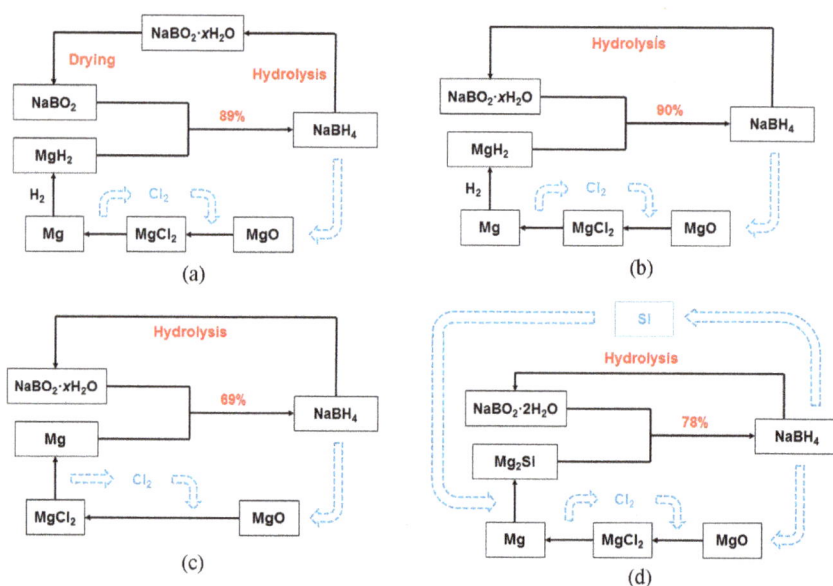

Scheme 1. Flowchart of (**a**) $NaBH_4$ regeneration via the reaction between MgH_2 and $NaBO_2$. (**b**) $NaBH_4$ regeneration via the reaction between MgH_2 and $NaBO_2 \cdot xH_2O$. (**c**) $NaBH_4$ regeneration via the reaction between Mg and $NaBO_2 \cdot xH_2O$. (**d**) $NaBH_4$ regeneration via the reaction between Mg_2Si and $NaBO_2 \cdot 2H_2O$. Numbers indicate the yield of $NaBH_4$.

2. NaBH₄ Regeneration via the Reaction between Metal or Other Hydrides and NaBO₂

As the hydrolysis byproduct of $NaBH_4$, $NaBO_2$ is the main research object of $NaBH_4$ regeneration studies. Many approaches have been adopted to reduce $NaBO_2$ to $NaBH_4$ with different reducing agents. Among the reducing agent, MgH_2 is the most effective. Kojima et al. [66] reacted MgH_2 with $NaBO_2$ under 550 °C and 7 MPa hydrogen pressure to regenerate $NaBH_4$, and about 97% $NaBH_4$ yield was achieved, while the high reaction temperature and high hydrogen pressure leads to a high energy consumption. Therefore, the thermochemistry method was substituted by room temperature ball milling in this reaction. Hsueh et al. [67], Kong et al. [68] and Çakanyildirim et al. [69] used MgH_2 to react with $NaBO_2$ by ball milling under argon. All of their $NaBH_4$ yields were over 70%, which strongly indicated that ball milling is suitable for the reaction between MgH_2 and $NaBO_2$. Based on the thermodynamics calculation, we found the maximum energy efficiency of the cycle was 49.91% [70]. Recently, we found that the addition of hydrogen pressure and methanol could further increase the $NaBH_4$ yield by this method [71]. The highest $NaBH_4$ yield could be increased to 89%. In addition to the energy consumption, raw material is another issue that should be considered. Hydrogenation of Mg to produce MgH_2 is hard due to its sluggish kinetics, thus resulting in the high cost and high energy consumption in MgH_2 production. By modifying the hydrogenation of Mg using Mg-based alloy, the above issue can be partly solved. Following this observation, we tried to use Mg_3La hydrides to react with $NaBO_2$ for its advantage of room temperature hydrogenation and low hydrogen purity requirement and found that $NaBH_4$ could be produced (Figure 1a) [70]. However, introduction of other elements influences the regeneration reaction of MgH_2. Directly using Mg and H_2 in the regeneration may solve the MgH_2 production problem. Kojima et al. [66] tried to directly react Mg with $NaBO_2$ under hydrogen, but the yield was extremely low, which may have resulted from the produced MgO obstruction. To promote the yield, Kojima et al. [66] found that Si addition could remarkably increase the $NaBH_4$ yield and Liu et al. [72] found transition metals, like Ni, Fe, and Co, addition could also promote the $NaBH_4$ yield. However, both Si and transition metals keep their own elemental form after the reaction, indicating that such additions would reduce the absolute $NaBH_4$ yield. A pre-milling of the reactants was then found that could also promote the yield. Eom et al. [73] proposed a large-scaled method for reacting Mg with $NaBO_2$ to synthesize $NaBH_4$. After 1 h of ball milling of the reactants, about 69% yield was achieved under 600 °C and 5.5 MPa hydrogen pressure.

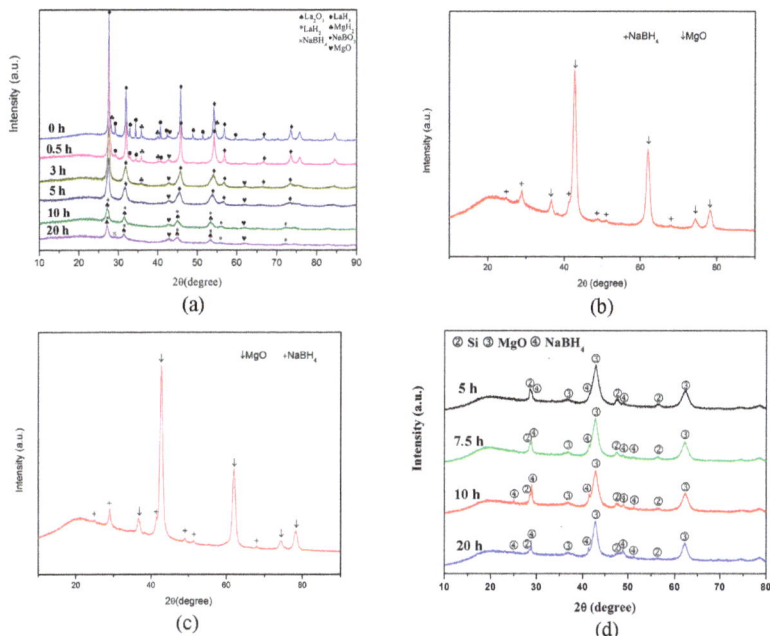

(a)

(b)

(c)

(d)

Figure 1. (a) XRD patterns of the $NaBO_2$–Mg_3La hydride mixture and the product after ball milling the $NaBO_2$–Mg_3La hydride mixture. (b) XRD pattern of products via ball-milling the mixture of $NaBO_2 \cdot 2H_2O$ and MgH_2 in 1:5.5 mol ratio for 15 h. (c) XRD pattern of products via ball-milling the mixture of $NaBO_2 \cdot 2H_2O$ and Mg in 1:5 mole ratio for 15 h. (d) XRD patterns of the products after ball milling Mg_2Si and $NaBO_2 \cdot 2H_2O$ mixtures (in 2:1 mol ratio).

For other reducing agents, the Gibbs free energy of the reaction using Ca is much lower than that of Mg. In addition, we found that the energy efficiency of the cycle using Ca is about 43%. For the experiment, Eom et al. tried to substitute Mg by Ca [73], but few $NaBH_4$ was regenerated. Another low cost and abundant metal reductant, Al, was studied by few researchers on $NaBH_4$ regeneration. The only work with respect to Al was reported by Liu et al. [74], expressing that Al could not react with $NaBO_2$ and H_2 to produce $NaBH_4$ because of the generated Al_2O_3. However, if $NaBO_2$ was exchanged to $Na_4B_2O_5$, the regeneration would succeed at 400 °C and 2.3 MPa pressure of of hydrogen.

3. $NaBH_4$ Regeneration via using $NaBO_2 \cdot xH_2O$ as Raw Materials

In $NaBH_4$ regeneration, many studies have focused on anhydrous $NaBO_2$ reducing. However, it should be noted that the direct hydrolysis byproduct is hydrated $NaBO_2$. For the $NaBH_4$ aqueous solution hydrolysis, the byproduct is $NaBO_2 \cdot 4H_2O$ [75], while for the solid $NaBH_4$ hydrolysis, the byproduct is $NaBO_2 \cdot 2H_2O$. Anhydrous $NaBO_2$ should be produced by drying hydrated $NaBO_2$ at 350 °C. If the drying process was omitted, more energy could be saved and the price can be lowered. The energy of the hydrated $NaBO_2$ and anhydrous $NaBO_2$ is shown in Scheme 2. Some studies thus worked on reducing hydrated $NaBO_2$ directly.

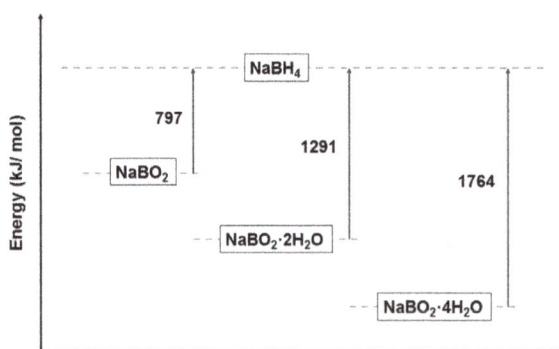

Scheme 2. Schematic energy diagram of the boron material for the recycling of $NaBO_2$ to $NaBH_4$, $NaBO_2 \cdot 2H_2O$ to $NaBH_4$, and $NaBO_2 \cdot 4H_2O$ to $NaBH_4$.

3.1. NaBH₄ Regeneration via the Reaction between MgH₂ and NaBO₂·xH₂O

For directly using hydrated $NaBO_2$ as the regeneration raw material, a thermochemistry method was tried. Liu et al. [76] reported that $NaBH_4$ can be regenerated by annealing Mg and $NaBO_2 \cdot 2H_2O$ under hydrogen atmosphere with only 12.3% yield. The low $NaBH_4$ yield may result from the obstruction of the thick generated MgO layer. However, they found that the coordinate water in $NaBO_2 \cdot 2H_2O$ was likely to be the hydrogen source. Considering the generated oxide layer, ball milling might be suitable to break the layer and continue the reaction. Therefore, we tried to used $NaBO_2 \cdot 4H_2O$ or $NaBO_2 \cdot 2H_2O$ to react with MgH_2 directly via ball milling to regenerate $NaBH_4$ [77]. $NaBH_4$ was successfully regenerated (Figure 1b):

$$NaBO_2 \cdot 2H_2O + 4MgH_2 \rightarrow NaBH_4 + 4MgO + 4H_2 \tag{2}$$

$$NaBO_2 \cdot 4H_2O + 6MgH_2 \rightarrow NaBH_4 + 6MgO + 8H_2 \tag{3}$$

The energy efficiency calculated in Section 2 could be improved by approximately 5.2%. Furthermore, a high $NaBH_4$ yield of 89.78% was achieved by this method, which is the highest compared with previous studies [67,69,78].

3.2. NaBH₄ Regeneration via the Reaction between Mg and NaBO₂·xH₂O

Hydrated $NaBO_2$ could be directly used in $NaBH_4$ regeneration, saving the energy consumption on the dehydration to produce anhydrous $NaBO_2$. However, production of MgH_2 from Mg requires high temperature to overcome the sluggish hydrogenation kinetics, resulting in the increased cost. In other words, the energy efficiency could be further promoted and the regeneration cost could be reduced, if the high-temperature hydrogenation process to produce MgH_2 can be avoided. According to Liu et al. [76], H in $NaBO_2 \cdot 2H_2O$ could transform to be the H of the regenerated $NaBH_4$. As a result, directly reacting Mg with hydrated $NaBO_2$ was possible to regenerate $NaBH_4$ and avoided the high-temperature hydrogenation process. We found that $NaBH_4$ could be produced by ball milling the $NaBO_2 \cdot 2H_2O$ and Mg mixture under argon (Figure 1c) [79] according to:

$$NaBO_2 \cdot 2H_2O + 4Mg \rightarrow NaBH_4 + 4MgO \tag{4}$$

It should be noted that the regenerated H of $NaBH_4$ was completely from the coordinate water. On the other hand, the reaction between $NaBO_2 \cdot 4H_2O$ and Mg could also generate $NaBH_4$ by ball milling:

$$NaBO_2 \cdot 4H_2O + 6Mg \rightarrow NaBH_4 + 6MgO + H_2 \tag{5}$$

Currently, the highest $NaBH_4$ yield of the reaction between Mg and $NaBO_2 \cdot 2H_2O$ is only 68.55%. The energy efficiency needs to be further promoted. Note that the cost of this method is 34-fold lower than the method using MgH_2 and $NaBO_2$ in terms of the raw materials required [79].

3.3. $NaBH_4$ Regeneration via the Reaction between Mg_2Si and $NaBO_2 \cdot 2H_2O$

Via ball milling hydrated $NaBO_2$ and Mg, $NaBH_4$ was regenerated and the energy efficiency was further increased. However, the highest $NaBH_4$ yield by this method was 68.55%, which did not reach the general yield of regenerated $NaBH_4$ (~76%) [67,68]. According to Kojima et al. [66], with Si added, the $NaBH_4$ yield was increased in the reaction between $NaBO_2$ and Mg under a hydrogen atmosphere. Therefore, Mg_2Si is possible to react with $NaBO_2 \cdot 2H_2O$ to regenerate $NaBH_4$ and improve the $NaBH_4$ yield. We have attempted the above idea in our previous study [80] and found that $NaBH_4$ was regenerated (Figure 1d) according to:

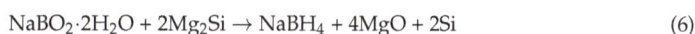

$$NaBO_2 \cdot 2H_2O + 2Mg_2Si \rightarrow NaBH_4 + 4MgO + 2Si \qquad (6)$$

The highest $NaBH_4$ yield was increased to 78% when the Mg_2Si and $NaBO_2 \cdot 2H_2O$ mixture was ball milled for 20 h. By using Mg_2Si as a reducing agent, the $NaBH_4$ yield was promoted and the H was still from the coordinate water in $NaBO_2 \cdot 2H_2O$. For the raw materials cost, this method is half of the commercial method and about 30-fold lower than the method using MgH_2 and $NaBO_2$ [80].

4. Mechanism of $NaBH_4$ Regeneration Using $NaBO_2 \cdot xH_2O$ as Raw Materials

The above three works [77,79,80] are new discoveries for direct regeneration of $NaBH_4$ from the hydrated $NaBO_2$ with high yield. Some common points were found in their mechanism studies. In all of the three works, a resonance at approximately -11.4 ppm was observed in the NMR spectra (Figure 2), which belongs to intermediate $[BH_3(OH)]^-$ [81]. Such an intermediate was likely to generate from $[BH(OH)_3]^-$ and $[BH_2(OH)_2]^-$. Conjecturing from the above intermediates, $[BH_4]^-$ was likely to generate from a step-by-step substitution process of $[OH]^-$ in $[B(OH)_4]^-$ by $[H]^-$. The $[H]^+$ in $NaBO_2 \cdot xH_2O$ thus transformed to $[H]^-$ in this process.

For the reaction between MgH_2 and $NaBO_2 \cdot 2H_2O$, the hydrogen transformation was realized by the substitution of the $[OH]^-$ in $NaBO_2 \cdot xH_2O$ by $[H]^-$ in MgH_2. For the reaction of Mg and $NaBO_2 \cdot 2H_2O$, $Mg(OH)_2$ and MgH_2 were generated as intermediates and the reactions can be written as:

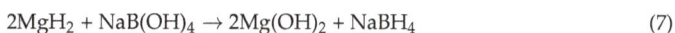

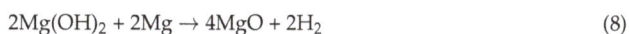

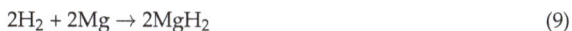

$$2MgH_2 + NaB(OH)_4 \rightarrow 2Mg(OH)_2 + NaBH_4 \qquad (7)$$

$$2Mg(OH)_2 + 2Mg \rightarrow 4MgO + 2H_2 \qquad (8)$$

$$2H_2 + 2Mg \rightarrow 2MgH_2 \qquad (9)$$

Since four moles of MgO were generated in this reaction (Equation (4)), it was a strong exergonic reaction. The reaction could be described as a substitution process of $[OH]^-$ through the $[H]^-$ from the produced intermediate MgH_2. During the substitution process, a side reaction may happen. $[B_3H_8]^-$ was generated and then may react with MgH_2 and Na^+ to form another part of $NaBH_4$ [82]. In the reaction between Mg_2Si and $NaBO_2 \cdot 2H_2O$, Si–H was found (Figure 2d). It was speculated that an intermediate consisting of Mg, O, Si, and H was generated. The $[OH]^-$ was transformed to $[H]^-$ through a Mg–O–Si–H intermediate. Therefore, though Si was generated with an elemental state after the reaction, Si played an important role in H^- formation. Consequently, the substitution process of $[OH]^-$ through $[H]^-$ was a direct process.

In conclusion, two forms of hydrogen molecules exist in the regeneration. They are H in $[OH]^-$ and H in $[H]^-$. When using MgH_2 as a reducing agent, H in MgH_2 directly substitutes $[OH]^-$ in $NaB(OH)_4$. This direct process contributes to the high $NaBH_4$ yield. In the situation of Mg, H in $[OH]^-$ first transfer to H in MgH_2. Then it substitutes the $[OH]^-$ in $NaB(OH)_4$ to form $NaBH_4$. The two-step reaction reduces the $NaBH_4$ yield. For the reaction between Mg_2Si and $NaBO_2 \cdot 2H_2O$, H in $[OH]^-$

first transfers to H in Si–H and then it transfers to NaBH$_4$. Although this process is also two steps, the more active Si–H benefits from the higher NaBH$_4$ yield. Therefore, all of the reactions are H transfer processes.

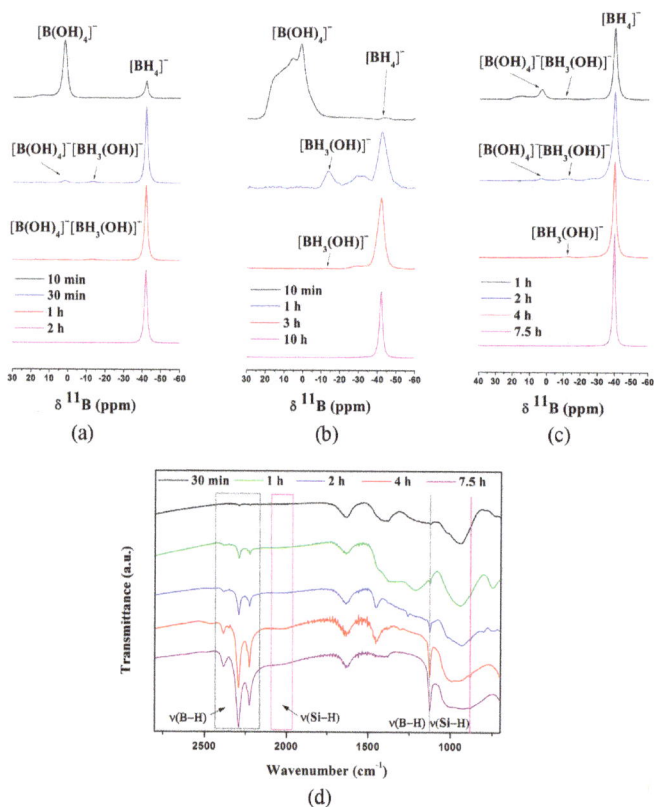

Figure 2. (**a**) Solid-state ^{11}B NMR spectra of products after ball milling MgH$_2$ and NaBO$_2$·2H$_2$O mixtures (in 5.0:1 mol ratio). (**b**) Solid-state ^{11}B NMR spectra of products after ball milling Mg and NaBO$_2$·2H$_2$O mixtures (in 5.0:1 mole ratio). (**c**) ^{11}B MAS NMR spectra of the products after ball milling Mg$_2$Si and NaBO$_2$·2H$_2$O mixtures (in 2:1 mol ratio). (**d**) FT-IR spectra of the products after ball milling Mg$_2$Si and NaBO$_2$·2H$_2$O mixtures (in 2:1 mol ratio).

5. Hydrolysis Property of Regenerated NaBH$_4$ Using NaBO$_2$·xH$_2$O as Raw Materials

Hydrolysis is the main application of the regenerated NaBH$_4$. By the catalysis of CoCl$_2$ [83], NaBH$_4$ could fast hydrolyze with stoichiometry H$_2$O. It was found that the regenerated NaBH$_4$ from NaBO$_2$·xH$_2$O had an excellent hydrolysis property, which was similar to the commercial NaBH$_4$. According to Figure 3, the highest system hydrogen capacity (containing water and catalyst) was 6.75 wt %, which was the highest compared with previous studies [67,69,78]. It was produced by the reaction between MgH$_2$ and NaBO$_2$·xH$_2$O. A system hydrogen capacity of 6.33 wt % and 6.3 wt % could also be obtained. Furthermore, the hydrolysis byproduct was indexed to be NaBO$_2$·2H$_2$O (inset, Figure 3), which was the raw material of our regeneration. As a result, it was demonstrated that a complete cycle of NaBH$_4$ hydrolysis could be achieved by existing works, which was suitable for sustainable application.

Inorganics **2018**, *6*, 10

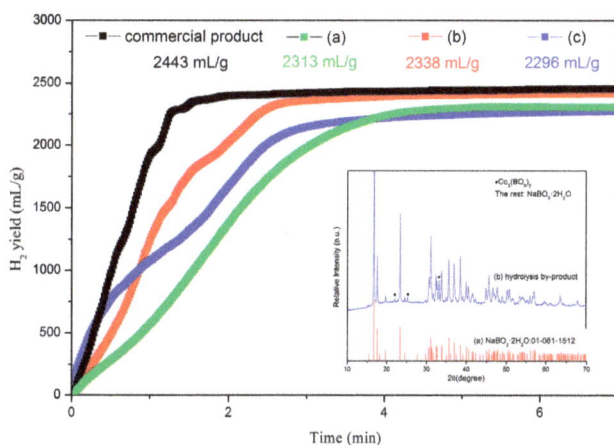

Figure 3. Hydrolysis curves of (**a**) the regeneration product (MgH$_2$ and NaBO$_2 \cdot x$H$_2$O). (**b**) The regeneration product (Mg and NaBO$_2 \cdot x$H$_2$O). (**c**) The purified product (Mg$_2$Si and NaBO$_2 \cdot 2$H$_2$O) and the commercial NaBH$_4$ in 5 wt % CoCl$_2$ aqueous solution. Inset: XRD patterns of (**a**) a standard PDF card of NaBO$_2 \cdot 2$H$_2$O and (**b**) the hydrolysis byproduct.

6. Summary and Perspective

Application of NaBH$_4$ hydrolysis is limited by its effective regeneration. NaBH$_4$ synthesis and regeneration thus become attractive research topics, especially for the recycling of byproduct NaBO$_2$. For the anhydrous NaBO$_2$ recycling, MgH$_2$ has the best reducing result. However, its high cost, resulting from the high hydrogenation temperature of Mg, limits the application of such methods. For the hydrolysis byproduct hydrated NaBO$_2$, it can also be reduced by MgH$_2$, Mg, or Mg$_2$Si via ball milling, and the highest NaBH$_4$ yield reaches 90%. This process using hydrated NaBO$_2$ exhibits significant advantages, whereby the dehydration process at 350 °C to obtain anhydrous NaBO$_2$ can be omitted and, more importantly, the water included can react with chemicals like Mg and Mg$_2$Si to provide hydrogen instead of using MgH$_2$. As a result, low cost metal (such as Mg, Ca, or Al) becomes possible to be the reducing agent for the NaBH$_4$ regeneration reaction via ball milling, because the [H]$^+$ in the hydrated NaBO$_2$ may directly transform to the [H]$^-$ in the hydrated NaBH$_4$. These reactions could operate without extra hydrogen inputs, which provides the possibility of a low-cost and sustainable regeneration. Furthermore, this strategy may also be promoted to other areas, such as LiBH$_4$ production.

Acknowledgments: This work was supported by the Fund for Innovative Research Groups of the National Natural Science Foundation of China (no. NSFC51621001), the National Natural Science Foundation of China Projects (nos. 51431001 and 51771075), and by the Project Supported by Natural Science Foundation of Guangdong Province of China (nos. 2016A030312011 and 2014A030311004). The Project Supported by Guangdong Province Universities and Colleges Pearl River Scholar Funded Scheme (2014) is also acknowledged.

Conflicts of Interest: The authors declare no conflict of interest.

References

1. Zhang, X.; Wei, Z.; Guo, Q.; Tian, H. Kinetics of sodium borohydride hydrolysis catalyzed via carbon nanosheets supported Zr/Co. *J. Power Sources* **2013**, *231*, 190–196. [CrossRef]
2. Awad, A.S.; El-Asmar, E.; Tayeh, T.; Mauvy, F.; Nakhl, M.; Zakhour, M.; Bobet, J.L. Effect of carbons (G and CFs), TM (Ni, Fe and Al) and oxides (Nb$_2$O$_5$ and V$_2$O$_5$) on hydrogen generation from ball milled Mg-based hydrolysis reaction for fuel cell. *Energy* **2016**, *95*, 175–186. [CrossRef]

3. Brack, P.; Dann, S.E.; Wijayantha, K.G.U. Heterogeneous and homogenous catalysts for hydrogen generation by hydrolysis of aqueous sodium borohydride (NaBH$_4$) solutions. *Energy Sci. Eng.* **2015**, *3*, 174–188. [CrossRef]

4. Luo, Y.-C.; Liu, Y.-H.; Hung, Y.; Liu, X.-Y.; Mou, C.-Y. Mesoporous silica supported cobalt catalysts for hydrogen generation in hydrolysis of ammonia borane. *Int. J. Hydrog. Energy* **2013**, *38*, 7280–7290. [CrossRef]

5. Ding, Q.; Meng, F.; English, C.R.; Cabán-Acevedo, M.; Shearer, M.J.; Liang, D.; Daniel, A.S.; Hamers, R.J.; Jin, S. Efficient photoelectrochemical hydrogen generation using heterostructures of Si and chemically exfoliated metallic MoS$_2$. *J. Am. Chem. Soc.* **2014**, *136*, 8504–8507. [CrossRef] [PubMed]

6. Ladomenou, K.; Natali, M.; Iengo, E.; Charalampidis, G.; Scandola, F.; Coutsolelos, A.G. Photochemical hydrogen generation with porphyrin-based systems. *Coord. Chem. Rev.* **2015**, *304*, 38–54. [CrossRef]

7. Huang, M.; Ouyang, L.; Wang, H.; Liu, J.; Zhu, M. Hydrogen generation by hydrolysis of MgH$_2$ and enhanced kinetics performance of ammonium chloride introducing. *Int. J. Hydrog. Energy* **2015**, *40*, 6145–6150. [CrossRef]

8. Bulut, A.; Yurderi, M.; Ertas, İ.E.; Celebi, M.; Kaya, M.; Zahmakiran, M. Carbon dispersed copper-cobalt alloy nanoparticles: A cost-effective heterogeneous catalyst with exceptional performance in the hydrolytic dehydrogenation of ammonia-borane. *Appl. Catal. B* **2016**, *180*, 121–129. [CrossRef]

9. Dudoladov, A.O.; Buryakovskaya, O.A.; Vlaskin, M.S.; Zhuk, A.Z.; Shkolnikov, E.I. Generation of hydrogen by aluminium oxidation in aquaeous solutions at low temperatures. *Int. J. Hydrog. Energy* **2016**, *41*, 2230–2237. [CrossRef]

10. Huynh, K.; Napolitano, K.; Wang, R.; Jessop, P.G.; Davis, B.R. Indirect hydrolysis of sodium borohydride: Isolation and crystallographic characterization of methanolysis and hydrolysis by-products. *Int. J. Hydrog. Energy* **2013**, *38*, 5775–5782. [CrossRef]

11. Jiang, J.; Materna, K.L.; Hedström, S.; Yang, K.R.; Crabtree, R.H.; Batista, V.S.; Brudvig, G.W. Antimony Complexes for Electrocatalysis: Activity of a Main-Group Element in Proton Reduction. *Angew. Chem. Int. Ed.* **2017**, *56*, 9111–9115. [CrossRef] [PubMed]

12. Yilmaz, C.; Kanoglu, M. Thermodynamic evaluation of geothermal energy powered hydrogen production by PEM water electrolysis. *Energy* **2014**, *69*, 592–602. [CrossRef]

13. Wang, M.; Wang, Z.; Gong, X.; Guo, Z. The intensification technologies to water electrolysis for hydrogen production—A review. *Renew. Sustain. Energy Rev.* **2014**, *29*, 573–588. [CrossRef]

14. Cardoso, D.S.P.; Amaral, L.; Santos, D.M.F.; Šljukić, B.; Sequeira, C.A.C.; Macciò, D.; Saccone, A. Enhancement of hydrogen evolution in alkaline water electrolysis by using nickel-rare earth alloys. *Int. J. Hydrog. Energy* **2015**, *40*, 4295–4302. [CrossRef]

15. Kibria, M.G.; Qiao, R.; Yang, W.; Boukahil, I.; Kong, X.; Chowdhury, F.A.; Trudeau, M.L.; Ji, W.; Guo, H.; Himpsel, F.J.; et al. Atomic-Scale Origin of Long-Term Stability and High Performance of *p*-GaN Nanowire Arrays for Photocatalytic Overall Pure Water Splitting. *Adv. Mater.* **2016**, *28*, 8388–8397. [CrossRef] [PubMed]

16. Ma, G.; Chen, S.; Kuang, Y.; Akiyama, S.; Hisatomi, T.; Nakabayashi, M.; Shibata, N.; Katayama, M.; Minegishi, T.; Domen, K. Visible light-driven Z-scheme water splitting using oxysulfide H$_2$ evolution photocatalysts. *J. Phys. Chem. Lett.* **2016**, *7*, 3892–3896. [CrossRef] [PubMed]

17. Senthil, V.; Badapanda, T.; Chithambararaj, A.; Chandra Bose, A.; Panigrahi, S. Impedance spectroscopy and photocatalysis water splitting for hydrogen production with cerium modified SrBi$_2$Ta$_2$O$_9$ ferroelectrics. *Int. J. Hydrog. Energy* **2016**, *41*, 22856–22865. [CrossRef]

18. Gao, L.; Li, Y.; Ren, J.; Wang, S.; Wang, R.; Fu, G.; Hu, Y. Passivation of defect states in anatase TiO$_2$ hollow spheres with Mg doping: Realizing efficient photocatalytic overall water splitting. *Appl. Catal. B* **2017**, *202*, 127–133. [CrossRef]

19. Schlapbach, L.; Zuttel, A. Hydrogen-storage materials for mobile applications. *Nature* **2001**, *414*, 353–358. [CrossRef] [PubMed]

20. Principi, G.; Agresti, F.; Maddalena, A.; Lo Russo, S. The problem of solid state hydrogen storage. *Energy* **2009**, *34*, 2087–2091. [CrossRef]

21. Sakintuna, B.; Lamari-Darkrim, F.; Hirscher, M. Metal hydride materials for solid hydrogen storage: A review. *Int. J. Hydrog. Energy* **2007**, *32*, 1121–1140. [CrossRef]

22. Lim, K.L.; Kazemian, H.; Yaakob, Z.; Daud, W.R.W. Solid-state Materials and Methods for Hydrogen Storage: A Critical Review. *Chem. Eng. Technol.* **2010**, *33*, 213–226. [CrossRef]

23. Khafidz, N.Z.A.K.; Yaakob, Z.; Lim, K.L.; Timmiati, S.N. The kinetics of lightweight solid-state hydrogen storage materials: A review. *Int. J. Hydrog. Energy* **2016**, *41*, 13131–13151. [CrossRef]

24. Amendola, S.C.; Sharp-Goldman, S.L.; Janjua, M.S.; Kelly, M.T.; Petillo, P.J.; Binder, M. An ultrasafe hydrogen generator: Aqueous, alkaline borohydride solutions and Ru catalyst. *J. Power Sources* **2000**, *85*, 186–189. [CrossRef]

25. Cao, N.; Hu, K.; Luo, W.; Cheng, G. RuCu nanoparticles supported on graphene: A highly efficient catalyst for hydrolysis of ammonia borane. *J. Alloys Compd.* **2014**, *590*, 241–246. [CrossRef]

26. Ouyang, L.Z.; Dong, H.W.; Peng, C.H.; Sun, L.X.; Zhu, M. A new type of Mg-based metal hydride with promising hydrogen storage properties. *Int. J. Hydrog. Energy* **2007**, *32*, 3929–3935. [CrossRef]

27. Liu, Y.; Du, H.; Zhang, X.; Yang, Y.; Gao, M.; Pan, H. Superior catalytic activity derived from a two-dimensional Ti$_3$C$_2$ precursor towards the hydrogen storage reaction of magnesium hydride. *Chem. Commun.* **2016**, *52*, 705–708. [CrossRef] [PubMed]

28. Ouyang, L.Z.; Cao, Z.J.; Wang, H.; Liu, J.W.; Sun, D.L.; Zhang, Q.A.; Zhu, M. Enhanced dehydriding thermodynamics and kinetics in Mg(In)–MgF$_2$ composite directly synthesized by plasma milling. *J. Alloys Compd.* **2014**, *586*, 113–117. [CrossRef]

29. Ouyang, L.Z.; Huang, J.M.; Fang, C.J.; Zhang, Q.A.; Sun, D.L.; Zhu, M. The controllable hydrolysis rate for LaMg$_{12}$ hydride. *Int. J. Hydrog. Energy* **2012**, *37*, 12358–12364. [CrossRef]

30. Ouyang, L.; Ma, M.; Huang, M.; Duan, R.; Wang, H.; Sun, L.; Zhu, M. Enhanced Hydrogen Generation Properties of MgH$_2$-Based Hydrides by Breaking the Magnesium Hydroxide Passivation Layer. *Energies* **2015**, *8*, 4237–4252. [CrossRef]

31. Varin, R.A.; Jang, M.; Czujko, T.; Wronski, Z.S. The effect of ball milling under hydrogen and argon on the desorption properties of MgH$_2$ covered with a layer of Mg(OH)$_2$. *J. Alloys Compd.* **2010**, *493*, L29–L32. [CrossRef]

32. Kadri, A.; Jia, Y.; Chen, Z.; Yao, X. Catalytically Enhanced Hydrogen Sorption in Mg-MgH$_2$ by Coupling Vanadium-Based Catalyst and Carbon Nanotubes. *Materials* **2015**, *8*, 3491–3507. [CrossRef]

33. Hiraki, T.; Hiroi, S.; Akashi, T.; Okinaka, N.; Akiyama, T. Chemical equilibrium analysis for hydrolysis of magnesium hydride to generate hydrogen. *Int. J. Hydrog. Energy* **2012**, *37*, 12114–12119. [CrossRef]

34. Tayeh, T.; Awad, A.S.; Nakhl, M.; Zakhour, M.; Silvain, J.F.; Bobet, J.L. Production of hydrogen from magnesium hydrides hydrolysis. *Int. J. Hydrog. Energy* **2014**, *39*, 3109–3117. [CrossRef]

35. Fernandes, R.; Patel, N.; Paris, A.; Calliari, L.; Miotello, A. Improved H$_2$ production rate by hydrolysis of Ammonia Borane using quaternary alloy catalysts. *Int. J. Hydrog. Energy* **2013**, *38*, 3313–3322. [CrossRef]

36. Kantürk Figen, A.; Coşkuner, B. A novel perspective for hydrogen generation from ammonia borane (NH$_3$BH$_3$) with Co–B catalysts: "Ultrasonic Hydrolysis". *Int. J. Hydrog. Energy* **2013**, *38*, 2824–2835. [CrossRef]

37. Rakap, M. PVP-stabilized Ru–Rh nanoparticles as highly efficient catalysts for hydrogen generation from hydrolysis of ammonia borane. *J. Alloys Compd.* **2015**, *649*, 1025–1030. [CrossRef]

38. Yang, X.J.; Li, L.L.; Sang, W.L.; Zhao, J.L.; Wang, X.X.; Yu, C.; Zhang, X.H.; Tang, C.C. Boron nitride supported Ni nanoparticles as catalysts for hydrogen generation from hydrolysis of ammonia borane. *J. Alloys Compd.* **2017**, *693*, 642–649. [CrossRef]

39. Durak, H.; Gulcan, M.; Zahmakiran, M.; Ozkar, S.; Kaya, M. Hydroxyapatite-nanosphere supported ruthenium(0) nanoparticle catalyst for hydrogen generation from ammonia-borane solution: Kinetic studies for nanoparticle formation and hydrogen evolution. *Rsc. Adv.* **2014**, *4*, 28947–28955. [CrossRef]

40. Garralda, M.A.; Mendicute-Fierro, C.; Rodriguez-Dieguez, A.; Seco, J.M.; Ubide, C.; Zumeta, I. Efficient hydridoirida-beta-diketone-catalyzed hydrolysis of ammonia- or amine-boranes for hydrogen generation in air. *Dalton Trans.* **2013**, *42*, 11652–11660. [CrossRef] [PubMed]

41. Barakat, N.A.M. Catalytic and photo hydrolysis of ammonia borane complex using Pd-doped Co nanofibers. *Appl. Catal. A* **2013**, *451*, 21–27. [CrossRef]

42. Jiang, H.-L.; Xu, Q. Catalytic hydrolysis of ammonia borane for chemical hydrogen storage. *Catal. Today* **2011**, *170*, 56–63. [CrossRef]

43. Li, H.; Yang, Q.; Chen, X.; Shore, S.G. Ammonia borane, past as prolog. *J. Organomet. Chem.* **2014**, *751*, 60–66. [CrossRef]

44. Summerscales, O.T.; Gordon, J.C. Regeneration of ammonia borane from spent fuel materials. *Dalton Trans.* **2013**, *42*, 10075–10084. [CrossRef] [PubMed]

45. Sutton, A.D.; Davis, B.L.; Bhattacharyya, K.X.; Ellis, B.D.; Gordon, J.C.; Power, P.P. Recycle of tin thiolate compounds relevant to ammonia-borane regeneration. *Chem. Commun.* **2010**, *46*, 148–149. [CrossRef] [PubMed]

46. Zhou, Y.; Fang, C.; Fang, Y.; Zhu, F.; Liu, H.; Ge, H. Hydrogen generation mechanism of spontaneous hydrolysis: A sight from ab initio calculation. *Int. J. Hydrog. Energy* **2016**, *41*, 22668–22676. [CrossRef]
47. Retnamma, R.; Novais, A.Q.; Rangel, C.M. Kinetics of hydrolysis of sodium borohydride for hydrogen production in fuel cell applications: A review. *Int. J. Hydrog. Energy* **2011**, *36*, 9772–9790. [CrossRef]
48. Zhong, H.; Wang, H.; Liu, J.W.; Sun, D.L.; Fang, F.; Zhang, Q.A.; Ouyang, L.Z.; Zhu, M. Enhanced hydrolysis properties and energy efficiency of MgH2-base hydrides. *J. Alloys Compd.* **2016**, *680*, 419–426. [CrossRef]
49. Nunes, H.X.; Ferreira, M.J.F.; Rangel, C.M.; Pinto, A.M.F.R. Hydrogen generation and storage by aqueous sodium borohydride (NaBH$_4$) hydrolysis for small portable fuel cells (H$_2$ – PEMFC). *Int. J. Hydrog. Energy* **2016**, *41*, 15426–15432. [CrossRef]
50. Chinnappan, A.; Puguan, J.M.C.; Chung, W.-J.; Kim, H. Hydrogen generation from the hydrolysis of sodium borohydride using chemically modified multiwalled carbon nanotubes with pyridinium based ionic liquid and decorated with highly dispersed Mn nanoparticles. *J. Power Sources* **2015**, *293*, 429–436. [CrossRef]
51. Lin, K.-Y.A.; Chang, H.-A. Efficient hydrogen production from NaBH$_4$ hydrolysis catalyzed by a magnetic cobalt/carbon composite derived from a zeolitic imidazolate framework. *Chem. Eng. J.* **2016**, *296*, 243–251. [CrossRef]
52. Demirci, U.B.; Miele, P. Cobalt in NaBH$_4$ hydrolysis. *Phys. Chem. Chem. Phys.* **2010**, *12*, 14651–14665. [CrossRef] [PubMed]
53. Zhu, J.; Li, R.; Niu, W.; Wu, Y.; Gou, X. Fast hydrogen generation from NaBH$_4$ hydrolysis catalyzed by carbon aerogels supported cobalt nanoparticles. *Int. J. Hydrog. Energy* **2013**, *38*, 10864–10870. [CrossRef]
54. Wang, M.C.; Ouyang, L.Z.; Liu, J.W.; Wang, H.; Zhu, M. Hydrogen generation from sodium borohydride hydrolysis accelerated by zinc chloride without catalyst: A kinetic study. *J. Alloys Compd.* **2017**, *717*, 48–54. [CrossRef]
55. Program, U.S.D.o.E.H. Go/No-Go Recommendation for Sodium Borohydride for On-Board Vehicular Hydrogen Storage. Available online: http://www.hydrogen.energy.gov/pdfs/42220.pdf (accessed on 1 July 2017).
56. Demirci, U.B.; Akdim, O.; Miele, P. Ten-year efforts and a no-go recommendation for sodium borohydride for on-board automotive hydrogen storage. *Int. J. Hydrog. Energy* **2009**, *34*, 2638–2645. [CrossRef]
57. Schlesinger, H.I.; Brown, H.C.; Finholt, A.E. The Preparation of Sodium Borohydride by the High Temperature Reaction of Sodium Hydride with Borate Esters1. *J. Am. Chem. Soc.* **1953**, *75*, 205–209. [CrossRef]
58. Johnson, W.C.; Isenberg, S. Hydrogen Compounds of Silicon. I. The Preparation of Mono- and Disilane. *J. Am. Chem. Soc.* **1935**, *57*, 1349–1353. [CrossRef]
59. Li, Z.P.; Morigazaki, N.; Liu, B.H.; Suda, S. Preparation of sodium borohydride by the reaction of MgH$_2$ with dehydrated borax through ball milling at room temperature. *J. Alloys Compd.* **2003**, *349*, 232–236. [CrossRef]
60. Çakanyildirim, Ç.; Gürü, M. The Production of NaBH$_4$ from Its Elements by Mechano-chemical Reaction and Usage in Hydrogen Recycle. *Energy Sources Part A* **2011**, *33*, 1912–1920. [CrossRef]
61. Bilen, M.; Gürü, M.; Çakanyıldırım, Ç. Role of NaCl in NaBH$_4$ production and its hydrolysis. *Energ Convers. Manag.* **2013**, *72*, 134–140. [CrossRef]
62. Marrero-Alfonso, E.Y.; Gray, J.R.; Davis, T.A.; Matthews, M.A. Minimizing water utilization in hydrolysis of sodium borohydride: The role of sodium metaborate hydrates. *Int. J. Hydrog. Energy* **2007**, *32*, 4723–4730. [CrossRef]
63. Ved, A.S.; Miley, G.H.; Seetaraman, T.S. Recycling Sodium Metaborate to Sodium Borohydride Using Wind-Solar Energy System for Direct Borohydride Fuel Cell. In Proceedings of the ASME 2010 8th International Fuel Cell Science, Engineering and Technology, Brooklyn, NY, USA, 14–16 June 2010; pp. 139–141.
64. Sanli, A.E.; Kayacan, İ.; Uysal, B.Z.; Aksu, M.L. Recovery of borohydride from metaborate solution using a silver catalyst for application of direct rechargable borohydride/peroxide fuel cells. *J. Power Sources* **2010**, *195*, 2604–2607. [CrossRef]
65. McLafferty, J.; Colominas, S.; Macdonald, D.D. Attempts to cathodically reduce boron oxides to borohydride in aqueous solution. *Electrochim. Acta* **2010**, *56*, 108–114. [CrossRef]
66. Kojima, Y.; Haga, T. Recycling process of sodium metaborate to sodium borohydride. *Int. J. Hydrog. Energy* **2003**, *28*, 989–993. [CrossRef]
67. Hsueh, C.-L.; Liu, C.-H.; Chen, B.-H.; Chen, C.-Y.; Kuo, Y.-C.; Hwang, K.-J.; Ku, J.-R. Regeneration of spent-NaBH4 back to NaBH4 by using high-energy ball milling. *Int. J. Hydrog. Energy* **2009**, *34*, 1717–1725. [CrossRef]

68. Kong, L.; Cui, X.; Jin, H.; Wu, J.; Du, H.; Xiong, T. Mechanochemical Synthesis of Sodium Borohydride by Recycling Sodium Metaborate. *Energy Fuels* **2009**, *23*, 5049–5054. [CrossRef]

69. Çakanyıldırım, Ç.; Gürü, M. Processing of NaBH₄ from NaBO₂ with MgH₂ by ball milling and usage as hydrogen carrier. *Renew. Energy* **2010**, *35*, 1895–1899. [CrossRef]

70. Ouyang, L.Z.; Zhong, H.; Li, Z.M.; Cao, Z.J.; Wang, H.; Liu, J.W.; Zhu, X.K.; Zhu, M. Low-cost method for sodium borohydride regeneration and the energy efficiency of its hydrolysis and regeneration process. *J. Power Sources* **2014**, *269*, 768–772. [CrossRef]

71. Lang, C.; Jia, Y.; Liu, J.; Wang, H.; Ouyang, L.; Zhu, M.; Yao, X. NaBH₄ regeneration from NaBO₂ by high-energy ball milling and its plausible mechanism. *Int. J. Hydrog. Energy* **2017**, *42*, 13127–13135. [CrossRef]

72. Liu, B. Kinetic characteristics of sodium borohydride formation when sodium meta-borate reacts with magnesium and hydrogen. *Int. J. Hydrog. Energy* **2008**, *33*, 1323–1328. [CrossRef]

73. Eom, K.; Cho, E.; Kim, M.; Oh, S.; Nam, S.-W.; Kwon, H. Thermochemical production of sodium borohydride from sodium metaborate in a scaled-up reactor. *Int. J. Hydrog. Energy* **2013**, *38*, 2804–2809. [CrossRef]

74. Liu, B.H.; Li, Z.P.; Zhu, J.K.; Morigasaki, N.; Suda, S. Sodium Borohydride Synthesis by Reaction of Na₂O Contained Sodium Borate with Al and Hydrogen. *Energy Fuels* **2007**, *21*, 1707–1711. [CrossRef]

75. Beaird, A.M.; Li, P.; Marsh, H.S.; Al-Saidi, W.A.; Johnson, J.K.; Matthews, M.A.; Williams, C.T. Thermal Dehydration and Vibrational Spectra of Hydrated Sodium Metaborates. *Ind. Eng. Chem. Res.* **2011**, *50*, 7746–7752. [CrossRef]

76. Liu, B.H.; Li, Z.P.; Zhu, J.K. Sodium borohydride formation when Mg reacts with hydrous sodium borates under hydrogen. *J. Alloys Compd.* **2009**, *476*, L16–L20. [CrossRef]

77. Chen, W.; Ouyang, L.Z.; Liu, J.W.; Yao, X.D.; Wang, H.; Liu, Z.W.; Zhu, M. Hydrolysis and regeneration of sodium borohydride (NaBH₄)—A combination of hydrogen production and storage. *J. Power Sources* **2017**, *359*, 400–407. [CrossRef]

78. Zhang, H.; Zheng, S.; Fang, F.; Chen, G.; Sang, G.; Sun, D. Synthesis of NaBH₄ based on a solid-state reaction under Ar atmosphere. *J. Alloys Compd.* **2009**, *484*, 352–355. [CrossRef]

79. Ouyang, L.; Chen, W.; Liu, J.; Felderhoff, M.; Wang, H.; Zhu, M. Enhancing the Regeneration Process of Consumed NaBH₄ for Hydrogen Storage. *Adv. Energy Mater.* **2017**, *7*, 1700299. [CrossRef]

80. Zhong, H.; Ouyang, L.Z.; Ye, J.S.; Liu, J.W.; Wang, H.; Yao, X.D.; Zhu, M. An one-step approach towards hydrogen production and storage through regeneration of NaBH4. *Energy Storage Mater.* **2017**, *7*, 222–228. [CrossRef]

81. Andrieux, J.; Demirci, U.B.; Hannauer, J.; Gervais, C.; Goutaudier, C.; Miele, P. Spontaneous hydrolysis of sodium borohydride in harsh conditions. *Int. J. Hydrog. Energy* **2011**, *36*, 224–233. [CrossRef]

82. Chong, M.; Karkamkar, A.; Autrey, T.; Orimo, S.-I.; Jalisatgi, S.; Jensen, C.M. Reversible dehydrogenation of magnesium borohydride to magnesium triborane in the solid state under moderate conditions. *Chem. Commun.* **2011**, *47*, 1330–1332. [CrossRef] [PubMed]

83. Dai, H.-B.; Ma, G.-L.; Kang, X.-D.; Wang, P. Hydrogen generation from coupling reactions of sodium borohydride and aluminum powder with aqueous solution of cobalt chloride. *Catal. Today* **2011**, *170*, 50–55. [CrossRef]

inorganics

MDPI

Article

Hydrogen Sorption in Erbium Borohydride Composite Mixtures with LiBH$_4$ and/or LiH

Michael Heere [1], Seyed Hosein Payandeh GharibDoust [2], Matteo Brighi [3], Christoph Frommen [1], Magnus H. Sørby [1], Radovan Černý [3], Torben R. Jensen [2] and Bjørn C. Hauback [1,*]

[1] Physics Department, Institute for Energy Technology, NO-2027 Kjeller, Norway; heere.michael@gmail.com (M.H.); Christoph.Frommen@ife.no (C.F.); Magnus.Sorby@ife.no (M.H.S.)

[2] Interdisciplinary Nanoscience Center (iNANO) and Department of Chemistry, University of Århus, Langelandsgade 140, DK-8000 Århus C, Denmark; hosein.payandeh@inano.au.dk (S.H.P.G.); trj@chem.au.dk (T.R.J.)

[3] Laboratory of Crystallography, Department of Quantum Matter Physics, University of Geneva, Quai Ernest-Ansermet 24, Ch-1211 Geneva, Switzerland; Matteo.Brighi@unige.ch (M.B.); Radovan.Cerny@unige.ch (R.Č.)

* Correspondence: Bjorn.Hauback@ife.no; Tel.: +47-9740-8844

Academic Editor: Hiroshi Kageyama
Received: 27 March 2017; Accepted: 20 April 2017; Published: 26 April 2017

Abstract: Rare earth (RE) metal borohydrides have recently been receiving attention as possible hydrogen storage materials and solid-state Li-ion conductors. In this paper, the decomposition and reabsorption of Er(BH$_4$)$_3$ in composite mixtures with LiBH$_4$ and/or LiH were investigated. The composite of 3LiBH$_4$ + Er(BH$_4$)$_3$ + 3LiH has a theoretical hydrogen storage capacity of 9 wt %, nevertheless, only 6 wt % hydrogen are accessible due to the formation of thermally stable LiH. Hydrogen sorption measurements in a Sieverts-type apparatus revealed that during three desorption-absorption cycles of 3LiBH$_4$ + Er(BH$_4$)$_3$ + 3LiH, the composite desorbed 4.2, 3.7 and 3.5 wt % H for the first, second and third cycle, respectively, and thus showed good rehydrogenation behavior. In situ synchrotron radiation powder X-ray diffraction (SR-PXD) after ball milling of Er(BH$_4$)$_3$ + 6LiH resulted in the formation of LiBH$_4$, revealing that metathesis reactions occurred during milling in these systems. Impedance spectroscopy of absorbed Er(BH$_4$)$_3$ + 6LiH showed an exceptional high hysteresis of 40–60 K for the transition between the high and low temperature phases of LiBH$_4$, indicating that the high temperature phase of LiBH$_4$ is stabilized in the composite.

Keywords: borohydride; rare earth element; hydrogen storage; decomposition; solid state electrolyte

---

1. Introduction

One of the most promising candidates for solid state hydrogen storage applications is LiBH$_4$, which has a theoretical capacity of 18.5 wt % H$_2$. The practical hydrogen amount that can be desorbed is 13.8 wt %, due to the formation of very stable LiH during decomposition, but LiBH$_4$ is still among the materials with the highest hydrogen capacity considered for solid state storage [1,2]. However, it requires very tough conditions for rehydrogenation [3] and suffers from capacity loss on cycling due to the formation of higher boranes [4]. Another class of materials to be considered is rare earth (RE) borohydrides, with hydrogen capacities varying between 9.0 wt % for Y(BH$_4$)$_3$ and 5.5 wt % for Yb(BH$_4$)$_3$ [5–19]. Li-containing RE borohydrides are also considered as solid state electrolytes for new battery applications, due to their high Li-ion conductivities [20–23]. RE borohydrides have shown novel properties such as luminescence, and a magnetocaloric effect has also been recently published [24–26].

Gennari et al. investigated the composite mixtures of $6LiBH_4$ and $RECl_3$ with RE = Ce, Gd and found a decrease in decomposition temperature of $LiBH_4$ due to an in situ formed REH_2 phase [17]. Additionally, the decomposed composite showed the ability to reabsorb 20% of the initial hydrogen content. Frommen et al. investigated the composite mixtures of $LiBH_4$, LiH and $RE(BH_4)_3$ with RE = La, Er [27]. It was reported that the desorption temperature of $LiBH_4$ in the mixture with RE = Er decreased by about 100 °C compared to pure $LiBH_4$. Reabsorption at 340 °C under 100 bar H_2 yielded a hydrogen uptake of 66% after desorption against vacuum, and 80% after desorption against 5 bar H_2, respectively, of the initially release of 3 wt % H_2.

The established technique to synthesize RE borohydrides is the solid state metathesis approach, as it has the advantage of direct synthesis of a RE borohydride without complicated in vacuo manipulations or possible decomposition [28,29]. The RE chloride and an alkali metal borohydride (i.e., Li, Na, K) are mixed in a one-step mechanochemical synthesis to form the RE borohydride and the alkali metal chloride via a metathesis reaction. $LiBH_4$ has proven to be an efficient precursor, but also other metal borohydrides e.g., $NaBH_4$, have been successfully used [30–32]. As an alternative method, the solvent-based synthesis of borohydrides has been used for over 50 years with the advantage of removing the alkali metal chloride (i.e., LiCl) [28], which is necessary to increase the hydrogen capacity. The halide-free synthesis in this paper is based on a mechanochemical synthesis [33] and wet-chemical extraction, as described by Hagemann et al. [34,35].

This work presents two new composite materials: $Er(BH_4)_3 + 6LiH$ and $3LiBH_4 + Er(BH_4)_3 + 3LiH$. The decomposition and reabsorption of hydrogen have been studied in detail by in situ synchrotron radiation powder X-ray diffraction (SR-PXD). Desorption-absorption cycling measurements were conducted in a Sieverts-type apparatus. The rehydrogenation properties of the $3LiBH_4 + Er(BH_4)_3 + 3LiH$ system were investigated during three desorption-absorption cycles. One cycle was studied for the $Er(BH_4)_3 + 6LiH$ system. The latter was also investigated by impedance spectroscopy. Finally, thermal gravimetric (TG) and differential scanning calorimetry (DSC), as well as mass spectrometry (MS) were conducted.

2. Results & Discussion

2.1. Er(BH₄)₃ + 6LiH (S1)

Figure 1 shows the in situ synchrotron radiation powder X-ray diffraction (SR-PXD) data of $Er(BH_4)_3 + 6LiH$ (S1) from room temperature (RT) to 280 °C. At RT, Bragg peaks corresponding to LiH, orthorhombic-$LiBH_4$ (o-$LiBH_4$) and ErH_2, as well as minor peaks from $Er(BH_4)_3$, are present, suggesting the reaction pathway of Equation (1) for the ball milling reaction.

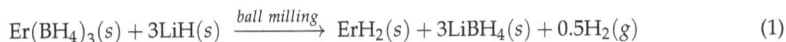

$$Er(BH_4)_3(s) + 3LiH(s) \xrightarrow{\text{ball milling}} ErH_2(s) + 3LiBH_4(s) + 0.5H_2(g) \qquad (1)$$

The small amount of $Er(BH_4)_3$ present (Table 1) indicates that Equation (1) has almost gone to completion during milling. The in situ SR-PXD data show the phase transition of o- to hexagonal-$LiBH_4$ (h-$LiBH_4$) at 100 °C. The Bragg peaks of $Er(BH_4)_3$ start to decrease simultaneously and are gone at 240 °C. ErH_2 is present in significant amounts (44.1(4) wt %) directly after ball milling. Its lattice parameters increase more than expected from the thermal expansion from 186 °C, indicating that it is further hydrogenated, induced by the hydrogen pressure in the in situ SR-PXD measurements setup, with a likely intermediate state of $ErH_{2+\delta}$ with ($0 \leq \delta \leq 1$) as shown in Figure S1. Bragg peaks from $LiBH_4$ disappear at 270 °C, caused by its melting. Above 280 °C, the sample oxidizes, as evident from strong Bragg peaks of Er_2O_3 (not shown), and further analysis of the decomposition process was therefore not possible.

Figure 2a shows the differential scanning calorimetry (DSC) data of S1. The phase transition from o- to h-$LiBH_4$ occurs at 105 °C and its melting at 260 °C (peak temperatures), which is in good agreement with the SR-PXD data. A rather minor exothermic DSC event (dashed circle in Figure 2a) can be seen in analogy to the increase in the lattice parameter observed in SR-PXD starting at 186 °C, corresponding to the reaction from ErH_2 towards ErH_3. The decomposition of the Li borohydride appears at 326 °C. There is a minor endothermic peak after the major endothermic event, which can be assigned to desorption of ErH_3 to ErH_2 (a similar decomposition temperature is observed for pure ErH_3, as shown in Figure S2). Thermogravimetric (TG) data for S1 are also presented in Figure 2 and show mass loss of 3.4 wt % between RT and 400 °C, including a 30-min isotherm.

The powder X-ray diffraction (PXD) pattern of one complete desorption absorption cycle in the Sieverts apparatus is shown in Figure 2b. The main product after desorption at 400 °C and 5–10 bar H_2 is ErH_3, formed by additional hydrogen absorption by ErH_2 upon cooling. Minor Bragg peaks of ErB_4 are also observed. Reabsorption at 340 °C under 100 bar H_2 for 12 h resulted in the formation of $LiBH_4$, as well as increased intensity of ErH_3 peaks.

Table 1. Compositions and synthesis method of the investigated samples. The products after ball milling are given with lattice parameters and fractions from Rietveld refinements of diffraction data at room temperature (RT). Refinements are shown in Table S1. Sample S2 was stored in a glove box for nine months before measurements were taken. Estimated standard deviations are given in parentheses.

Sample	Reactants	Method	Products	Lattice Parameters (Å)	Refined Fractions (wt %)
			$Er(BH_4)_3$	$a = 10.884(3)$	1.1(2)
S1	$Er(BH_4)_3 + 6LiH$	1 h ball milling	$LiBH_4$	$a = 7.174(7)$ $b = 4.426(5)$ $c = 6.807(9)$	29.9(3)
			LiH	$a = 4.0851(6)$	24.8(2)
			ErH_2	$a = 5.1458(4)$	44.1(4)
			$Er(BH_4)_3$	$a = 10.8134(3)$	29.7(1)
S2	$Er(BH_4)_3 + 3LiBH_4 + 3LiH$	1 h ball milling	$LiBH_4$	$a = 7.172(5)$ $b = 4.430(3)$ $c = 6.804(6)$	38.2(16)
			LiH	$a = 4.0851$	0.0(9)
			ErH_2	$a = 5.1823(3)$	32.1(1)

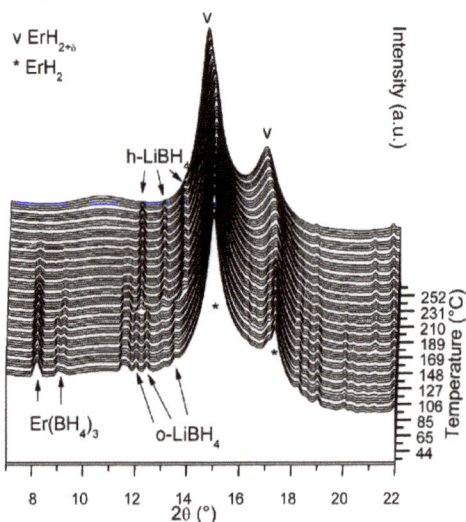

Figure 1. In situ SR-PXD data of sample S1 during heating from RT to 280 °C (5 °C·min^{-1}) under 1 bar H_2. Arrows mark the Bragg peaks of $Er(BH_4)_3$, o- and h-$LiBH_4$. $\lambda = 0.77787$ Å.

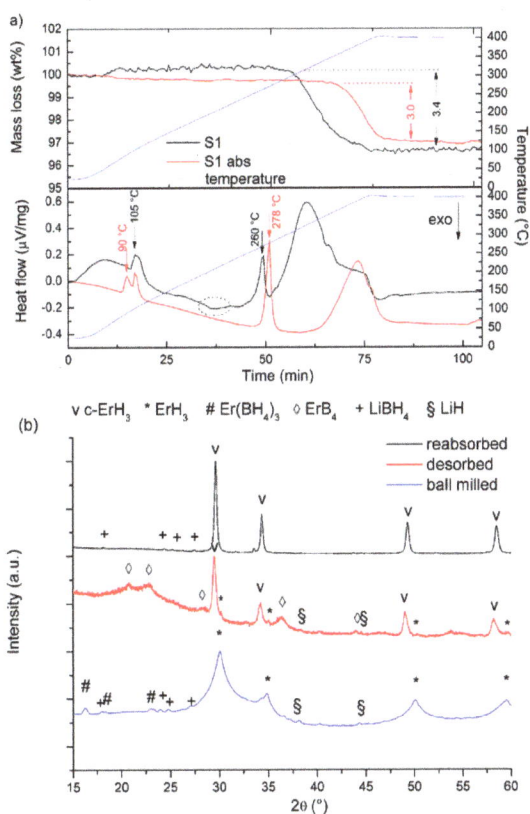

Figure 2. (a) TG-DSC data of ball milled S1 (black curves) and reabsorbed S1 (S1 abs) (red curves), showing TG data (top) and DSC signal (bottom). *X*-axis in time (min) as the heating from RT to 400 °C proceeded with a 30-min isotherm. Temperature (°C) is given on the secondary *y*-axis to the right; (b) Ex situ PXD data of S1 showing one absorption-desorption cycle with ball milled (bottom), desorbed (middle), and reabsorbed (top) composite. λ = 1.54056 Å.

The thermogravimetric and differential scanning calorimetry (TG-DSC) data for the decomposition of reabsorbed S1 is shown in Figure 2a (red curves). TG data show a mass loss of 3.0 wt %, corresponding to 88% of the initial released hydrogen. An increase in onset temperature of hydrogen release can be observed ($\Delta T = 70$ °C). In the DSC data, two peaks occur at rather low temperatures. The first peak is at 90 °C, but the process behind it is not completely understood (see Li-ion conductivity section). The second one at 105 °C corresponds to the *o*- to *h*-LiBH$_4$ transition. The melting of LiBH$_4$ occurs at 278 °C. The last endothermic event coincides with the mass loss and occurs at 390 °C, which is probably a superposition of two events. The first is proposed to be the decomposition of LiBH$_4$, while the second is possibly the reduction of ErH$_3$ to ErH$_2$, see Figure S2.

In general, the desorption-absorption cycle shown in this section follows the reaction pathways outlined in Equations (2) and (3).

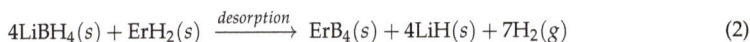

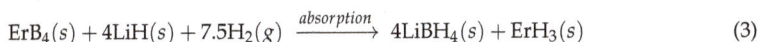

$$4\text{LiBH}_4(s) + \text{ErH}_2(s) \xrightarrow{\text{desorption}} \text{ErB}_4(s) + 4\text{LiH}(s) + 7\text{H}_2(g) \tag{2}$$

$$\text{ErB}_4(s) + 4\text{LiH}(s) + 7.5\text{H}_2(g) \xrightarrow{\text{absorption}} 4\text{LiBH}_4(s) + \text{ErH}_3(s) \tag{3}$$

2.2. $3LiBH_4 + Er(BH_4)_3 + 3LiH$ (S2)

Figure 3a shows the ex situ lab PXD patterns of $3LiBH_4 + Er(BH_4)_3 + 3LiH$ (S2) in the as-milled, desorbed, and reabsorbed states. The Bragg peaks of the reactants $LiBH_4$, $Er(BH_4)_3$ and LiH, are observed after ball milling. In addition, ErH_2 is formed in minor amounts, showing that a reaction has already started during the milling process. The products after desorption at 400 °C and 5–10 bar H_2 are mainly ErB_4 and several Er-hydrides. Cubic (c-) ErH_3 (*Fm-3m*) is the main phase, but small amounts of ErH_2, ErH and trigonal (t-) ErH_3 (*P-3c1*) are also present. As suggested above, ErH_3 appeared during cooling under hydrogen pressure. Minor Bragg peaks corresponding to ErB_{12} are detected as well. Reabsorption at 340 °C under 100 bar H_2 for 12 h led to the complete consumption of the Er borides. Trivalent Er-hydride was found after reabsorption in both the cubic and trigonal modifications.

Interestingly, the Bragg peaks of ErB_4 show less intensity in the ex situ PXD pattern of desorbed S1 in Figure 2b than in desorbed S2, although the conditions were the same. This observation let us to a hypothesis that the excess of $LiBH_4$ somehow increases the crystallinity of the material. This is supported by the observations in another recently published study, where the decomposition of pure $Er(BH_4)_3$ resulted in a completely amorphous material [36]. The crystallinity was improved by addition of LiH and by reabsorption. The products were ErH_3 and $LiBH_4$, and the final material was crystalline. Whether this observation is caused by amorphous Er–B–H species, which simply decrease when $LiBH_4$ is formed, or if the process is triggered directly by $LiBH_4/ErH_3$ formation, cannot be concluded.

TG-DSC data of S2, stored for three months in a glove box, is presented in Figure 3b. TG data show a mass loss of 3.9 wt % up to 400 °C, and DSC data show three endothermic events, two sharp and one broad. The two sharp events are the phase transition and melting of $LiBH_4$ occurring at 111 and 271 °C, respectively. These events are not related to any mass loss in the TG data. The third endothermic peak is broad, and therefore probably a superposition of the decomposition of $Er(BH_4)_3$ and $LiBH_4$ with peak temperatures at 336 and 391 °C, respectively. Figure 3bii shows H_2 ($m/z = 2$) desorption between 275 and 400 °C. Minor amounts of B_2H_6 ($m/z = 26$) are also seen in Figure 3bi, as well as $S(CH_3)_2$ ($m/z = 62$) between 75 and 125 °C.

Figure 4a shows in situ SR-PXD desorption data of S2, which was stored for nine months in a glove box. The RT data show significant Bragg peaks from ErH_2. Compared to the rather minor peaks of ErH_2 directly after ball milling in Figure 3a, it becomes clear that ErH_2 is formed during the storage in Ar. This observation suggests that the reaction in Equation (1) even occurs at ambient temperature for S2. Equation (1) is an intermediate reaction, showing the formation of $LiBH_4$, initiated during the milling of $Er(BH_4)_3$ and LiH, as well as the formation of ErH_2. Bragg peaks of the other phases appear for the as-milled S2 (Figure 3a) at RT. The atomic positions for Er in $Er(BH_4)_3$ were refined (Figure S3 and Table S1) and are within the standard deviations of those reported for Y in $Y(BH_4)_3$ [5]. The in situ SR-PXD desorption starts with a heating ramp from RT to 400 °C with a heating rate of 5 °C·min^{-1} and a 30-min isotherm at 400 °C. The gap in the data is due to a lost connection to the detector for 15 min in the temperature interval of 30–80 °C. The phase transition of $LiBH_4$ occurs at 100 °C. The Bragg peaks of $Er(BH_4)_3$ decrease and are gone at 195 °C, which is 55 °C lower compared to the in situ SR-PXD desorption data of S1. In any case, these observations are in agreement with earlier reports, where the halide free yttrium borohydride turns into "an X-ray amorphous solid" without melting at ~200 °C [35]. Therefore, we assume an amorphisation, as all Bragg peaks of $Er(BH_4)_3$ disappear and the DSC data indicate no melting. The amorphisation temperatures in S1 and S2 are, however, 50–100 °C lower than the reported amorphisation temperature of pure $Er(BH_4)_3$ [36]. The melting of $LiBH_4$ can be observed by the in situ SR-PXD data at 295 °C. The hydrogenation of the bivalent to trivalent erbium hydride occurs simultaneously, which is evident from a shift to lower 2θ angles for the Er hydride peaks (Figure 4a and Figure S1) and in good agreement with the in situ SR-PXD desorption data of S1. The Bragg peaks of ErH_3 increase in intensity until the heating is stopped after 30 min at 400 °C, concluding that the timeframe of the in situ SR-PXD desorption measurement for S2 does not allow for the formation of ErB_4, which is the suggested desorption product, see Equation (2).

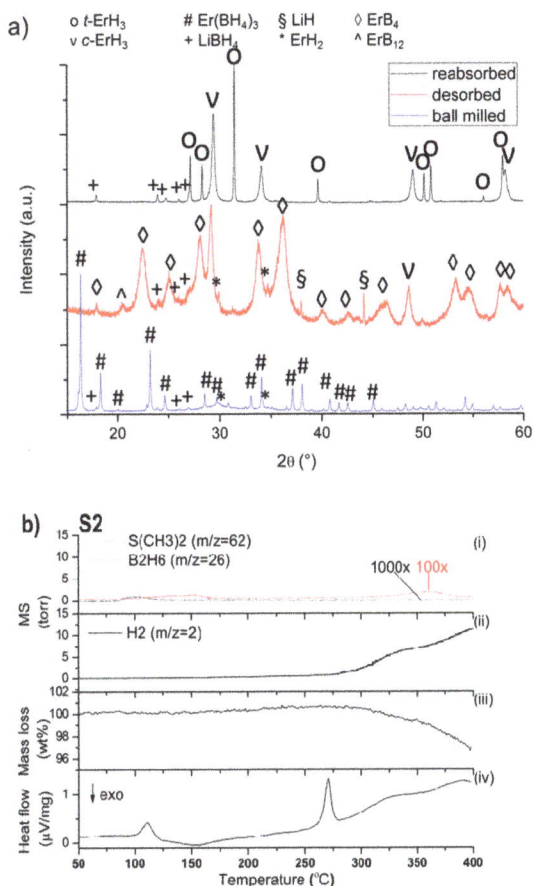

Figure 3. (a) Ex situ PXD data of S2 at RT showing one absorption-desorption cycle with ball milled (bottom), desorbed (middle), and reabsorbed (top) composite. $\lambda = 1.54056$ Å; (b) MS-TG-DSC of ball milled S2 from RT to 400 °C (heating rate 5 °C·min^{-1}) showing: (i) release of S(CH$_3$)$_2$ and B$_2$H$_6$; (ii) H$_2$; (iii) TG data; and (iv) DSC data.

The starting point of the in situ SR-PXD reabsorption is the ex-situ desorbed S2, see Figure 3a (middle, red curve). The in situ SR-PXD reabsorption of desorbed S2 was performed with a heating ramp of 5 °C·min^{-1} from RT to 340 °C under 93 bar H$_2$. From RT to 261 °C, the intensities of all phases appear constant. The connection to the detector was lost at 261 °C for 1 h until 42 min into the isothermal regime at 340 °C. During the connection loss, the Bragg peaks of c-ErH$_3$ increase moderately, while those of ErB$_4$ decrease slightly. This trend continues during the additional 2-h isotherm shown in Figure 4b. The minor changes in intensity indicate that it is again not possible to follow a complete hydrogenation during the timeframe of the in situ measurement.

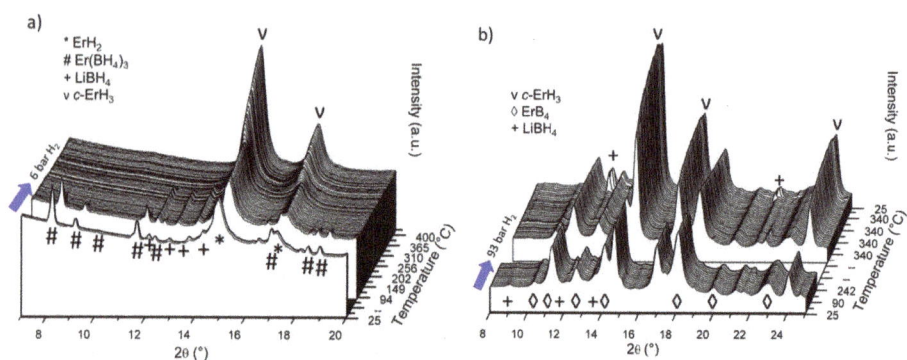

Figure 4. In situ SR-PXD data showing (**a**) desorption of S2 during heating from RT to 400 °C (5 °C·min^{-1}) under 6 bar H$_2$. A detector connection loss is the reason for the gaps in intensity (heating continued). λ = 0.77787 Å; (**b**) reabsorption of Sieverts apparatus desorbed S2 during heating from RT to 340 °C (5 °C·min^{-1}) under 93 bar H$_2$ followed by a 2-h isotherm, and cooling to RT. λ = 0.77787 Å.

Three desorption-absorption cycles were measured for S2 in a Sieverts-type apparatus. The volumetric desorption data after 5 h at 400 °C and 5–10 bar H$_2$ is illustrated in Figure 5. The first, second and third desorption show a hydrogen release of 4.2 wt %, 3.7 wt % and 3.5 wt %, respectively. During the first desorption of S2, Er(BH$_4$)$_3$ is decomposed to Er-hydride according to the ex situ PXD data, which resulted in a slower hydrogen release at the beginning of the first desorption (black curve in Figure 5). As Er(BH$_4$)$_3$ is not formed on rehydrogenation, this step is not repeated after the first desorption, thus resulting in a faster desorption for the second/third reabsorption.

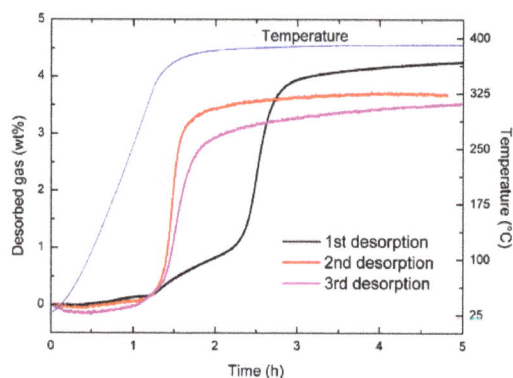

Figure 5. Hydrogen release in the first three desorption during cycling of S2 in a Sieverts-type apparatus. The hydrogen pressure is 5–10 bar.

The TG-DSC data of S2 after the first reabsorption is shown in Figure 6. The DSC signal is similar to the ball milled sample in Figure 3b. The phase transition of LiBH$_4$ shifts to a slightly lower temperature, 103 °C (111 °C for ball milled S2). The melting of LiBH$_4$ still occurs at 270 °C. The third minor endothermic peak at 322 °C in Figure 6 cannot be explained. Although the set temperature for the isotherm is 400 °C, the fourth peak was observed at a peak temperature of 403 °C (due to a slight temperature overshoot) in Figure 6. It is suggested to be the decomposition of LiBH$_4$, and hence slightly higher than for S2 after ball milling. The mass loss starts simultaneously with the fourth endothermic DSC event, also suggesting the decomposition of LiBH$_4$. The fifth peak in the isothermal regime after 7 min at 400 °C is suggested to be the reduction of ErH_3 to ErH_2 as the mass loss continues through this event (Figure S2). The TG data in Figure 6 show a mass loss of 4.5 wt % for reabsorbed S2. Although it was reported that hydrogen pressure has a beneficial effect on desorption behavior, the TG and Sieverts data for the reabsorbed S2 show a weight loss of 4.5 and 3.7 wt %, respectively. The total hydrogen content of S2 is 9.0 wt %, but only 6.0 wt % can be obtained when forming LiH, meaning that our TG data shows a hydrogen uptake of 75%, compared to the nominal maximum value.

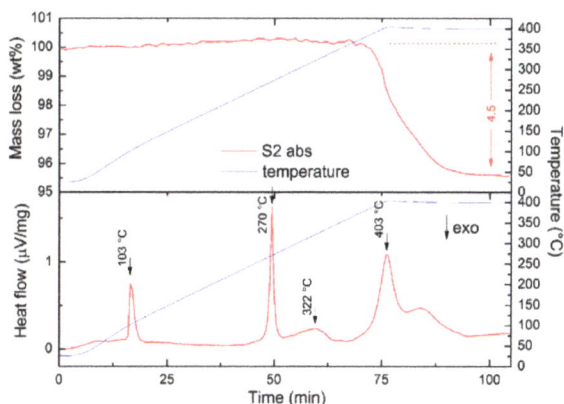

Figure 6. TG (top) and DSC (bottom) data of S2 after the first reabsorption. The *x*-axis is in time (min) as the measurement proceeds isothermally for 30 min after the heating ramp. Temperature (°C) is shown on the *y*-axis to the right.

To summarize, we observe from the PXD and SR-PXD data an increasing amount of ErB_4 during thermal decomposition and an increasing amount of $ErH_{2+\delta}$ during reabsorption for S2. This is in good agreement with the following reaction pathway, with an assumed start of decomposition similar to reaction products in Equation (1)

$$ErH_3(s) + 4LiBH_4(s) \leftrightarrow ErB_4(s) + 4LiH(s) + 7.5H_2(g) \tag{4}$$

The theoretical storage capacity of Equation (4) of 5.1 wt % H was not achieved, which is possibly caused by the uncompleted decomposition reactions, as residuals of ErH_3 have been found in decomposed samples.

The comparison of S2 to the recently investigated $6LiBH_4 + ErCl_3 + 3LiH$ composite by Frommen et al. [27] shows that the two systems behave very similar with respect to rehydrogenation. During the first reabsorption, 80% to 85% of the original hydrogen content could be reabsorbed for both systems at similar conditions. The weight loss for our system shows 4.5 wt % after one desorption-absorption cycle while 2.4 wt % was reported for the other composite [27], meaning that the absorption in our composite has almost double the hydrogen capacity after the first cycle.

There are some differences in the hydrogenation products. In our halide-free S2 composite, c-ErH_3 and t-ErH_3 are both formed during rehydrogenation with 27.4(2) wt % and 14.4(1) wt %, respectively. $LiBH_4$ is the major reabsorption product with refined 58.1(2) wt %, suggesting, as reported recently, that LiH plays a crucial role in the formation of new $LiBH_4$ at the employed conditions [36]. In Reference [27] only c-ErH_3 was observed after rehydrogenation, but no t-ErH_3, similar to our absorbed S1 sample.

2.3. Li-Ion Conductivity

The Li-ion conductivity of absorbed S1 was measured with alternating current (AC) impedance spectroscopy from RT to 140 °C (Figure 7). Further details to the experimental setup are given in

Reference [37] and the references therein. The measurement was motivated by the hypothesis that the high temperature phase of $LiBH_4$ was stabilized in the composite, thus shifting the phase transition to lower temperatures. Evidently, the absorbed S1 showed an endothermic DSC event as low as 90 °C, which we first considered to be the phase transition in $LiBH_4$ (see DSC data, red curve in Figure 2a).

Figure 7 shows the Li^+ ionic conductivity for absorbed S1, and the values are systematically lower than for pure $LiBH_4$ as expected, since $LiBH_4$ only makes up 21 wt % in the present sample (The low conductivity values for the samples can be explained by minor amounts of $LiBH_4$ after rehydrogenation. For absorbed S1 the content is approx. 2.5 molar equivalent of $LiBH_4$ corresponding to 21 wt %. This value was calculated from the mass loss shown by TG data of absorbed S1). However, upon heating above 110 °C, the conductivities reach about four orders of magnitude higher than the RT modification, which is in good agreement with the conductivity enhancement in pure $LiBH_4$ [38]. The conductivity in absorbed S1 (Figure 7) shows a slightly steeper slope than pure $LiBH_4$ [38], which manifested in the absence of the "conductivity jump" in our data. We first suggested that this observation was caused by the initial ball milling of S1. In a recent publication, it was shown that ball milled $LiBH_4$-LiI samples have a higher conductivity before annealing caused by the formation of a "defect rich microstructure," which influences the conductivity positively [39]. However, the sample chosen for impedance spectroscopy was heated to 400 °C during the first decomposition and then reabsorbed at 340 °C. This heat treatment should heal all defects and disregard their influence [39]. In summary, the high temperature phase transition is observed at 110 °C, which is in agreement with pure $LiBH_4$ [38,40]. From the Li-ion conductivity measurements, the DSC event at 90 °C for absorbed S1 cannot be explained by the high temperature phase transition of $LiBH_4$.

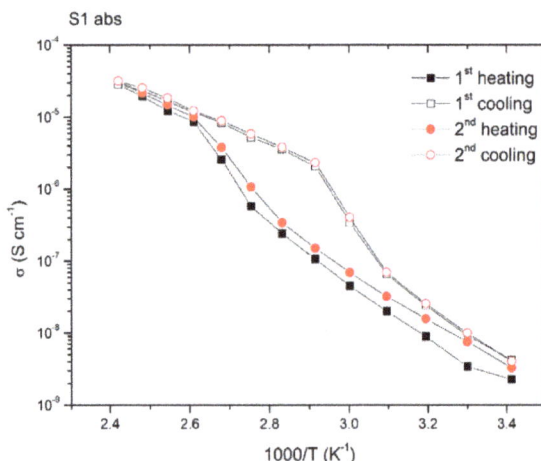

Figure 7. Li-ion conductivity measured by AC impedance spectroscopy for S1 abs. Conductivity was measured every 10 °C from RT to 140 °C and back down to RT. See text for further discussion.

However, a rather wide hysteresis is observed in Figure 7, which is in the range of 40–60 K. That is in strong contrast to the hysteresis of pure and ball milled $LiBH_4$, which is only 4 K [41], but rather close to the $LiBH_4$-LiCl system in which the hysteresis is about 20–40 K [40]. The presence of lithium halides can be ruled out, as it would significantly lower the phase transition temperature [40]. It has been observed [42] that nanoconfined $LiBH_4$ in mesoporous silica scaffolds shows a remarkable Li-conductivity in the temperature range of RT-140 °C. According to Blanchard et al. [42], the high conductivity is a consequence of two different fractions of $LiBH_4$, a bulk $LiBH_4$ fraction and a thin (1.0 nm) interfacial layer of $LiBH_4$. Assuming a morphological change in the grain size and in the inter-grains arrangement, it could be assumed that the formation of the abovementioned layer between

LiBH$_4$ and ErH$_3$ after several cycles of hydrogenation has happened. This can explain the inertia of the system to undergo the transition to the RT phase. The ionic conductivity contribution of this layer is dominant with respect to the bulk LiBH$_4$, however, it is not visible by diffraction due to its nanoscopic nature. This effect of nanoconfined LiBH$_4$ could explain the rather low endothermic DSC event, which were discussed above, as similar events were also reported in Reference [42].

SEM images (see Figures S4–S6) were collected, but their resolution is not sufficient to observe a 1.0-nm thin interfacial layer. Further measurements with TEM are necessary to conclude with certainty that an interfacial layer of LiBH$_4$ has formed. Due to the paramagnetic properties of erbium, it was not possible to conduct NMR.

3. Materials and Methods

ErCl$_3$ (99.9%), LiBH$_4$ (>95%), LiH (>95%) and dimethyl sulfide (S(CH$_3$)$_2$, anhydrous, 99.9%) were purchased from Sigma Aldrich, St. Louis, MO, USA and used as received.

The synthesis of Er(BH$_4$)$_3$ has been described in Reference [36]. The composite mixtures were ball milled using a Fritsch Pulverisette 6 planetary mill (Fritsch, Idar-Oberstein, Germany) employing an 80-mL tungsten carbide-coated steel vial and balls. A ball to powder ration of 40:1 was used. All sample descriptions of the composite mixtures are given in Table 1, including their composition and synthesis method as well as refined lattice parameters and phase fractions obtained by Rietveld refinements.

The samples were stored and handled in an MBraun glove box (MBraun Inertgas-Systeme GmBH, Garching, Germany) fitted with a recirculation system and oxygen/humidity sensors with H$_2$O/O$_2$ levels below 1 ppm. All procedures outside of the glove box were performed using in vacuo or Schlenk line techniques under a purified Ar atmosphere. An in-house manufactured Sieverts-type apparatus [43] was used for hydrogen desorption–absorption cycling experiments. Desorption was performed using a temperature ramp of 5 °C·min^{-1} from room temperature to 400 °C under 3 bar H$_2$, followed by a 12 h isothermal step. Absorption was conducted for 12 h at 340 °C and 100 bar H$_2$.

Powder X-ray diffraction (PXD) data were collected using a Rigaku SmartLab diffractometer (Rigaku, Tokyo, Japan). The samples were measured within rotating glass capillaries, with an inner diameter of 0.5 mm and sealed with silicone grease. All measurements were completed in Debye-Scherrer geometry using CuKα radiation with λ = 1.54056 Å, over the scattering angles $2\theta = 10°$–$70°$.

In situ synchrotron radiation powder X-ray diffraction (SR-PXD) was executed at Swiss Norwegian Beam Lines, BM01 [44], at the European Synchrotron Radiation Facility (ESRF), Grenoble, France. The cycling experiments were conducted using a heating/cooling rate of 5 °C·min^{-1} with a typical H$_2$ pressure of ~100 bar for absorption and ~3 bar for desorption. Hydrogen pressure was employed for all volumetric and in situ SR-PXD experiments, as similar conditions were reported to increase gas desorption compared to desorption under vacuum [17,27,45]. Data were collected using a Pilatus 2 M detector (DECTRIS Ltd., Baden-Daettwil, Switzerland) and a sample-to-detector distance of 146 mm at wavelengths of 0.77787 Å. Wavelength and sample-to-detector distance were calibrated from an NIST LaB$_6$ standard. Exposure time was set to 30 s, giving a temperature resolution of 2.5 K/pattern. The sample was contained in a single-crystal sapphire tube (inner diameter 0.8 mm) connected to the cell with Vespel ferrules and Swagelok fittings. Hydrogen was introduced and removed from the cell with an in-house built computer-controlled gas rig. The cell was rotated by 10° to improve powder averaging during each exposure. The single-crystal reflections from the sapphire tubes were masked out manually in Fit2D [46]. Rietveld refinements were performed with GSAS and Expgui software [47,48]. Thompson-Cox-Hastings pseudo-Voigt functions with three Gaussian and one Lorentzian parameter were used to model the Bragg peak profiles [49]. The background was fitted with a shifted Chebyshev polynomial with up to 36 terms. The atomic positions for the Er(BH$_4$)$_3$ were taken from Y(BH$_4$)$_3$ [5] and refined with the data shown in Figure S3 at room temperature. BH$_4$ units were treated as rigid bodies with B–H distances of 1.13 Å.

Simultaneous thermogravimetric and differential scanning calorimetry (TG-DSC) experiments were carried out with a Netzsch STA 449 F3 Jupiter instrument (Netzsch, Bavaria, Germany). In some measurements, mass spectrometry (MS) was performed simultaneously by connecting a Hiden Analytical HPR-20 QMS (Hiden, Warrington, UK) to the TG-DSC. The samples were loaded in aluminum crucibles (~5–10 mg) and were heated up to 400–500 °C with a heating rate of 5 °C·min^{-1}. An argon flow of 70 mL·min^{-1} was used as a protection gas and purge gas. All given temperatures are peak temperatures, unless stated otherwise.

Li-ion conductivity was measured by impedance spectroscopy, employing an HP 4192A FL impedance analyzer (Keysight Technologies, Santa Rosa, CA, USA). The frequency range was 5 Hz to 10 MHz with a signal amplitude of 60 mV. The temperature was varied by 10 °C for each measurement from RT to 140 °C, and back to RT. Three sets were performed for each measurement in order to improve reproducibility. A 3-ton mechanical axial press was used for pressing pellets with a typical diameter of 6.35 mm and a thickness of 0.5–1.0 mm. The samples were placed in a BDS 1200 Novocontrol sample cell (Novocontrol Technologies GmbH Co. KG, Montabaur, Germany) under an argon atmosphere between two gold ion-blocking electrodes. A single parallel resistor(R)-constant phase element (CPE) was used as an equivalent circuit at low temperatures, where the noise is higher. At higher temperatures, additional effects were taken into account, such as the polarization of the electrode interfaces (modeled with a single CPE) and, eventually, a second R-CPE parallel circuit due to grain boundary contributions. A Hitachi S-4800 Scanning electron microscope (SEM) (Hitachi, Tokyo, Japan) equipped with a Noran System Six energy dispersive spectrometer (EDS) was employed for investigating the absorbed S1 after the conductivity measurements.

4. Conclusions

This work characterized two new composite materials consisting of Er(BH$_4$)$_3$, LiBH$_4$ and/or LiH. The composites react during ball milling, storage as well as decomposition in a two-step reaction where Er-hydrides and LiBH$_4$ are formed in the first step. ErB$_4$ is formed in a second step during thermal decomposition. The composites can be cycled between the first and second decomposition steps by applying hydrogen pressure. Volumetric measurements show a hydrogen capacity of 88% after the first cycle of the initially released hydrogen content and seem stable after the third cycle, with 95% rehydrogenation compared to the second cycle. With a weight loss of 4.5 wt % after the first desorption–absorption cycle, the hydrogen capacity had almost doubled compared to earlier investigated systems, which included LiCl.

Li-ion conductivity measurements of absorbed Er(BH$_4$)$_3$ + 6LiH showed an exceptional high hysteresis of 40–60 K for the transition between the high and low temperature phases of LiBH$_4$, which may be a good starting point to investigate further, for all solid state Li-ion batteries.

Supplementary Materials: The following are available online at www.mdpi.com/2304-6740/5/2/31/s1. Figure S1: SR-PXD data of thermal desorption of S1 and S2, Figure S2: TG-DSC data of pure ErH$_3$ between 25 and 1000 °C. Figure S3: SR-PXD Rietveld refinement and difference plot of S2, Figures S4–S6: SEM image of the absorbed S1. Figure S7: SR-PXD Rietveld refinement and difference plot of S1, Figure S8: PXD Rietveld refinement and difference plot of S2, Table S1: Atomic positions and displacement factor refined for Er and B in Er(BH$_4$)$_3$.

Acknowledgments: The research leading to these results has received funding from the People Program (Marie Curie Actions) of the European Union's Seventh Framework Program FP7/2007–2013/ under REA grant agreement n° 607040 (Marie Curie ITN ECOSTORE) and is thankfully acknowledged. The authors acknowledge the skillful assistance from the staff of the Swiss-Norwegian Beamline, at the European Synchrotron Radiation Facility, Grenoble, France.

Author Contributions: Michael Heere was involved in all stages of the reported work, from planning and conducting of most experiments to the analysis of the data. Michael Heere wrote the first draft of the manuscript, implemented feedback from the co-authors and submitted the work. Michael Heere will include the requests of the reviewers and accompany the whole submission process to its completion. Seyed Hosein Payandeh GharibDoust was involved in all stages of the reported work, from planning and helping with experiments to the analysis of the data. He contributed with the synthesis method and helped in various experiments started from MS-TG-DSC, PXD, SR-PXD including the help with analysis of the data. Magnus H. Sørby and Christoph Frommen acted as co-supervisors and main co-reader/writer. They helped with Rietveld refinement and with data anaylsis from

Inorganics **2017**, *5*, 31

synchrotron radiation powder X-ray diffraction experiments. Matteo Brighi and Radovan Černý conducted the impedance spectroscopy experiments, analysed the data and added valuable sections in the discussion of the results. Torben R. Jensen and Bjørn C. Hauback acted as main supervisor for this work. They contributed to initiate the work, followed up the experiments in Aarhus and Kjeller, respectively, and helped with the preparation of the manuscript.

Conflicts of Interest: The authors declare no conflict of interest.

References

1. Züttel, A.; Rentsch, S.; Fischer, P.; Wenger, P.; Sudan, P.; Mauron, P.H.; Emmenegger, C.H. Hydrogen storage properties of LiBH$_4$. *J. Alloys Compd.* **2003**, *356–357*, 515–520. [CrossRef]
2. Züttel, A.; Borgschulte, A.; Orimo, S.I. Tetrahydroborates as new hydrogen storage materials. *Scr. Mater.* **2007**, *56*, 823–828. [CrossRef]
3. Mauron, P.; Buchter, F.; Friedrichs, O.; Remhof, A.; Bielmann, M.; Zwicky, C.N.; Züttel, A. Stability and Reversibility of LiBH$_4$. *J. Phys. Chem. B* **2008**, *112*, 906–910. [CrossRef] [PubMed]
4. Pitt, M.P.; Paskevicius, M.; Brown, D.H.; Sheppard, D.A.; Buckley, C.E. Thermal Stability of Li$_2$B$_{12}$H$_{12}$ and its Role in the Decomposition of LiBH$_4$. *J. Am. Chem. Soc.* **2013**, *135*, 6930–6941. [CrossRef] [PubMed]
5. Sato, T.; Miwa, K.; Nakamori, Y.; Ohoyama, K.; Li, H.W.; Noritake, T.; Aoki, M.; Towata, S.I.; Orimo, S.I. Experimental and computational studies on solvent-free rare-earth metal borohydrides R(BH$_4$)$_3$ (R = Y, Dy, and Gd). *Phys. Rev. B* **2008**, *77*, 104–114. [CrossRef]
6. Olsen, J.E.; Frommen, C.; Sørby, M.H.; Hauback, B.C. Crystal structures and properties of solvent-free LiYb(BH$_4$)$_{4-x}$Cl$_x$, Yb(BH$_4$)$_3$ and Yb(BH$_4$)$_{2-x}$Cl$_x$. *RSC Adv.* **2013**, *3*, 10764–10774. [CrossRef]
7. Jaroń, T.; Grochala, W. Y(BH$_4$)$_3$—An old-new ternary hydrogen store aka learning from a multitude of failures. *Dalton Trans.* **2010**, *39*, 160–166. [CrossRef] [PubMed]
8. Frommen, C.; Aliouane, N.; Deledda, S.; Fonneløp, J.E.; Grove, H.; Lieutenant, K.; Llamas-Jansa, I.; Sartori, S.; Sørby, M.H.; Hauback, B.C. Crystal structure, polymorphism, and thermal properties of yttrium borohydride Y(BH$_4$)$_3$. *J. Alloys Compd.* **2010**, *496*, 710–716. [CrossRef]
9. Frommen, C.; Sørby, M.H.; Ravindran, P.; Vajeeston, P.; Fjellvåg, H.; Hauback, B.C. Synthesis, crystal structure, and thermal properties of the first mixed-metal and anion-substituted rare earth borohydride LiCe(BH$_4$)$_3$Cl. *J. Phys. Chem. C* **2011**, *115*, 23591–23602. [CrossRef]
10. Ravnsbæk, D.B.; Filinchuk, Y.; Černý, R.; Ley, M.B.; Haase, D.; Jakobsen, H.J.; Skibsted, J.; Jensen, T.R. Thermal polymorphism and decomposition of Y(BH$_4$)$_3$. *Inorg. Chem.* **2010**, *49*, 3801–3809. [CrossRef] [PubMed]
11. Ley, M.B.; Jepsen, L.H.; Lee, Y.S.; Cho, Y.W.; von Colbe, J.M.B.; Dornheim, M.; Rokni, M.; Jensen, J.O.; Sloth, M.; Filinchuk, Y. Complex hydrides for hydrogen storage—New perspectives. *Mater. Today* **2014**, *17*, 122–128. [CrossRef]
12. Yan, Y.; Li, H.W.; Sato, T.; Umeda, N.; Miwa, K.; Towata, S.; Orimo, S. Dehydriding and rehydriding Properties of yttrium borohydride Y(BH$_4$)$_3$ prepared by liquid-phase synthesis. *Int. J. Hydrogen Energy* **2009**, *34*, 5732–5736. [CrossRef]
13. Jaron, T.; Kozminski, W.; Grochala, W. Phase transition induced improvement in H$_2$ desorption kinetics: The case of the high-temperature form of Y(BH$_4$)$_3$. *Phys. Chem. Chem. Phys.* **2011**, *13*, 8847–8851. [CrossRef] [PubMed]
14. Gennari, F.C.; Esquivel, M.R. Synthesis and dehydriding process of crystalline Ce(BH$_4$)$_3$. *J. Alloys Compd.* **2009**, *485*, L47–L51. [CrossRef]
15. Li, H.W.; Yan, Y.; Orimo, S.I.; Züttel, A.; Jensen, C.M. Recent Progress in Metal Borohydrides for Hydrogen Storage. *Energies* **2011**, *4*, 185–214. [CrossRef]
16. Zhang, B.J.; Liu, B.H.; Li, Z.P. Destabilization of LiBH$_4$ by (Ce, La)(Cl, F)$_3$ for hydrogen storage. *J. Alloys Compd.* **2011**, *509*, 751–757. [CrossRef]
17. Gennari, F.; Albanesi, L.F.; Puszkiel, J.; Larochette, P.A. Reversible hydrogen storage from 6LiBH$_4$–MCl$_3$ (M = Ce, Gd) composites by in-situ formation of MH$_2$. *Int. J. Hydrogen Energy* **2011**, *36*, 563–570. [CrossRef]
18. Paskevicius, M.; Jepsen, L.H.; Schouwink, P.; Černý, R.; Ravnsbæk, D.B.; Filinchuk, Y.; Dornheim, M.; Besenbacher, F.; Jensen, T.R. Metal borohydrides and derivatives—Synthesis, structure and properties. *Chem. Soc. Rev.* **2017**, *46*, 1565–1634. [CrossRef] [PubMed]

19. Callini, E.; Atakli, Z.Ö.K.; Hauback, B.C.; Orimo, S.; Jensen, C.; Dornheim, M.; Grant, D.; Cho, Y.W.; Chen, P.; Hjörvarsson, B.; et al. Complex and liquid hydrides for energy storage. *Appl. Phys. A* **2016**, *122*, 353. [CrossRef]

20. Ley, M.B.; Boulineau, S.; Janot, R.; Filinchuk, Y.; Jensen, T.R. New Li Ion Conductors and Solid State Hydrogen Storage Materials: LiM(BH$_4$)$_3$Cl, M = La, Gd. *J. Phys. Chem. C* **2012**, *116*, 21267–21276. [CrossRef]

21. Skripov, A.V.; Soloninin, A.V.; Ley, M.B.; Jensen, T.R.; Filinchuk, Y. Nuclear Magnetic Resonance Studies of BH$_4$ Reorientations and Li Diffusion in LiLa(BH$_4$)$_3$Cl. *J. Phys. Chem. C* **2013**, *117*, 14965–14972. [CrossRef]

22. Ley, M.B.; Ravnsbæk, D.B.; Filinchuk, Y.; Lee, Y.S.; Janot, R.l.; Cho, Y.W.; Skibsted, J.; Jensen, T.R. LiCe(BH$_4$)$_3$Cl, a new lithium-ion conductor and hydrogen storage material with isolated tetranuclear anionic clusters. *Chem. Mater.* **2012**, *24*, 1654–1663. [CrossRef]

23. Roedern, E.; Lee, Y.S.; Ley, M.B.; Park, K.; Cho, Y.W.; Skibsted, J.; Jensen, T.R. Solid state synthesis, structural characterization and ionic conductivity of bimetallic alkali-metal yttrium borohydrides MY(BH$_4$)$_4$ (M = Li and Na). *J. Mater. Chem. A* **2016**, *4*, 8793–8802. [CrossRef]

24. Marks, S.; Heck, J.G.; Habicht, M.H.; Oña-Burgos, P.; Feldmann, C.; Roesky, P.W. [Ln(BH$_4$)$_2$(THF)$_2$](Ln = Eu, Yb)—A Highly Luminescent Material. Synthesis, Properties, Reactivity, and NMR Studies. *J. Am. Chem. Soc.* **2012**, *134*, 16983–16986. [PubMed]

25. Schouwink, P.; Ley, M.B.; Tissot, A.; Hagemann, H.; Jensen, T.R.; Smrčok, L'.; Černý, R. Structure and properties of complex hydride perovskite materials. *Nat. Commun.* **2014**, *5*, 5706. [CrossRef] [PubMed]

26. Schouwink, P.; Didelot, E.; Lee, Y.S.; Mazet, T.; Černý, R. Structural and magnetocaloric properties of novel gadolinium borohydrides. *J. Alloys Compd.* **2016**, *664*, 378–384. [CrossRef]

27. Frommen, C.; Heere, M.; Riktor, M.D.; Sørby, M.H.; Hauback, B.C. Hydrogen storage properties of rare earth (RE) borohydrides (RE = La, Er) in composite mixtures with LiBH$_4$ and LiH. *J. Alloys Compd.* **2015**, *645*, S155–S159. [CrossRef]

28. James, B.; Wallbridge, M. Metal tetrahydroborates. *Prog. Inorg. Chem.* **1970**, *11*, 99–231.

29. Visseaux, M.; Bonnet, F. Borohydride complexes of rare earths, and their applications in various organic transformations. *Coord. Chem. Rev.* **2011**, *255*, 374–420. [CrossRef]

30. Yang, C.H.; Tsai, W.T.; Chang, J.K. Hydrogen desorption behavior of vanadium borohydride synthesized by modified mechano-chemical process. *Int. J. Hydrogen Energy* **2011**, *36*, 4993–4999. [CrossRef]

31. Korablov, D.; Ravnsbæk, D.B.; Ban, V.; Filinchuk, Y.; Besenbacher, F.; Jensen, T.R. Investigation of MBH$_4$–VCl$_2$, M = Li, Na or K. *Int. J. Hydrogen Energy* **2013**, *38*, 8376–8383. [CrossRef]

32. Gennari, F.C. Mechanochemical synthesis of erbium borohydride: Polymorphism, thermal decomposition and hydrogen storage. *J. Alloys Compd.* **2013**, *581*, 192–195. [CrossRef]

33. Olsen, J.E.; Frommen, C.; Jensen, T.R.; Riktor, M.D.; Sørby, M.H.; Hauback, B.C. Structure and thermal properties of composites with RE-borohydrides (RE = La, Ce, Pr, Nd, Sm, Eu, Gd, Tb, Er, Yb or Lu) and LiBH$_4$. *RSC Adv.* **2014**, *4*, 1570–1582. [CrossRef]

34. Hagemann, H.; Černý, R. Synthetic approaches to inorganic borohydrides. *Dalton Trans.* **2010**, *39*, 6006–6012. [CrossRef] [PubMed]

35. Ley, M.B.; Paskevicius, M.; Schouwink, P.; Richter, B.; Sheppard, D.A.; Buckley, C.E.; Jensen, T.R. Novel solvates M(BH$_4$)$_3$S(CH$_3$)$_2$ and properties of halide-free M(BH$_4$)$_3$ (M = Y or Gd). *Dalton Trans.* **2014**, *43*, 13333–13342. [CrossRef] [PubMed]

36. Heere, M.; Payandeh GharibDoust, S.H.; Frommen, C.; Humphries, T.D.; Ley, M.B.; Sørby, M.H.; Jensen, T.R.; Hauback, B.C. The influence of LiH on the rehydrogenation behavior of halide free rare earth (RE) borohydrides (RE = Pr, Er). *Phys. Chem. Chem. Phys.* **2016**, *18*, 24387–24395. [CrossRef] [PubMed]

37. Brighi, M.; Schouwink, P.; Sadikin, Y.; Černý, R. Fast ion conduction in garnet-type metal borohydrides Li$_3$K$_3$Ce$_2$(BH$_4$)$_{12}$ and Li$_3$K$_3$La$_2$(BH$_4$)$_{12}$. *J. Alloys Compd.* **2016**, *662*, 388–395. [CrossRef]

38. Maekawa, H.; Matsuo, M.; Takamura, H.; Ando, M.; Noda, Y.; Karahashi, T.; Orimo, S.I. Halide-Stabilized LiBH$_4$, a Room-Temperature Lithium Fast-Ion Conductor. *J. Am. Chem. Soc.* **2009**, *131*, 894–895. [CrossRef] [PubMed]

39. Sveinbjörnsson, D.; Myrdal, J.S.G.; Blanchard, D.; Bentzen, J.J.; Hirata, T.; Mogensen, M.B.; Norby, P.; Orimo, S.I.; Vegge, T. Effect of Heat Treatment on the Lithium Ion Conduction of the LiBH$_4$–LiI Solid Solution. *J. Phys. Chem. C* **2013**, *117*, 3249–3257. [CrossRef]

40. Matsuo, M.; Takamura, H.; Maekawa, H.; Li, H.W.; Orimo, S.I. Stabilization of lithium superionic conduction phase and enhancement of conductivity of $LiBH_4$ by $LiCl$ addition. *Appl. Phys. Lett.* **2009**, *94*, 084103. [CrossRef]

41. Nakamori, Y.; Orimo, S.I.; Tsutaoka, T. Dehydriding reaction of metal hydrides and alkali borohydrides enhanced by microwave irradiation. *Appl. Phys. Lett.* **2006**, *88*, 112104. [CrossRef]

42. Blanchard, D.; Nale, A.; Sveinbjörnsson, D.; Eggenhuisen, T.M.; Verkuijlen, M.H.; Vegge, T.; Kentgens, A.P.; Jongh, P.E. Nanoconfined $LiBH_4$ as a fast lithium ion conductor. *Adv. Funct. Mater.* **2015**, *25*, 184–192. [CrossRef]

43. Brinks, H.; Fossdal, A.; Bowman, R.; Hauback, B.C. Pressure–composition isotherms of $TbNiAlH_x$. *J. Alloys Compd.* **2006**, *417*, 92–95. [CrossRef]

44. Dyadkin, V.; Pattison, P.; Dmitriev, V.; Chernyshov, D. A new multipurpose diffractometer PILATUS@SNBL. *J. Synchrotron Radiat.* **2016**, *23*, 825–829. [CrossRef] [PubMed]

45. Shim, J.H.; Lim, J.H.; Rather, S.; Lee, Y.S.; Reed, D.; Kim, Y.; Book, D.; Cho, Y.W. Effect of Hydrogen Back Pressure on Dehydrogenation Behavior of $LiBH_4$-Based Reactive Hydride Composites. *J. Phys. Chem. Lett.* **2010**, *1*, 59–63. [CrossRef]

46. Hammersley, A. *FIT2D: An Introduction and Overview*; European Synchrotron Radiation Facility Internal Report ESRF97HA02T; European Synchrotron Radiation Facility: Grenoble, France, 1997.

47. Larson, A.; Von Dreele, R. *General Structure Analysis System (GSAS)*; Report LAUR 86-748; Los Alamos National Laboratory: Los Alamos, NM, USA, 2000.

48. Toby, B.H. *EXPGUI*, a graphical user interface for *GSAS*. *J. Appl. Crystallogr.* **2001**, *34*, 210–213. [CrossRef]

49. Thompson, P.; Cox, D.; Hastings, J. Rietveld refinement of Debye–Scherrer synchrotron X-ray data from Al_2O_3. *J. Appl. Crystallogr.* **1987**, *20*, 79–83. [CrossRef]

inorganics

MDPI

Article

Thermodynamic Properties and Reversible Hydrogenation of LiBH$_4$–Mg$_2$FeH$_6$ Composite Materials

Guanqiao Li [1,*], Motoaki Matsuo [2], Shigeyuki Takagi [1], Anna-Lisa Chaudhary [3], Toyoto Sato [1], Martin Dornheim [3] and Shin-ichi Orimo [1,4]

[1] Institute for Materials Research, Tohoku University, Sendai 980-8577, Japan; shigeyuki.takagi@imr.tohoku.ac.jp (S.T.); toyoto@imr.tohoku.ac.jp (T.S.); orimo@imr.tohoku.ac.jp (S.O.)
[2] Department of Nanotechnology for Sustainable Energy, School of Science and Technology, Kwansei Gakuin University, Sanda 669-1337, Japan; matsuo35@kwansei.ac.jp
[3] Department of Nanotechnology, Institute of Materials Research, Helmholtz-Zentrum Geesthacht, D-21502 Geesthacht, Germany; anna-lisa.chaudhary@hzg.de (A.-L.C.); martin.dornheim@hzg.de (M.D.)
[4] WPI-Advanced Institute for Materials Research (WPI-AIMR), Tohoku University, Sendai 980-8577, Japan
* Correspondence: likk@imr.tohoku.ac.jp

Received: 8 October 2017; Accepted: 2 November 2017; Published: 16 November 2017

Abstract: In previous studies, complex hydrides LiBH$_4$ and Mg$_2$FeH$_6$ have been reported to undergo simultaneous dehydrogenation when ball-milled as composite materials $(1 - x)$LiBH$_4$ + xMg$_2$FeH$_6$. The simultaneous hydrogen release led to a decrease of the dehydrogenation temperature by as much as 150 K when compared to that of LiBH$_4$. It also led to the modified dehydrogenation properties of Mg$_2$FeH$_6$. The simultaneous dehydrogenation behavior between stoichiometric ratios of LiBH$_4$ and Mg$_2$FeH$_6$ is not yet understood. Therefore, in the present work, we used the molar ratio $x = 0.25, 0.5,$ and 0.75, and studied the isothermal dehydrogenation processes via pressure–composition–isothermal (PCT) measurements. The results indicated that the same stoichiometric reaction occurred in all of these composite materials, and $x = 0.5$ was the molar ratio between LiBH$_4$ and Mg$_2$FeH$_6$ in the reaction. Due to the optimal composition ratio, the composite material exhibited enhanced rehydrogenation and reversibility properties: the temperature and pressure of 673 K and 20 MPa of H$_2$, respectively, for the full rehydrogenation of $x = 0.5$ composite, were much lower than those required for the partial rehydrogenation of LiBH$_4$. Moreover, the $x = 0.5$ composite could be reversibly hydrogenated for more than four cycles without degradation of its H$_2$ capacity.

Keywords: complex hydride; composite material; hydrogen storage

1. Introduction

Boron-based complex hydrides MBH$_4$ (M = Li, Na, and K), which consist of an M^+ cation and a [BH$_4$]$^-$ complex anion, have high gravimetric H$_2$ densities (7.7–18.4 mass %); therefore, these materials have the potential to be used as hydrogen storage materials [1–3]. However, MBH$_4$ are thermodynamically stable and the hydrogenation is difficult to achieve under mild temperatures and pressures. For example, when undergoing the following Reaction (1), the dehydrogenation temperature of LiBH$_4$ at 0.1 MPa H$_2$ has been estimated to be 683 K on the basis of enthalpy ΔH and entropy ΔS changes of 66.6 kJ/(mole of H$_2$) and 97.4 J/K(mole of H$_2$), respectively [4]. In practice, significant dehydrogenation only occurs at temperatures greater than 700 K. In addition, partial rehydrogenation of LiBH$_4$ requires a much higher pressure and temperature of 35 MPa and 873 K.

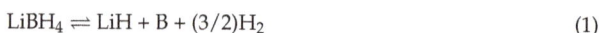

$$\text{LiBH}_4 \rightleftharpoons \text{LiH} + \text{B} + (3/2)\text{H}_2 \tag{1}$$

Combining borohydrides with metallic hydrides (e.g., MgH_2) or complex hydrides (e.g., Mg_2NiH_4) has been proven to be effective at modifying the dehydrogenation properties of MBH_4. The reactions between MBH_4 and the partner hydrides in these composite materials can produce the stable boride MgB_2 or $MgNi_{2.5}B_2$ that results in a decrease of the dehydrogenation temperature of $LiBH_4$ [5–11]. Therefore, in a previous study, we have tried to bring down the dehydrogenation temperature of $LiBH_4$ by combining it with a transition-metal-based complex hydride, Mg_2FeH_6. We ball-milled the two complex hydrides over a large composition range: $(1 − x)LiBH_4 + xMg_2FeH_6$ $(0.25 \leq x \leq 0.9)$ [12–15]. The dehydrogenation properties were investigated using a dynamic measurement: thermogravimetry–mass spectrometry (TG–MS). It was observed that although the thermodynamic stabilities of $LiBH_4$ and Mg_2FeH_6 differed substantially, they underwent dehydrogenation simultaneously when heated up during the TG–MS analysis, by which the dehydrogenation temperature of the composite materials was lowered by, at most, 150 K when compared to that of $LiBH_4$. In addition to the dehydrogenation properties of $LiBH_4$, those of Mg_2FeH_6 were also modified; Mg_2FeH_6 no longer underwent independent dehydrogenation, and the temperature of the simultaneous dehydrogenation shifted closer to that of pure Mg_2FeH_6, both continuously and with an increasing ratio x.

The dehydrogenation process of $(1 − x)LiBH_4 + xMg_2FeH_6$ is special when compared to the processes of other composite materials, e.g., $2LiBH_4 + MgH_2$ [16]. The dehydrogenation behavior of MgH_2 was unaffected by its combination with $LiBH_4$; furthermore, a stoichiometric reaction between $LiBH_4$ and MgH_2 existed, the dehydrogenation temperature of which would not be affected by the composition ratio in the composite materials. However, in the case of $(1 − x)LiBH_4 + xMg_2FeH_6$, the properties of both $LiBH_4$ and Mg_2FeH_6 were modified and the stoichiometric reaction between these two complex hydrides at a specific composition ratio was not yet understood. We considered that the stoichiometric reaction was hidden in the continuous hydrogen releasing events in the dynamic TG–MS measurements, where the thermodynamic and kinetic factors both affected the dehydrogenation process.

Besides our study, the research on $LiBH_4 + Mg_2FeH_6$ composite materials has been conducted on $LiBH_4$-rich compositions only. The reaction processes investigated by various dynamic measurement methods (for example, differential scanning calorimetry (DSC)) has been explained as Mg_2FeH_6 dehydrogenating to form elemental Mg and Fe, followed by the Mg and Fe reacting with $LiBH_4$ and thereby destabilizing it thermodynamically. Because of the absence of studies on any Mg_2FeH_6-rich compositions, the simultaneous dehydrogenation was not discovered [17,18]. In addition, Ghaani et al. used an isothermal method (i.e., the pressure–composition–isothermal (PCT) measurement) to study the dehydrogenation process of $2LiBH_4 + Mg_2FeH_6$. Here, it was found that $LiBH_4$ and Mg_2FeH_6 react with each other; however, unreacted $LiBH_4$ was apparent using this composition ratio [19]. Whether the reaction between $LiBH_4$ and Mg_2FeH_6 changes with variations in composition ratio is unclear, and whether an optimal composition ratio exists at which a stoichiometric reaction occurs without independent dehydrogenation of the parent complex hydrides remains unknown.

In this study, to confirm the stoichiometric reaction between Mg_2FeH_6 and $LiBH_4$, and the optimal composition ratios for it, we used PCT measurements to evaluate the isothermal dehydrogenation processes of $(1 − x)LiBH_4 + xMg_2FeH_6$. Among a large range of composition ratios, the focus was on three in the present study: $x = 0.25$, 0.5, and 0.75. TG–MS measurements done in our previous study revealed that $x = 0.5$ was a critical composition. Composites with $x \geq 0.5$ exhibited a single dehydrogenation event, whereas composites with $x < 0.5$ underwent multiple events involving both simultaneous dehydrogenation of $LiBH_4$ and Mg_2FeH_6, and individual dehydrogenation of $LiBH_4$. We considered that the optimal ratio for the stoichiometric reaction could be deduced from these representative compositions. After deciding on the optimal ratio, its effect on the modification of the reversible hydrogenation properties of the composite materials were investigated.

2. Results

2.1. Optimal Composition Ratio for the Stoichiometric Reaction between LiBH₄ and Mg₂FeH₆

The PCT results obtained at 643 K for the composites ($x = 0.25$, 0.5, and 0.75; temperatures were slightly different within each measurement) are shown in Figure 1. Although only one dehydrogenation event was observed for the $x = 0.5$ and 0.75 composites in the dynamic TG–MS measurements, the PCT results revealed two and three thermodynamically independent reactions, respectively. For $x = 0.25$, three independent reactions were observed, similar to the TG–MS results in which multiple dehydrogenation events appeared. The first and second dehydrogenation reactions of all of these compositions exhibited the same equilibrium pressure of 1.85 MPa and 0.85 MPa, which indicates that these compositions have undergone similar reaction pathways.

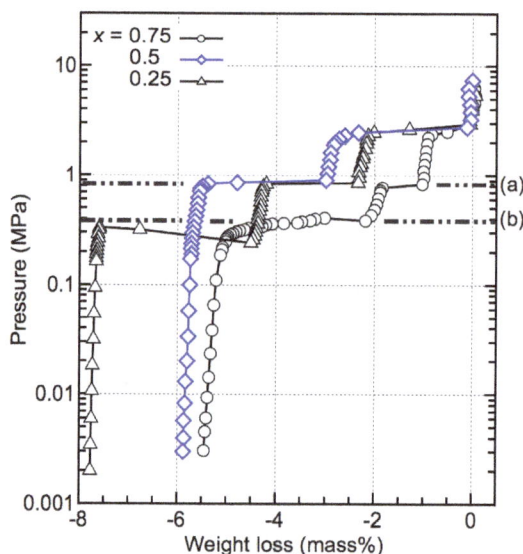

Figure 1. Pressure–composition–isothermal (PCT) results for $(1 - x)\text{LiBH}_4 + x\text{Mg}_2\text{FeH}_6$ ($x = 0.25$, 0.5, and 0.75) at 643 K. The temperatures are slightly different within each measurement. The first and second reactions in all the compositions exhibit the same equilibrium pressures of 1.85 MPa and 0.85 MPa, respectively. Line (a) represents the equilibrium pressure of the dehydrogenation of MgH_2, according to Refs. [20,21]; line (b) represents the equilibrium pressure of the dehydrogenation of Mg_2FeH_6, according to Refs. [22–24].

To determine the reaction pathway, the materials were quenched under H_2 pressure after each thermodynamically independent reaction and each phase was identified using X-ray diffraction (XRD). The XRD results for the $x = 0.5$ composite after each reaction are shown in Figure 2. After the first reaction at 1.85 MPa H_2, the diffraction peaks of Mg_2FeH_6 were no longer apparent, whilst the intensity of the diffraction peak of Fe increased. Furthermore, peaks attributable to MgH_2 appeared, accompanied by several small broad peaks of an unclear phase. The results indicate that Mg_2FeH_6 was fully dehydrogenated. After the second reaction at 0.85 MPa H_2, the diffraction peaks of MgH_2 transitioned to those of Mg. The small broad unclear peaks were not prominent but were still present, indicating that this phase did not participate in the second reaction. The presence of LiBH_4 was not confirmed by XRD due to its weak diffraction intensity compared to the other reactants as well as the amorphisation after ball-milling. Because direct dehydrogenation of Mg_2FeH_6 cannot produce MgH_2;

we considered that $LiBH_4$ reacted with Mg_2FeH_6 in the first reaction. Also, we used scanning electron microscopy (SEM) to characterize the presence of $LiBH_4$ before and after the first reaction.

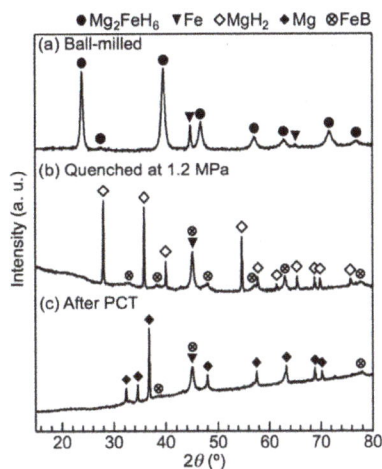

Figure 2. X-ray diffraction (XRD) profiles of the x = 0.5 composite during the pressure–composition–isothermal (PCT) measurement at 643 K: (**a**) the ball-milled material before the PCT measurement; (**b**) the material quenched at 1.2 MPa, which is the pressure between the first isothermal reaction at 1.85 MPa and the second one at 0.85 MPa; and (**c**) the material after the PCT measurement. As a result of Fe in the material, the baseline tilted at high angles. The strongest diffraction peak of LiH overlapped with that of Fe at 44°.

The backscattering images (BSE) and energy-dispersive X-ray spectroscopy (EDX) analysis results obtained before the PCT measurements and after the first reaction are shown in Figure 3. Before the PCT measurement, the composite material was a uniform mixture; separate $LiBH_4$ and Mg_2FeH_6 particles were not observed at about 100-nm scale. After the first isothermal reaction at 1.85 MPa, the composite became an obvious mixture. A large blank area was observed where no B, Fe, or Mg were detected, suggesting that this area was possibly composed of the dehydrogenation product of $LiBH_4$: LiH. These results complement the XRD results, indicating that both $LiBH_4$ and Mg_2FeH_6 underwent dehydrogenation in the first reaction. Element mapping of Fe and B showed significant overlap, and the particle size was of the order of several nanometers, indicating the possible formation of iron boride. A 1:1 reaction ratio between $LiBH_4$ and Mg_2FeH_6 in the first isothermal reaction suggests that FeB is a possible iron boride product. The recognized diffraction patterns in the XRD results can be attributed to FeB (orthorhombic, *Cmcm*), but the very broad peaks made the phase identification difficult. Because of the overlapping of the strongest diffraction peaks of Fe (cubic, *Im-3m*) and FeB, it is hard to rule out the existence of Fe. Therefore, we still keep it in the reaction of Equation (2). The formation of FeB is crucial for the reaction between $LiBH_4$ and Mg_2FeH_6. The estimated enthalpy change ΔH and entropy change ΔS for the reaction ($LiBH_4$ + Mg_2FeH_6 → LiH + $2MgH_2$ + FeB + $5/2H_2$) is 64 kJ/(mol of H_2) and 125 J/K(mole of H_2), respectively. According to the theoretical value, the equilibrium pressure at 643 K would be 2.1 MPa [23,25–27]. It is very close to the experimental data 1.85 MPa.

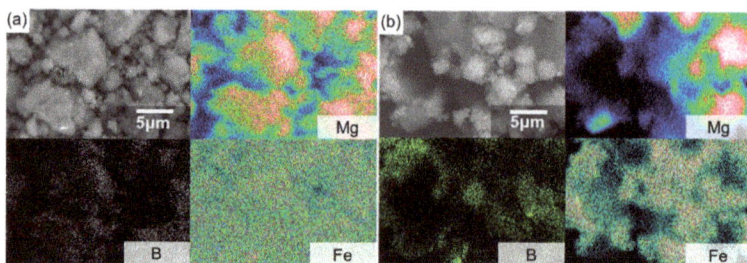

Figure 3. Backscattered electron (BSE) images and energy-dispersive X-ray spectroscopy (EDX) analyses of the $x = 0.5$ composite: (**a**) the material after ball milling was a uniform mixture of $LiBH_4$ and Mg_2FeH_6; (**b**) the material quenched at 1.2 MPa during the PCT measurement exhibited black areas, possibly LiH. These analyses indicate that $LiBH_4$ dehydrogenated simultaneously with Mg_2FeH_6 during the first isothermal reaction at 1.85 MPa.

On the basis of the phase identification results and the equilibrium pressure of the dehydrogenation reaction of MgH_2 calculated from previously reported thermodynamic data [20,21], Mg_2FeH_6 and $LiBH_4$ clearly reacted during the first reaction (Equation (2)), whereas the second reaction (Equation (3)) was the dehydrogenation of MgH_2.

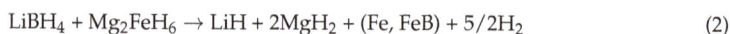

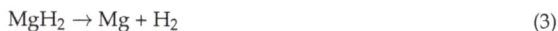

$$LiBH_4 + Mg_2FeH_6 \rightarrow LiH + 2MgH_2 + (Fe, FeB) + 5/2H_2 \qquad (2)$$

$$MgH_2 \rightarrow Mg + H_2 \qquad (3)$$

The theoretical weight loss due to hydrogen release of Reactions (2) and (3) is 3.8 mass % and 3.0 mass %, respectively. This correlates well with the dehydrogenation of MgH_2 in the second reaction very well. However, the actual weight loss of the first reaction was only 3 mass %, and this deviation could be due to the residual Fe present from Mg_2FeH_6 synthesis.

As shown in Figure 4a, for $x = 0.25$, the phases changes at the first and second reaction were almost the same as those of $x = 0.5$, except for the presented diffraction peaks of $LiBH_4$ after the second reaction. The theoretical weight loss of Reactions (2) and (3) in $x = 0.25$ would be 2.9 mass % and 2.3 mass %, respectively. The actual weight loss of 2.5 mass % and 2.0 mass % at the first and second reaction, respectively, correlate with the theoretical value. The results indicate that the same stoichiometric reaction between $LiBH_4$ and Mg_2FeH_6 occurred in $x = 0.25$, and the amount of $LiBH_4$ in this composition is excessive for the stoichiometric reaction. An additional reaction with an incubation process occurred after the dehydrogenation of MgH_2, and diffraction peaks of MgB_2 appeared after this reaction. Therefore, the final isothermal reaction appears to be between $LiBH_4$ and Mg, as shown in Equation (4).

$$2LiBH_4 + Mg \rightarrow 2LiH + MgB_2 + 3H_2 \qquad (4)$$

In the case of $x = 0.75$, after the dehydrogenation of MgH_2, diffraction peaks of Mg_2FeH_6 were still observed and finally disappeared after the last reaction, as shown in Figure 4b. Therefore, the last reaction can be attributed to the dehydrogenation of any residual Mg_2FeH_6:

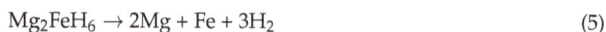

$$Mg_2FeH_6 \rightarrow 2Mg + Fe + 3H_2 \qquad (5)$$

The equilibrium pressure of the dehydrogenation of Mg_2FeH_6, calculated using reference data, also supports this interpretation [22,23]. The same stoichiometric reaction between $LiBH_4$ and Mg_2FeH_6, and the dehydrogenation of MgH_2, also occurred in $x = 0.75$. They contributed 1.4 mass % and 1.1 mass % weight loss in theoretical, and 1.1 mass % and 1.1 mass % in practice, respectively.

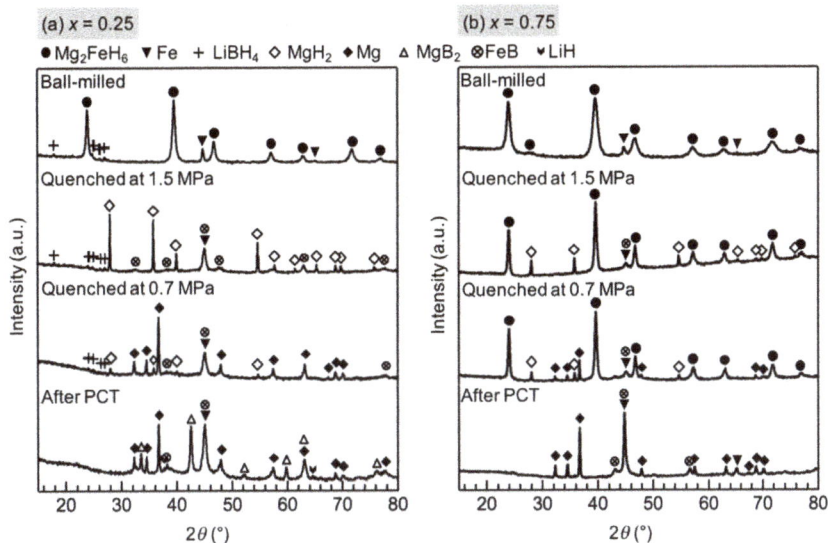

Figure 4. XRD profiles of (**a**) *x* = 0.25 and (**b**) *x* = 0.75 composite during the PCT measurement at 643 K: the ball-milled material before the PCT measurement; the material quenched at 1.5 MPa, which is the pressure after the first isothermal reaction at 1.85 MPa; the material quenched at 0.7 MPa, which is the pressure after the second isothermal reaction at 0.85 MPa; and after the PCT measurement. The strongest diffraction peak of LiH overlapped with that of Fe at 44°.

2.2. Reversible Hydrogenation Due to the Optimal Composition Ratio

After attempting rehydrogenation at different temperatures and pressures, we confirmed that full rehydrogenation was achievable at 673 K and 20 MPa H_2. The XRD profiles of the *x* = 0.5 composites before and after the rehydrogenation reaction at 673 K and 20 MPa H_2 are shown in Figure 5. Before rehydrogenation, the material released H_2 via a PCT program such that only Mg, LiH, Fe, and a small amount of FeB remained. After rehydrogenation, the XRD pattern showed the diffraction peaks of well-defined crystalline Mg_2FeH_6, indicating that Mg_2FeH_6 had been fully rehydrogenated. The weak diffraction peaks of α-Fe are attributed to the Fe located in the cores of the Mg_2FeH_6 grains, which, as reported previously [22,28], cannot be avoided. The diffraction peaks of Mg_2FeH_6 were sharp due to the fact that the heat treatment reduced the effect of the ball milling. $LiBH_4$ was not observed from XRD after ball-milling nor was its presence detected after rehydrogenation. Therefore, we used infrared (IR) spectroscopy to detect the presence of $LiBH_4$. As shown in Figure 6, the asymmetric stretching vibration mode and the bending mode of BH_4 appeared in the spectra of the rehydrogenated materials at approximately 2329 cm^{-1} and 1234–1095 cm^{-1}, respectively. The IR pattern correlated well with the data reported for pure $LiBH_4$ [29], demonstrating that $LiBH_4$ was present in the rehydrogenated material. The asymmetric stretching vibration of $[FeH_6]^{4-}$ at 1785 cm^{-1} was also seen [30].

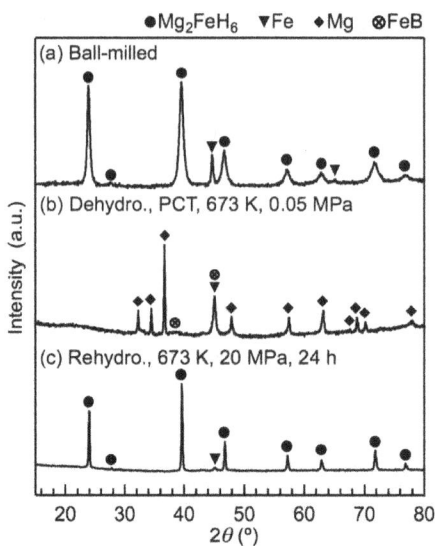

Figure 5. XRD profiles of the *x* = 0.5 composite before and after rehydrogenation at 673 K and 20 MPa H₂: (**a**) the material immediately after ball milling; (**b**) the material dehydrogenated at 673 K via a PCT measurement, and (**c**) the material rehydrogenated at 673 K and 20 MPa H₂.

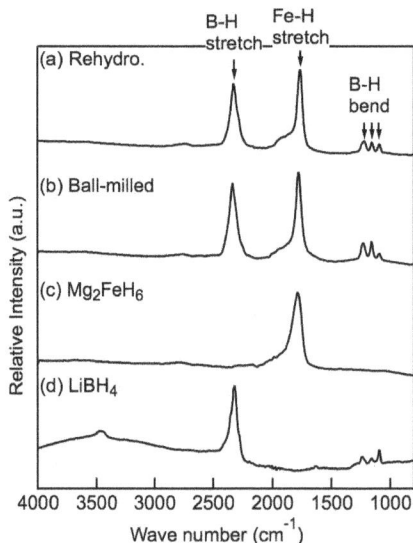

Figure 6. Infrared (IR) spectra of the *x* = 0.5 composite: (**a**) directly after ball-milling, (**b**) after rehydrogenation. The IR spectra of (**c**) pure Mg_2FeH_6 and (**d**) pure $LiBH_4$ are shown as references. The B–H vibration stretching at 2329 cm^{-1} and the bending vibration at 1234–1095 cm^{-1} demonstrate that $LiBH_4$ in the composite materials was successfully rehydrogenated.

The PCT plots of the dehydrogenation part during the reversible hydrogenation tests at 673 K are shown in Figure 7. The equilibrium pressure of the reaction between $LiBH_4$ and Mg_2FeH_6 at 6.8 MPa and that of the dehydrogenation reaction of MgH_2 at 1.8 MPa remained stable, indicating that the properties of the composite materials did not degrade after cycling. The weight loss during each cycle did not decrease, suggesting that the composite material was fully rehydrogenated.

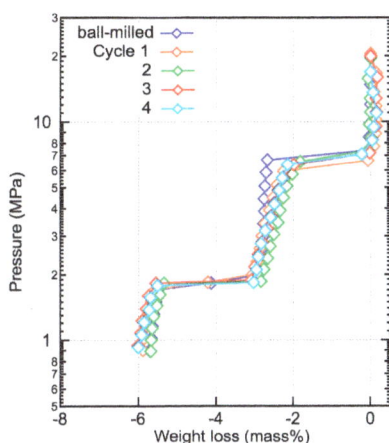

Figure 7. De/rehydrogenation reversibility of the $x = 0.5$ composite. The PCT plots were constructed for the dehydrogenation process after rehydrogenation in each cycle. After four cycles, the H_2 capacity of the materials remained the same.

3. Discussion

According to the PCT results, $(1 - x)LiBH_4 + xMg_2FeH_6$ ($x = 0.25$, 0.5, and 0.75) shared the same reaction between $LiBH_4$ and Mg_2FeH_6. This reaction was stoichiometric and its thermodynamic properties did not alter when excessive Mg_2FeH_6 or $LiBH_4$ were present. The optimal composition for this stoichiometric reaction was $x = 0.5$, which can be explained by the formation of FeB. The kinetics of FeB formation was slow. For example, the reaction between $LiBH_4$ and Mg_2FeH_6 in the $x = 0.5$ composite required more than 5 h to achieve a total hydrogen release of 3 mass %. The slow kinetics also caused the dehydrogenation process observed in the TG–MS and PCT measurements to exhibit different features. During the PCT measurements, the reaction between $LiBH_4$ and Mg_2FeH_6 can be separated from the dehydrogenation of MgH_2 for $x = 0.5$ and 0.75 because of the long reacting time. On the other hand, the high heating rate of 5 K·min^{-1} during the TG measurements in our previous report was not sufficient to incubate the reaction at its thermodynamically equilibrium temperature. According to theoretical data, it should be below 512 K (the equilibrium termperature at 0.1 MPa H_2) [23,25–27], but in our previous work, the dehydrogenation of the composite materials was observed beyond 630 K. At this temperature, MgH_2 was so unstable that the independent dehydrogenation of MgH_2 no longer occurred. The thermodynamic property of Mg_2FeH_6 is similar to that of MgH_2 [27]; thus, in the Mg_2FeH_6-rich compositions, independent dehydrogenation of Mg_2FeH_6 also no longer occurred and only one event appeared in the TG measurements for $x \geq 0.5$, with which all of the hydrogen-containing species dehydrogenated simultaneously. $LiBH_4$ is stable at the temperature of the simultaneous dehydrogenation. Therefore, for the $x < 0.5$ composite, the residue $LiBH_4$ reacted with Mg or dehydrogenated independently after the simultaneous dehydrogenation in which several dehydrogenation events were observed. The activation energy of the reaction between $LiBH_4$ and Mg_2FeH_6 should be different at the equilibrium reaction pressure and in the dynamic measurement. The detailed investigation will be carried out in a further study.

Ghaani et al. reported a composite reaction similar to that shown in Equation (2) in the $LiBH_4$-rich composition [13,19]. Our results demonstrate that the reaction between $LiBH_4$ and Mg_2FeH_6 was stoichiometric and did not degrade when excess $LiBH_4$ or Mg_2FeH_6 was present. Also, an optimal composition ratio existed: $x = 0.5$, with which the independent dehydrogenation of the parent complex hydrides was avoided.

At the PCT experiment temperature of 643 K, the equilibrium pressure of the reaction between $LiBH_4$ and Mg_2FeH_6 is 38 times or 4 times higher than that of the dehydrogenation reaction of pure $LiBH_4$ or Mg_2FeH_6. This result indicates that both $LiBH_4$ and Mg_2FeH_6 were thermodynamically destabilized in the composite. In other composite materials, such as that between $LiBH_4$ and MgH_2, only $LiBH_4$ could be destabilized [3,20]. Another merit is that the H_2 capacity and the dehydrogenation process remained stable even under an Ar atmosphere. For example, for $x = 0.5$ composite, all the H_2 (6 mass %) was released in the temperature range 580 K to 630 K in the TG–MS experiments. However, an initial backpressure of at least 0.5 MPa H_2 is needed for the incubation of the composite reaction between $LiBH_4$ and MgH_2 [31,32]. Also, the dehydrogenation process of the composite material of $LiBH_4$ and Mg_2NiH_4 separated into more than three reactions and lasted from room temperature to beyond 673 K [6].

Full rehydrogenation of $x = 0.5$ composite requires much lower temperatures and pressures than those required in the partial rehydrogenation of $LiBH_4$ (pressures as high as 35 MPa H_2 and temperatures as high as 823 K [33]). Because previous studies on $LiBH_4$-rich compositions did not achieve full rehydrogenation, we consider that the optimal ratio contributed to the good reversible hydrogenation performance of the composite material [17,19,34].

4. Materials and Methods

4.1. Synthesis of $(1 - x)LiBH_4 + xMg_2FeH_6$

The synthesis of $(1 - x)LiBH_4 + xMg_2FeH_6$ ($x = 0.25$, 0.5, and 0.75) occurred via ball milling of commercially available $LiBH_4$ (purity $\geq$ 90%, Sigma-Aldrich, St. Louis, MO, USA) with laboratory-synthesized Mg_2FeH_6 (purity $\geq$ 90%) in an Ar atmosphere. The detailed synthesis method is described in the previous paper [12].

4.2. Pressure–Composition–Isothermal (PCT) Measurements

PCT measurements were conducted using a Sievert-type apparatus (Suzuki Shokan Co. Ltd., Tokyo, Japan, High-Pressure System Co., Saitama, Japan). The composite materials were heated to the designated temperature under a high H_2 pressure (>10 MPa) to prevent decomposition before the measurements. During the dehydrogenation measurements, H_2 was released in increments smaller than 0.01 MPa. The pressure change was kept stable between 15 min and 5 h to allow equilibration of the reaction at each step. Hydrogen weight loss was calculated using the function related to the pressure; volume of the apparatus, container, and material; and the temperatures at each part of the apparatus. The investigated temperature range was from 623 K to 673 K. The investigation of the rehydrogenation and reversible hydrogenation property of the optimized composition was performed in two steps. First, the dehydrogenation reaction was measured via PCT measurements using the same process described previously. Second, 20 MPa H_2 was added to the container, which was subsequently maintained for 20–24 h at the same temperature. This two-step process was repeated four times.

4.3. Phase Identification

Phase identification was performed using powder XRD (PANalytical X'Pert-Pro, Almelo, The Netherlands, Cu Kα radiation, λ = 1.5405 Å) at room temperature. The microstructure and element distributions of the synthesized samples and the dehydrogenated products were recorded via SEM (JSM-6009, JEOL Ltd., Tokyo, Japan) using an instrument equipped for EDS (EX-54175JMH, JEOL Ltd., Tokyo, Japan). During the synthesis and measurement processes, the samples were always handled under Ar or vacuum to avoid contamination by air or water.

5. Conclusions

The isothermal dehydrogenation processes of $(1 - x)\text{LiBH}_4 + x\text{Mg}_2\text{FeH}_6$ ($x = 0.25$, 0.5, and 0.75) composite materials were studied using PCT measurements. The PCT results suggest that a stoichiometric reaction between LiBH_4 and Mg_2FeH_6 occurred at all of the investigated composition ratios, i.e., $\text{LiBH}_4 + \text{Mg}_2\text{FeH}_6 \rightarrow \text{LiH} + 2\text{MgH}_2 + (\text{Fe, FeB}) + 5/2\text{H}_2$. Both LiBH_4 and Mg_2FeH_6 were thermodynamically destabilized by this reaction. $x = 0.5$ is the optimal ratio for the composite material, which is also the reacting ratio of the two complex hydrides in the stoichiometric reaction. With the optimal ratio, the independent dehydrogenation of the excessive LiBH_4 or Mg_2FeH_6 was avoided. During the dynamic TG–MS experiments, only one dehydrogenation event was observed for the $x = 0.5$ and 0.75 composites. The difference between the PCT and TG–MS results is explained by the slow kinetics of the reaction between LiBH_4 and Mg_2FeH_6. Beside the dehydrogenation property, the optimal ratio also contributed to the enhanced reversible hydrogenation properties of the composite materials. The $x = 0.5$ composites can be de/rehydrogenated completely at 673 K and 20 MPa H_2 for at least four cycles without the loss of H_2 capacity.

Acknowledgments: This research has been financed by the Grant-in-Aid for Young Scientists (B) (17K14830), the Grant-in-Aid for Research Fellow of Japan Society for the Promotion of Science (15J10604), the JSPS KAKENHI Grant (25220911), and the German Federal Government under the European ERA-NET CONCERT Japan scheme via the iTHEUS project (grant CONCERT-EN-015). The authors would like to acknowledge Ms. N. Warifune for providing technical support.

Author Contributions: Guanqiao Li, Motoaki Matsuo, and Shin-ichi Orimo conceived and designed the experiments; Guanqiao Li and Anna-Lisa Chaudhary performed the experiments; Toyoto Sato and Shigeyuki Takagi helped analyze the data; Anna-Lisa Chaudhary and Martin Dornheim gave advice concerning the analysis of the data; and Guanqiao Li wrote the paper.

Conflicts of Interest: The authors declare no conflict of interest.

References

1. Paskevicius, M.; Jepsen, L.H.; Schouwink, P.; Cerny, R.; Ravnsbaek, D.B.; Filinchuk, Y.; Dornheim, M.; Besenbacher, F.; Jensen, T.R. Metal Borohydrides and Derivatives—Synthesis, Structure and Properties. *Chem. Soc. Rev.* **2017**, *46*, 1565–1634. [CrossRef] [PubMed]
2. Callini, E.; Atakli, Z.O.K.; Hauback, B.C.; Orimo, S.; Jensen, C.; Dornheim, M.; Grant, D.; Cho, Y.W.; Chen, P.; Hjorvarsson, B.; et al. Complex and Liquid Hydrides for Energy Storage. *Appl. Phys. A* **2016**, *122*, 353. [CrossRef]
3. Orimo, S.; Nakamori, Y.; Eliseo, J.R.; Zuttel, A.; Jensen, C.M. Complex Hydrides for Hydrogen Storage. *Chem. Rev.* **2007**, *107*, 4111–4132. [CrossRef] [PubMed]
4. Li, H.W.; Yan, Y.G.; Orimo, S.; Züttel, A.; Jensen, C.M. Recent Progress in Metal Borohydrides for Hydrogen Storage. *Energies* **2011**, *4*, 185–214. [CrossRef]
5. Bösenberg, U.; Doppiu, S.; Mosegaard, L.; Barkhordarian, G.; Eigen, N.; Borgschulte, A.; Jensen, T.R.; Cerenius, Y.; Gutfleisch, O.; Klassen, T.; et al. Hydrogen Sorption Properties of MgH_2–LiBH_4 Composites. *Acta Mater.* **2007**, *55*, 3951–3958. [CrossRef]
6. Vajo, J.J.; Li, W.; Liu, P. Thermodynamic and Kinetic Destabilization in LiBH_4–Mg_2NiH_4: Promise for Borohydride-Based Hydrogen Storage. *Chem. Commun.* **2010**, *46*, 6687–6689. [CrossRef] [PubMed]
7. Javadian, P.; Zlotea, C.; Ghimbeu, C.M.; Latroche, M.; Jensen, T.R. Hydrogen Storage Properties of Nanoconfined LiBH_4–Mg_2NiH_4 Reactive Hydride Composites. *J. Phys. Chem. C* **2015**, *119*, 5819–5826. [CrossRef]
8. Yan, Y.G.; Li, H.W.; Maekawa, H.; Miwa, K.; Towata, S.; Orimo, S. Formation of Intermediate Compound $\text{Li}_2\text{B}_{12}\text{H}_{12}$ during The Dehydrogenation Process of The LiBH_4–MgH_2 System. *J. Phys. Chem. C* **2011**, *115*, 19419–19423. [CrossRef]
9. Bergemann, N.; Pistidda, C.; Milanese, C.; Emmler, T.; Karimi, F.; Chaudhary, A.L.; Chierotti, M.R.; Klassen, T.; Dornheim, M. $\text{Ca(BH}_4)_2$–Mg_2NiH_4: On The Pathway to A $\text{Ca(BH}_4)_2$ System with A Reversible Hydrogen Cycle. *Chem. Commun.* **2016**, *52*, 4836–4839. [CrossRef] [PubMed]

10. Bosenberg, U.; Kim, J.W.; Gosslar, D.; Eigen, N.; Jensen, T.R.; von Colbe, J.M.B.; Zhou, Y.; Dahms, M.; Kim, D.H.; Gunther, R.; et al. Role of Additives in LiBH$_4$–MgH$_2$ Reactive Hydride Composites for Sorption Kinetics. *Acta Mater.* **2010**, *58*, 3381–3389. [CrossRef]

11. Yang, J.; Sudik, A.; Wolverton, C. Destabilizing LiBH$_4$ with A Metal (*M* = Mg, Al, Ti, V, Cr, or Sc) or Metal Hydride (MH$_2$, MgH$_2$, TiH$_2$, or CaH$_2$). *J. Phys. Chem. C* **2007**, *111*, 19134–19140. [CrossRef]

12. Li, G.; Matsuo, M.; Deledda, S.; Sato, R.; Hauback, B.C.; Orimo, S. Dehydriding Property of LiBH$_4$ Combined with Mg$_2$FeH$_6$. *Mater. Trans.* **2013**, *54*, 1532–1534. [CrossRef]

13. Li, G.; Matsuo, M.; Aoki, K.; Ikeshoji, T.; Orimo, S. Dehydriding Process and Hydrogen-Deuterium Exchange of LiBH$_4$–Mg$_2$FeD$_6$ Composites. *Energies* **2015**, *8*, 5459–5466. [CrossRef]

14. Chaudhary, A.-L.; Li, G.; Matsuo, M.; Orimo, S.; Deledda, S.; Sørby, M.H.; Hauback, B.C.; Pistidda, C.; Klassen, T.; Dornheim, M. Simultaneous Desorption Behavior of *M* Borohydrides and Mg$_2$FeH$_6$ Reactive Hydride Composites (*M* = Mg, then Li, Na, K, Ca). *Appl. Phys. Lett.* **2015**, *107*, 073905. [CrossRef]

15. Li, G.; Matsuo, M.; Deledda, S.; Hauback, B.C.; Orimo, S. Dehydriding Property of NaBH$_4$ Combined with Mg$_2$FeH$_6$. *Mater. Trans.* **2014**, *55*, 1141–1143. [CrossRef]

16. Pinkerton, F.E.; Meyer, M.S.; Meisner, G.P.; Balogh, M.P.; Vajo, J.J. Phase Boundaries and Reversibility of LiBH$_4$–MgH$_2$ Hydrogen Storage Material. *J. Phys. Chem. C* **2007**, *111*, 12881–12885. [CrossRef]

17. Deng, S.S.; Xiao, X.Z.; Han, L.Y.; Li, Y.; Li, S.Q.; Ge, H.W.; Wang, Q.D.; Chen, L.X. Hydrogen Storage Performance of 5LiBH$_4$ + Mg$_2$FeH$_6$ Composite System. *Int. J. Hydrog. Energy* **2012**, *37*, 6733–6740. [CrossRef]

18. Langmi, H.W.; McGrady, G.S.; Newhouse, R.; Rönnebro, E. Mg$_2$FeH$_6$–LiBH$_4$ and Mg$_2$FeH$_6$–LiNH$_2$ Composite Materials for Hydrogen Storage. *Int. J. Hydrog. Energy* **2012**, *37*, 6694–6699. [CrossRef]

19. Ghaani, M.R.; Catti, M.; Nale, A. Thermodynamics of Dehydrogenation of the 2LiBH$_4$–Mg$_2$FeH$_6$ Composite. *J. Phys. Chem. C* **2012**, *116*, 26694–26699. [CrossRef]

20. Bogdanovic, B.; Bohmhammel, K.; Christ, B.; Reiser, A.; Schlichte, K.; Vehlen, R.; Wolf, U. Thermodynamic Investigation of the Magnesium–Hydrogen System. *J. Alloys Compd.* **1999**, *282*, 84–92. [CrossRef]

21. Bohmhammel, K.; Wolf, U.; Wolf, G.; Konigsberger, E. Thermodynamic Optimization of the System Magnesium–Hydrogen. *Thermochim. Acta* **1999**, *337*, 195–199. [CrossRef]

22. Zhang, X.; Yang, R.; Qu, J.; Zhao, W.; Xie, L.; Tian, W.; Li, X. The Synthesis and Hydrogen Storage Properties of Pure Nanostructured Mg$_2$FeH$_6$. *Nanotechnology* **2010**, *21*, 095706. [CrossRef] [PubMed]

23. Puszkiel, J.A.; Larochette, P.A.; Gennari, F.C. Thermodynamic and Kinetic Studies of Mg–Fe–H after Mechanical Milling Followed by Sintering. *J. Alloys Compd.* **2008**, *463*, 134–142. [CrossRef]

24. Wang, Y.; Cheng, F.Y.; Li, C.S.; Tao, Z.L.; Chen, J. Preparation and Characterization of Nanocrystalline Mg$_2$FeH$_6$. *J. Alloys Compd.* **2010**, *508*, 554–558. [CrossRef]

25. Siegel, D.J.; Wolverton, C.; Ozoliņš, V. Thermodynamic Guidelines for the Prediction of Hydrogen Storage Reactions and Their Application to Destabilized Hydride Mixtures. *Phys. Rev. B* **2007**, *76*. [CrossRef]

26. Price, T.E.C.; Grant, D.M.; Telepeni, I.; Yu, X.B.; Walker, G.S. The Decomposition Pathways for LiBD$_4$–MgD$_2$ Multicomponent Systems Investigated by In Situ Neutron Diffraction. *J. Alloys Compd.* **2009**, *472*, 559–564. [CrossRef]

27. Miwa, K.; Takagi, S.; Matsuo, M.; Orimo, S. Thermodynamical Stability of Complex Transition Metal Hydrides M$_2$FeH$_6$. *J. Phys. Chem. C* **2013**, *117*, 8014–8019. [CrossRef]

28. Bogdanovic, B.; Reiser, A.; Schlichte, K.; Spliethoff, B.; Tesche, B. Thermodynamics and Dynamics of the Mg−Fe−H System and Its Potential for Thermochemical Thermal Energy Storage. *J. Alloys Compd.* **2002**, *345*, 77–89. [CrossRef]

29. Zavorotynska, O.; Corno, M.; Damin, A.; Spoto, G.; Ugliengo, P.; Baricco, M. Vibrational Properties of *M*BH$_4$ and *M*BF$_4$ Crystals (*M* = Li, Na, K): A Combined DFT, Infrared, and Raman Study. *J. Phys. Chem. C* **2011**, *115*, 18890–18900. [CrossRef]

30. Parker, S.F.; Williams, K.P.J.; Bortz, M.; Yvon, K. Inelastic Neutron Scattering, Infrared, and Raman Spectroscopic Studies of Mg$_2$FeH$_6$ and Mg$_2$FeD$_6$. *Inorg. Chem.* **1997**, *36*, 5218–5221. [CrossRef]

31. Kim, K.B.; Shim, J.H.; Park, S.H.; Choi, I.S.; Oh, K.H.; Cho, Y.W. Dehydrogenation Reaction Pathway of the LiBH$_4$−MgH$_2$ Composite under Various Pressure Conditions. *J. Phys. Chem. C* **2015**, *119*, 9714–9720. [CrossRef]

32. Bosenberg, U.; Ravnsbaek, D.B.; Hagemann, H.; D'Anna, V.; Minella, C.B.; Pistidda, C.; van Beek, W.; Jensen, T.R.; Bormann, R.; Dornheim, M. Pressure and Temperature Influence on the Desorption Pathway of the LiBH$_4$−MgH$_2$ Composite System. *J. Phys. Chem. C* **2010**, *114*, 15212–15217. [CrossRef]

33. Orimo, S.; Nakamori, Y.; Kitahara, G.; Miwa, K.; Ohba, N.; Towata, S.; Zuttel, A. Dehydriding and Rehydriding Reactions of LiBH$_4$. *J. Alloys Compd.* **2005**, *404*, 427–430. [CrossRef]
34. Gosselin, C.; Deledda, S.; Hauback, B.C.; Huot, J. Effect of Synthesis Route on the Hydrogen Storage Properties of 2MgH$_2$–Fe Compound Doped with LiBH$_4$. *J. Alloys Compd.* **2015**, *645*, S304–S307. [CrossRef]

![inorganics logo] *inorganics*

MDPI

Article

Hydrogen Storage Stability of Nanoconfined MgH$_2$ upon Cycling

Priscilla Huen [1], Mark Paskevicius [2], Bo Richter [1], Dorthe B. Ravnsbæk [3] and Torben R. Jensen [1,*]

[1] Center for Materials Crystallography, Interdisciplinary Nanoscience Center and Department of Chemistry, Aarhus University, Langelandsgade 140, 8000 Aarhus C, Denmark; priscilla.huen@inano.au.dk (P.H.); richter@chem.au.dk (B.R.)

[2] Department of Physics and Astronomy, Fuels and Energy Technology Institute, Curtin University, Kent Street, Bentley, WA 6102, Australia; M.Paskevicius@curtin.edu.au

[3] Department of Physics, Chemistry and Pharmacy, University of Southern Denmark, Campusvej 55, 5230 Odense M, Denmark; dbra@sdu.dk

* Correspondence: trj@chem.au.dk; Tel.: +45-8715-5939

Received: 4 July 2017; Accepted: 18 August 2017; Published: 23 August 2017

Abstract: It is of utmost importance to optimise and stabilise hydrogen storage capacity during multiple cycles of hydrogen release and uptake to realise a hydrogen-based energy system. Here, the direct solvent-based synthesis of magnesium hydride, MgH$_2$, from dibutyl magnesium, MgBu$_2$, in four different carbon aerogels with different porosities, i.e., pore sizes, $15 < D_{avg} < 26$ nm, surface area $800 < S_{BET} < 2100$ m^2/g, and total pore volume, $1.3 < V_{tot} < 2.5$ cm^3/g, is investigated. Three independent infiltrations of MgBu$_2$, each with three individual hydrogenations, are conducted for each scaffold. The volumetric and gravimetric loading of MgH$_2$ is in the range 17 to 20 vol % and 24 to 40 wt %, which is only slightly larger as compared to the first infiltration assigned to the large difference in molar volume of MgH$_2$ and MgBu$_2$. Despite the rigorous infiltration and sample preparation techniques, particular issues are highlighted relating to the presence of unwanted gaseous by-products, Mg/MgH$_2$ containment within the scaffold, and the purity of the carbon aerogel scaffold. The results presented provide a research path for future researchers to improve the nanoconfinement process for hydrogen storage applications.

Keywords: hydride; nanoconfinement; carbon scaffold

---

1. Introduction

The development of a cleaner and more sustainable energy system is urgently needed to meet our increasing energy demand, and to avoid global warming and environmental pollution due to increasing levels of carbon dioxide and other toxic gases. Hydrogen is considered a potential energy carrier, since it is an abundant, non-greenhouse gas and can be produced by the electrolysis of water [1–4]. However, gaseous hydrogen at ambient conditions has a low density of 0.082 g/L, which is a disadvantage for mobile applications, even with compression [5]. Therefore, the solid state storage of hydrogen in a metal hydride has been investigated [3,4,6,7]. Magnesium hydride, MgH$_2$, as one of the most extensively studied hydride materials, has a moderately high theoretical gravimetric H$_2$ density of ρ_m(MgH$_2$) = 7.6 wt % H$_2$, and a volumetric H$_2$ density of ρ_v(MgH$_2$) = 110 g H$_2$/L [8]. However, the practical application of an MgH$_2$-based system is hindered from the unfavourable thermodynamics and the typically slow kinetics of the hydrogen release and uptake [9,10].

To improve the hydrogen storage properties of MgH$_2$, nanoconfinement in porous materials can be considered [11–17]. Preparing nanosized MgH$_2$ from this bottom-up approach can reduce the hydrogen diffusion distance and increase the amount of hydrogen in the grain boundaries, leading to improved kinetics of hydrogenation/dehydrogenation [18]. Nanoconfinement has also been employed for other hydride materials (e.g., $NaAlH_4$, $LiBH_4$, and NH_3BH_3) and demonstrates an improvement in gas release properties [15,19–22]. Nanoconfined MgH_2 in mesoporous scaffolds can be prepared through an Mg melt infiltration process followed by hydrogenation, or by a direct synthesis route using a precursor (e.g., dibutyl magnesium, $MgBu_2$) [11–13,23–25]. The loading of MgH_2 in the scaffold is between 3.6 wt % and 22.0 wt % [19].

Previous work reveals that smaller pore sizes within resorcinol-formaldehyde carbon aerogel (CA) scaffolds lead to improved hydrogen release kinetics of nanoconfined MgH_2, by reducing the particle size and increasing the surface area of MgH_2 [11]. However, mainly the first hydrogen release cycle has been investigated up to now. Therefore, this present study includes multiple cycles of hydrogen release and uptake. Thermal treatment of the CA scaffold in a gas flow (often CO_2) can increase the surface area, up to >2000 m²/g, and the total pore volume to 2–3 mL/g, but has almost no effect on the pore size distribution. This procedure is often denoted scaffold "activation". Therefore, CA scaffolds are considered very customisable and may possess a wide range of porosity parameters. Previous investigations of sodium aluminium hydride, $NaAlH_4$, nanoconfined in activated scaffolds reveal that more material can be infiltrated onto an activated scaffold, i.e., there is a larger hydrogen storage capacity due to a larger pore volume, but these materials show slower kinetics for hydrogen release as compared to nonactivated scaffold [26]. Nanoconfined hydrides are mostly shown to exhibit improved kinetics of hydrogen release and uptake, but a change in thermodynamics is only observed when the scaffolds have pore sizes smaller than 2–3 nm [27].

There are a number of studies that have investigated the effect of nanoconfinement on the dehydrogenation properties of MgH_2, but there is little information about the reversible hydrogen storage capacity of nanoconfined MgH_2 upon cycling (hydrogen release and uptake). In addition, it has been discovered that butane gas is released (in conjunction with hydrogen) in the thermal treatment of nanoconfined MgH_2, which may refer to the incomplete hydrogenation of $MgBu_2$ after infiltration [28]. Here, we maximize the hydrogen storage capacity of nanoconfined MgH_2 through multiple infiltrations and use a variety of carbon aerogel scaffolds with different pore networks. The properties of nanoconfined MgH_2 samples are then compared with a focus on their hydrogen storage capacity after multiple hydrogen release and uptake cycles.

2. Results and Discussion

2.1. Porosity of the Nanoporous Scaffolds and Confinement of Magnesium Hydride

Magnesium hydride, MgH_2, was nanoconfined in four different carbon aerogel scaffolds with different texture properties as shown in Table 1. The porosities of the as-synthesised scaffolds X1 and X2 are similar except for the average pore sizes, D_{max}, of 16.6 ± 0.5 and 27.1 ± 2.7 nm, respectively. The surface area, S_{BET}, and total pore volume, V_{tot}, of the activated scaffolds CX1 and CX2 increase significantly after heat treatment in a flow of carbon dioxide, but D_{max} remains almost constant.

Table 1. Texture properties of the carbon aerogel scaffolds and amount of magnesium hydride present after three infiltrations.

Carbon Aerogels	S_{BET} (m²/g)	D_{avg} (nm)	V_{micro} (cm³/g)	V_{meso} (cm³/g)	V_{tot} (cm³/g)	MgH_2 (wt %) [a]	MgH_2 (vol %) [b]
X1	829 ± 16	16.6 ± 0.5	0.23 ± 0.01	1.13 ± 0.03	1.32 ± 0.04	24.8	17.3
X2	801 ± 16	27.1 ± 2.7	0.25 ± 0.01	1.11 ± 0.08	1.32 ± 0.10	24.3	16.7
CX1	1940 ± 131	14.7 ± 0.6	0.54 ± 0.07	1.85 ± 0.06	2.37 ± 0.12	37.1	17.1
CX2	1803 ± 30	25.0 ± 0.8	0.56 ± 0.02	1.89 ± 0.04	2.38 ± 0.05	40.3	19.6

[a] Calculated stoichiometrically from the uptake of di-*n*-butylmagnesium; [b] Calculated from the volume of the empty scaffold and the bulk density of MgH_2.

The procedures for the direct synthesis of nanoconfined magnesium hydride, MgH_2 utilised in this investigation are a new modification of a previously described approach using monoliths of carbon aerogel scaffold [11,12]. The aim of this investigation is to explore new approaches to prepare high hydrogen capacity materials based on nanoconfined magnesium hydride. A total of three dibutyl magnesium infiltrations, each with three hydrogenations, were conducted in order to increase the loading of MgH$_2$ in the porous scaffolds, with details provided in Table 1 and Table S1. The infiltrated amount of dibutyl magnesium, MgBu$_2$, is measured gravimetrically after mechanically removing excess dibutyl magnesium that was crystallised on the surface of the scaffolds. Scaffolds X2 and CX2 show decreasing amounts of infiltrated MgBu$_2$ for each consecutive cycle of infiltration, see Table S1, assigned to increasing amounts successfully infiltrated in each cycle. In contrast, the amount of infiltrated MgBu$_2$ in X1 and CX1 vary more so, possibly due to difficulties in efficiently removing MgBu$_2$ from the surface. A graphical presentation of the results from the infiltrations is presented in Figure 1. Dibutyl magnesium is assumed to be completely converted to MgH$_2$ following the reaction Scheme (1):

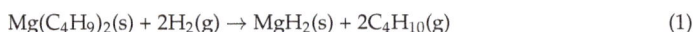

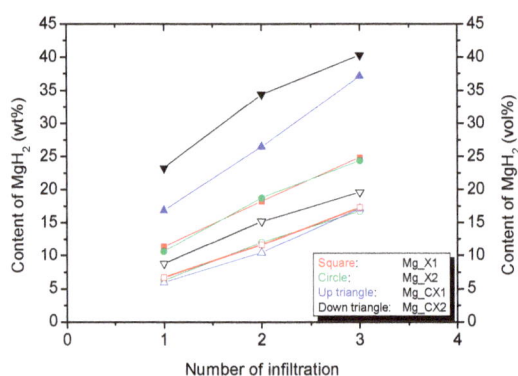

$$Mg(C_4H_9)_2(s) + 2H_2(g) \rightarrow MgH_2(s) + 2C_4H_{10}(g) \tag{1}$$

Figure 1. The cumulative gravimetric (solid symbols) and volumetric (open symbols) infiltration of magnesium hydride MgH$_2$ after each infiltration procedure in the four different carbon aerogel (CA) scaffolds.

The gravimetric and volumetric quantity of infiltrated magnesium hydride is calculated using the mass of scaffold, total pore volume, and bulk density of MgH$_2$. The volumetric loading of MgH$_2$ in the three scaffolds X1, X2, and CX1 are similar, ~17 vol %, whereas CX2 is slightly larger, ~20 vol %. However, the gravimetric hydride content varies more significantly, ~24 wt % for X1 and X2, ~37 wt % for CX1, and ~40 wt % for CX2. Recall that three independent infiltrations of MgBu$_2$ were conducted in this work, each with three individual hydrogenations. However, this work reveals that only a moderate increase in the infiltrated amount of MgH$_2$ is obtained after three infiltrations as compared to 12 vol % MgH$_2$ after one infiltration in a previous work [11]. That is mainly assigned to the large difference in molar volume of MgH$_2$ (18.2 cm^3/mol) and MgBu$_2$ (188.2 cm^3/mol). As such, MgBu$_2$ takes up a large volume after the infiltration, and only one-tenth of this volume is converted to MgH$_2$. This is similar to the utilisation of butyllithium for the direct synthesis of nanoconfined LiH, where loadings in the range of 12–17 wt % were obtained [29]. Secondly, MgH$_2$ may have a tendency to block the pores and stop further infiltration, which may hamper the full infiltration of the smaller pores.

2.2. Hydrogen Storage Capacity upon Cycling

Reversible hydrogen storage properties were investigated for five cycles of hydrogen release ($T = 355\,°C$, $t = 15$ h in vacuum) and uptake ($T = 355\,°C$, $t = 15$ h in $p(H_2) = 50$ bar), i.e., $\Delta p(H_2) = 50$ bar, denoted *condition 1*, for the four nanoconfined MgH$_2$ samples (see Figure 2). In the first decomposition, Mg_CX1 released 3.1 wt % H$_2$, which is slightly higher than the calculated hydrogen content of the sample based on the calculated quantity of MgH$_2$, 2.82 wt % (see Table 2). The observed hydrogen release from Mg_X1, 1.8 wt % H$_2$, is in accordance with the calculated value (1.88 wt % H$_2$). Samples Mg_X2 and Mg_CX2, with larger average pore sizes, release a lower quantity of gas, 1.3 and 2.2 wt % H$_2$, which corresponds to 68% and 71% of the calculated hydrogen content, respectively. For the following cycles, Table 2 and Figure 2 reveal a general stabilisation of the hydrogen storage capacity after the second desorption cycle.

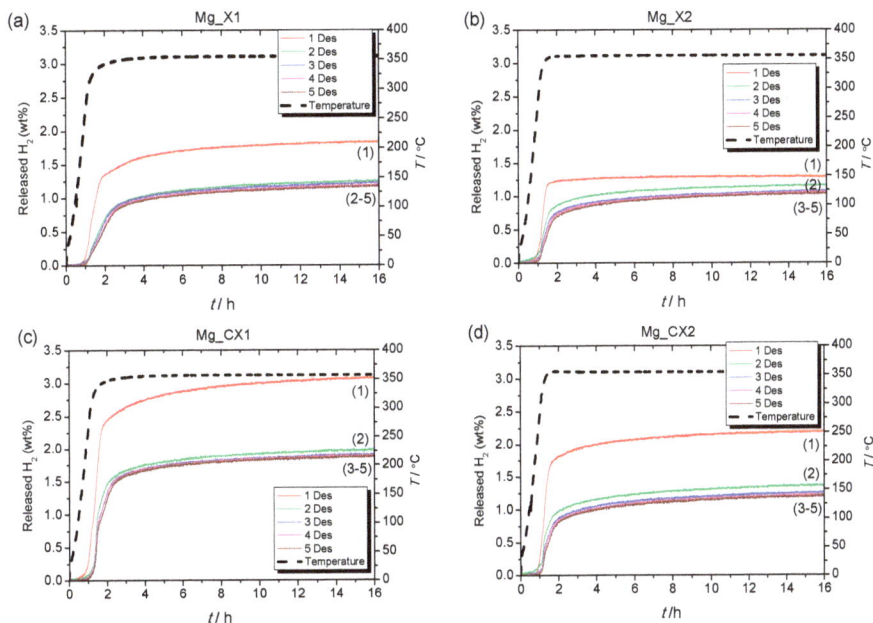

Figure 2. Sievert's measurements of (**a**) Mg_X1; (**b**) Mg_X2; (**c**) Mg_CX1; (**d**) Mg_CX2 under *condition* 1. Samples heated in vacuum from room temperature to 355 °C ($\Delta T/\Delta t = 5$ °C/min) for 15 h and reabsorbed for 15 h under $p(\text{H}_2) = 50$ bar.

Table 2. Calculated hydrogen content in infiltrated carbon aerogels and the hydrogen release measured by Sievert's method in desorption one (Des1) to five (Des5) using *condition* 1. The percentages in parentheses correspond to the retained hydrogen storage capacity compared to the initial values in Des1.

Sample	D_{max} (nm)	ρ_m (H$_2$)/(wt %)	Des1 (H$_2$ wt %)	Des2 (H$_2$ wt %)	Des3 (H$_2$ wt %)	Des4 (H$_2$ wt %)	Des5 (H$_2$ wt %)
Mg_X1	17	1.88	1.8 (100%)	1.3 (72%)	1.2 (67%)	1.2 (67%)	1.2 (67%)
Mg_X2	26	1.85	1.3 (100%)	1.2 (92%)	1.1 (85%)	1.1 (85%)	1.0 (77%)
Mg_CX1	15	2.82	3.1 (100%)	2.0 (67%)	1.9 (61%)	1.9 (61%)	1.9 (61%)
Mg_CX2	25	3.06	2.2 (100%)	1.4 (64%)	1.3 (59%)	1.2 (55%)	1.2 (55%)

For all four samples, the hydrogen release temperature is lower for the first cycle in comparison to further cycles. This may indicate that other reactions, besides the release of hydrogen, mainly occur in the first cycle. A thermal analysis using mass spectroscopy revealed that butane release occurs in addition to hydrogen release. This is unexpected due to the rigorous infiltration procedure, where a total of nine hydrogenation and evacuation steps are undertaken. In fact, butane is typically released at lower temperatures than hydrogen, generally in the range of 100 to 350 °C (this is further discussed later).

A similar investigation of the reversible hydrogen storage properties of the four nanoconfined samples was conducted using *condition* 2, i.e., the same temperature and time but a higher back-pressure for hydrogen release ($p(H_2)$ = 4–5 bar) and lower hydrogen pressure for uptake ($p(H_2)$ = 12 bar), i.e., $\Delta p(H_2)$ ~7.5 bar. Figure S1 shows a dramatic difference in the hydrogen release properties in comparison to Figure 2, where hydrogen desorption was conducted under vacuum and hydrogen absorption was conducted under 50 bar.

2.2.1. Thermodynamic Considerations

Conditions 1 and 2 for hydrogen release and uptake were selected so that *condition* 2 was just above/below the thermodynamic equilibrium pressure for hydrogen absorption/release of Mg/MgH$_2$ at 355 °C, i.e., $p_{eq}(H_2)$ = 6.4 bar [30], whereas *condition* 1 operates at a considerable "over-pressure". The hydrogen release data is presented in Figure 2 and Figure S1, respectively, showing dramatically different hydrogen release properties. Specifically, Figure S1 displays much lower gravimetric hydrogen release (i.e., 0.4 wt % vs. 1.8 wt % for the same sample and same cycle).

For *condition* 2, hydrogen is absorbed at $p(H_2)$ = 11–12 bar and desorbed at $p(H_2)$ < 5.2 bar, which is well above/below the thermodynamically limiting equilibrium pressure of $p_{eq}(H_2)$ = 6.4 bar [30]. Thus, from a thermodynamic point of view, *conditions* 1 and 2 should provide the same hydrogen storage properties, including hydrogen capacity. The hydrogen release profiles of *conditions* 1 and 2 (Figure 2 and Figure S1) are similar, which suggest that hydrogen release kinetics are similar and the majority of hydrogen release is within the first 3 h in all cases. However, the amount of hydrogen release is much lower in *condition* 2.

The very different pressures during hydrogenation, 50 or 12 bar for *conditions* 1 and 2, may lead to large differences in the degree of hydrogenation for several reasons: (i) Hydrogen is known to have slow diffusion in bulk Mg and MgH$_2$; (ii) The larger molar volume of magnesium hydride, $\rho_{mol}(MgH_2)$ = 18.15 cm^3/mol as compared to magnesium $\rho_{mol}(Mg)$ = 13.98 cm^3/mol may lead to core/shell formation during the hydrogenation of magnesium particles. Thus, a magnesium hydride layer may retard further hydrogenation; (iii) The material expansion of Mg to MgH$_2$ could lead to the blocking of the smaller pores in the scaffold, which may also retard further hydrogenation. A larger "over-pressure" as applied in *condition* 1 may limit the above mentioned drawbacks, (i) to (iii), and lead to complete hydrogenation of the samples.

2.2.2. Kinetics of Hydrogen Release of Nanoconfined MgH$_2$

For all the nanoconfined magnesium hydride samples, the majority of hydrogen is desorbed during heating from room temperature to 355 °C. Furthermore, in all cases, the first H$_2$ release profile is significantly different to the following ones, whereas the second is similar to the third, and then the H$_2$ release profiles become almost identical. This is clearly observed in Figure 2. For all four samples, the first decomposition has faster kinetics for hydrogen release and also a lower onset temperature. The initial 10 to 50% H$_2$ for the first cycle is released at a rate of 0.024, 0.030, 0.046, and 0.046 wt % H$_2$/min for the samples Mg_X1, Mg_X2, Mg_CX1 and Mg_CX2, respectively. The later hydrogen release profiles, cycle no. 2 to 5, consist of two regimes, see Figure S2. Initially, the hydrogen release rate appears to increase exponentially and then linearly at higher temperatures (see Figure S2). This suggests that the hydrogen release mechanism consists of more than one process, which is also observed for Mg$_{1-x}$Ti$_x$H$_2$ nanoparticles [31]. Here, we assume that the individual hydrogen release processes are independent and are due to differences in particle size, location in small or large pores or being located outside the scaffold, or consisting of Mg/MgH$_2$/MgBu$_2$ core–shell particles [32]. Assuming independent individual processes for hydrogen release, then the fastest process would occur at lower temperatures.

The data presented here for hydrogen release is not measured under isothermal conditions, which makes the kinetic analysis more challenging. The overall hydrogen release profile has a distorted sigmoidal shape, which cannot be modelled using Avrami-type kinetic equations, which have previously successfully been used to evaluate hydrogen release from Mg–Al–H, Mg–Cu–H, and Mg–Ni–H systems [33–35]. The first exponentially increasing hydrogen release does not match a power law, but the linear part of the profile can be fitted to a linear equation of the type, $\alpha(t) = b + kt$, where k is assigned an apparent rate constant. Apparent kinetic data is useful to compare similar samples in a more quantitative way. The degree of hydrogen release, $\alpha(t)$, from the normalised hydrogen release profiles (see Figure 3) also expresses the degree of magnesium formation. For the two as-synthesised scaffolds, the linear part of the curve is approximately in the range $0.3 < \alpha(t) < 0.6$. The apparent rate constants for these two samples, Mg_X1 and Mg_X2, are $k_1 = 1.33(4) \times 10^{-4}\,\text{s}^{-1}$ and $k_2 = 2.3(1) \times 10^{-4}\,\text{s}^{-1}$, respectively. The carbon dioxide activated sample, Mg_CX2, is somewhat similar, $0.37 < \alpha(t) < 0.62$, with $k_4 = 1.65(7) \times 10^{-4}\,\text{s}^{-1}$, whereas the linear hydrogen release profile occurs at a higher degree of formation, $0.50 < \alpha(t) < 0.75$, for Mg_CX1, with $k_3 = 1.23(2) \times 10^{-4}\,\text{s}^{-1}$. The linear regime for the hydrogen release rates have onsets in the temperature range 300 to ~330 °C and in some cases continue into the isothermal heating at $T = 355$ °C. We note that the calculated values for the apparent rate constants have the same order of magnitude as the values for bulk- and nickel-doped magnesium hydride, i.e., $1.0 < k < 5.3 \times 10^{-4}\,\text{s}^{-1}$, but at significantly higher temperatures, 370 to 390 °C [35].

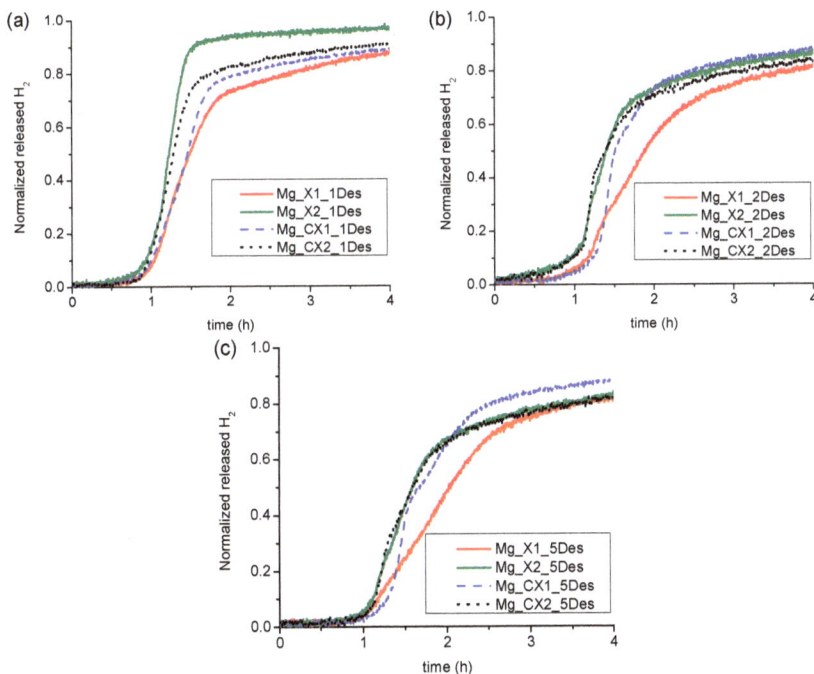

Figure 3. Normalized Sieverts gas release profiles of the four samples of nanoconfined magnesium hydride, (**a**) first desorption cycle; (**b**) second desorption cycle; and (**c**) fifth desorption cycle.

The first gas release with an exponential increasing rate is assigned to MgH$_2$ confined in the smaller pores, whereas hydrogen release at higher temperatures in the linear regime is assigned to MgH$_2$ confined in the larger cavities or outside the scaffold. Clearly, the rate of hydrogen release is lower for the larger particles as compared to the initial hydrogen release for the smaller in all cases, despite the significantly higher temperatures in the linear regime, which is illustrated in Figure S2. Accordingly, the four samples have similar apparent rate constants. However, the hydrogen storage capacities for the nanoconfined samples presented in Table 2 are significantly lower as compared to well-known magnesium hydride–metal oxide systems, which may also show fast kinetics, e.g., MgH_2–Nb_2O_5 [36,37]. However, this is due to a reduction of the metal and the formation of a solid solution, $Mg_xNb_{1-x}O$ [38].

2.3. Analysis of the Released Gases and Samples after Cycling

TGA-MS reveals that nanoconfined MgH_2 samples release hydrogen in accordance with Sievert's measurements (see Figure 4). However, there is also a significant quantity of butane gas that is also released, not just in the first cycle but also small but still detectable amounts on the fifth desorption cycle. However, after five desorption/absorption cycles under *condition* 1, the amount of butane released by Mg_X2 is about 100 times less compared to the as-prepared Mg_X2. It should again be reiterated that the sample preparation in this study was meticulous in pre-cycling a hydrogen reduction step three times in an attempt to completely transform the $MgBu_2$ precursor, but the release gas stream is still contaminated with butane. The conversion of the $MgBu_2$ precursor to MgH_2 was conducted at $T = 150\,°C$ during sample preparation. This treatment appears more efficient for scaffolds with larger pores, which release less butane. Scaffolds with smaller pores may more effectively contain and isolate $MgBu_2$, preventing it from hydrogenating during activation. This leads to butane release in the later hydrogenation cycles. In terms of hydrogen release, the temperature of maximum hydrogen release shifts to a higher temperature due to the particle growth of MgH_2, as revealed by powder X-ray diffraction (see Section 2.4).

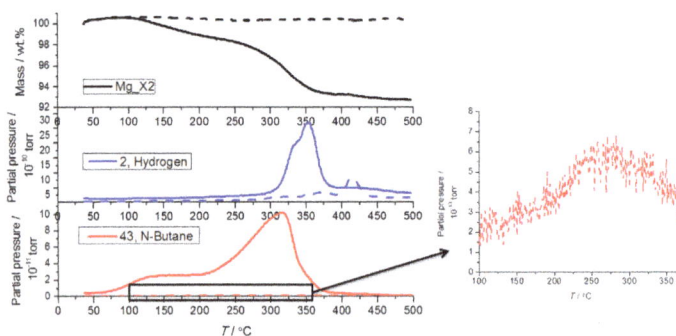

Figure 4. Thermogravimetric and mass spectroscopic analysis of the hydrogen and butane release from as-prepared Mg_X2 (solid line) and cycled Mg_X2 (dash line) during constant heating from room temperature (RT) to 500 °C ($\Delta T/\Delta t = 5\,°C/min$).

The minor increase in the measured mass at low temperature is caused by buoyancy. The total mass loss of the as-prepared Mg_X2 upon decomposition was 7.3 wt %, which is significantly higher than the calculated hydrogen content (1.85 wt %). Larger than expected mass loss is also observed for other samples. In addition to hydrogen and butane, other types of gas (e.g., observed as m/z ratio = 28, 36, and 38) are also released from the samples in the first decomposition (see Figure S3). The impurities may come from the organic solvent or from the scaffolds above 250 °C [11]. In the first decomposition cycles, impurities in the as-prepared samples vaporize. Thus, in further cycles, the gas stream is more pure hydrogen whilst other gases are absent and do not contribute to extra mass loss.

2.4. Comparison of As-Prepared and Cycled Nanoconfined MgH_2

The four nanoconfined magnesium hydride samples were examined by powder X-ray diffraction (PXD) before and after five cycles of hydrogen release and uptake. Figure 5 reveals that the as-prepared nanoconfined sample and the five-times cycled sample Mg_CX1 contain crystalline MgH_2 and MgO. Figure 5 also reveals an extreme difference in the diffraction peak width for MgH_2 in the two samples.

All the diffraction data was analysed quantitatively for the composition of the crystalline fraction of the sample and the average crystallite sizes using Rietveld refinement (see Table 3). In the as-prepared samples, the crystallite size of MgH_2 is significantly smaller than the average pore size of the scaffold. This is due to the relatively low temperature for conversion of $MgBu_2$ to MgH_2 (150 °C), and the fact that the molar volume of $MgBu_2$ is a factor ten larger than that of MgH_2. However, only 38% to 48% of the crystalline fraction is MgH_2; the major part is nanocrystalline MgO.

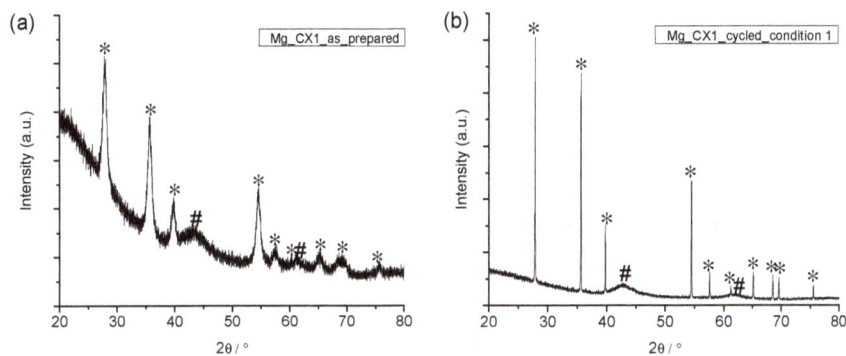

Figure 5. Powder X-ray diffraction (PXD) of Mg_CX1 (**a**) before; and (**b**) after five desorption/absorption cycles. * MgH_2; # MgO.

Table 3. Calculated average MgH_2 crystallite size and crystalline weight fraction from PXD as infiltrated and after five desorption cycles using *condition* 1. The remaining crystalline weight fraction is from MgO, which in all cases exists with ~1 nm crystallites.

Sample	D_{max} (nm)	As Infiltrated		After Five Cycles	
		MgH_2 Cryst. Size (nm)	MgH_2 Cryst. wt %	MgH_2 Cryst. Size (nm)	MgH_2 Cryst. wt %
Mg_X1	17	13	0.38	210	0.21
Mg_X2	26	10	0.40	248	0.19
Mg_CX1	15	8	0.48	300	0.35
Mg_CX2	25	13	0.23	95	0.12

For sample Mg_CX2, the distribution of MgH_2 and MgO is 23% and 77%, respectively. This decrease in active hydrogen storage material is in accordance with the decrease in hydrogen storage capacity measured by Sievert's method (see Figures 2 and 3). For all investigated samples, magnesium oxide is present as stable nanocrystallites (~1 nm). This can be ascribed to the fact that MgO is a much more refractory material, which does not take part in any reactions at temperatures used in the present study. The presence of oxygen is obviously a significant problem for the long term stability of nanoconfined MgH_2. The primary source of oxygen appears to be the "inert" carbon aerogel scaffold. It has been found that a carbon aerogel synthesised by a variety of routes has a significant oxygen content (C–O and C=O) [39]. Typically, the oxygen content is a few percent, with much higher oxygen content reported on the surface (~10%). Magnesium is an excellent oxygen scavenger, and the results here show that it strongly reacts with the oxygen within the carbon aerogel scaffold during synthesis and hydrogen cycling at an elevated temperature.

After five cycles of hydrogen release/uptake, the Bragg peaks of MgH_2 are much sharper, revealing an average crystallite size that is one order of magnitude or two orders of magnitude greater than in the as-prepared samples (Table 3). These average crystallite sizes are also much larger than the average pore sizes in the scaffolds, which demonstrates the high mobility of Mg/MgH_2 during cycling (hydrogen release and uptake) at 350 °C. Thus, Mg/MgH_2 tends to migrate or agglomerate in larger pore voids or outside of the scaffold. Particle growth contributes to increasing temperatures for hydrogen release due to hindered kinetics. Nanoparticles have a well-known tendency to grow to larger particles. Previous work demonstrates that sodium alanate, $NaAlH_4$, prefers to crystallise in the larger pores in CA scaffolds [26], and may also migrate out of the scaffold upon cycling [40].

The infiltrated scaffolds, before and after hydrogen cycling, were investigated by transmission electron microscopy (TEM) (see Figure 6). After infiltration, the MgH_2 is well-dispersed in the carbon scaffold (<25 nm). After five desorption/absorption cycles, MgH_2 particles appear to form larger agglomerations (~100 nm). However, it is difficult to determine if the agglomerates of MgH_2 are still within the scaffold or on the surface from the TEM data given that it is a transmission-based technique. Given the average carbon aerogel pore size of 25 nm, it seems likely that Mg/MgH_2 has migrated to the surface of the scaffold outside of the pore network.

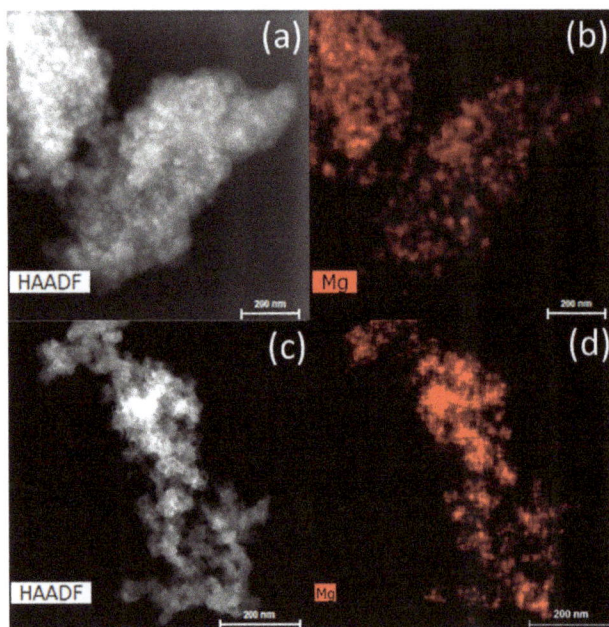

Figure 6. Scanning transmission electron microscope-high-angle annular dark-field (STEM-HAADF) images and elemental mapping of the as-prepared Mg_CX2 (**a,b**) and Mg_CX2 after five desorption/absorption cycles (**c,d**).

3. Materials and Methods

3.1. Synthesis of Carbon Scaffolds

Two batches (denoted X1 and X2) of resorcinol-formaldehyde carbon aerogel were synthesized as described previously [11,41]. Resorcinol (41.3 g, Sigma-Aldrich, Brøndby, Denmark, ≥99.0%) and formaldehyde (56.9 mL, 37 wt % in H_2O, stabilized by 10–15% methanol, Sigma-Aldrich) were added to deionized water (56.6 mL) under stirring. Sodium carbonate, Na_2CO_3 (65 mg, Sigma-Aldrich, 99.999%) was added to the synthesis of X1 (pH = 6.47) and 40 mg to that of X2 (pH = 6.20). The mixtures were kept in sealed containers at room temperature for 24 h, then at 50 °C for 24 h, and finally at 90 °C for 72 h. The depth of the solution in the sealed containers was less than 0.5 cm to ensure the homogeneity of the carbon aerogel. After cooling, the solid gels were immersed in an acetone bath to exchange all the water inside the pores. The solid gels were then cut into small pieces with average dimension

1 cm $\times$ 0.5 cm $\times$ 0.4 cm and pyrolysed at 800 °C ($\Delta T/\Delta t = 3$ °C/min) in N_2 for 6 h. A portion of both samples X1 and X2 underwent further heat treatment from room temperature (RT) to 950 °C ($\Delta T/\Delta t = 6$ °C/min) followed by an isothermal step at 950 °C for 5 h in a constant CO_2 flow in order to increase the surface area (S_{BET}) and total pore volume (V_{tot}) [42]. These samples are denoted CX1 and CX2. The average dimension of the monoliths decreased significantly to only 10–20% of their initial volume. All the synthesized carbon aerogels were degassed in vacuum at 350 °C for several hours and stored inside an argon-filled glovebox.

3.2. Direct Synthesis of Nanoconfined Magnesium Hydride

Monoliths of carbon aerogel with an average volume of 0.2 cm^3 were immersed in 1 M di-*n*-butylmagnesium, $Mg(CH_2CH_2CH_2CH_3)_2$, denoted $MgBu_2$ (~5 mL, in ether and hexanes, Sigma-Aldrich) for two days. The solvent was removed using Schlenk techniques and the monoliths were dried for several hours in an inert argon atmosphere. Excess white $MgBu_2$ on the surface of the black scaffold was removed mechanically. The amount of infiltrated dibutyl magnesium was determined from the weight gain of the monoliths before and after each infiltration. Afterwards, the infiltrated monoliths were placed in an autoclave (Swagelok, Esbjerg, Denmark) and heated to 150 °C ($\Delta T/\Delta t = 5$ °C/min) under $p(H_2) = 100$ bar and kept at 150 °C for 1 h to convert $MgBu_2$ to MgH_2 and butane. The autoclave was then evacuated and kept in dynamic vacuum for 30 min to remove the released butane gas. The hydrogenation and evacuation procedures were repeated two further times at 150 °C to ensure a high conversion of $MgBu_2$ to MgH_2. Finally, the samples were cooled to room temperature under hydrogen pressure. These $MgBu_2$ infiltration and consequent hydrogenation procedures were repeated three times (3×) for each of the four monolithic samples, and finally the prepared samples were hand ground into powder for further characterisation. The infiltrated volumetric quantity of hydrogen storage material, MgH_2, is calculated from the weight gain of the scaffold and the bulk densities $\rho(MgH_2) = 1.45$ g/cm^3 and $\rho(MgBu_2) = 0.736$ g/cm^3. Table S1 provide details about the amounts of $MgBu_2$ infiltrated in each procedure and the total amounts of magnesium hydride in each scaffold. The magnesium hydride-containing scaffolds are denoted Mg_X1, Mg_X2, etc. The samples were stored and handled inside an argon-filled glovebox with H_2O/O_2 levels below 1 ppm.

3.3. Characterisation

The porosity analysis was performed using a Nova 2200e surface area and pore size analyser (Quantachrome Instruments, Odelzhausen, Germany). The properties of the carbon aerogels were deduced from N_2 adsorption/desorption measurements at 77 K. The surface area (S_{BET}) was measured using the Brunauer–Emmett–Teller (BET) method, and the micropore volume (V_{micro}) was determined by the *t*-plot method [43,44]. The average pore size (D_{max}) and mesopore volume (V_{meso}) were recorded by the Barrett–Joyner–Halenda (BJH) method during desorption [45]. The total pore volume (V_{tot}) of the scaffold was obtained from the point at maximum $p/p_0 \sim 1$.

The thermal properties of nanoconfined MgH_2 before and after the desorption/absorption cycles were studied by thermogravimetric analysis (TGA) coupled with mass spectroscopy (MS). TGA was carried out using a STA 6000 (Perkin Elmer, Skovlunde, Denmark), and the evolved gases were detected by a HPR-20 QMS Mass Spectrometer (Hiden Analytical, Warrington, UK). A few milligrams of sample was placed in an aluminium crucible and heated ($\Delta T/\Delta t = 5$ °C/min) in an argon flow of 40 mL/min.

The stability of the hydrogen storage capacity of nanoconfined MgH_2 samples was investigated over five cycles of hydrogen release and uptake by Sievert's measurements using an in-house custom apparatus [30]. Approximately 100 mg of sample was sealed in an autoclave and studied for five desorption and absorption cycles under two different conditions. For *condition 1*, the samples were heated in vacuum from room temperature to 355 °C ($\Delta T/\Delta t = 5$ °C/min) and kept isothermal for 15 h during hydrogen release. Then, hydrogen absorption was conducted at $p(H_2) = 50$ bar for 15 h at 355 °C, i.e., $\Delta p(H_2)$ = 50 bar. The sample was then cooled to room temperature under the same hydrogen pressure. For *condition* 2, the samples were heated to 355 °C ($\Delta T/\Delta t$ = 5 °C/min) and kept at 355 °C for five cycles. Hydrogen release was conducted at $p(H_2)$ = 4–5 bar for 15 h at 355 °C and hydrogen absorption at $p(H_2)$ = 12 bar for 15 h at 355 °C, i.e., $\Delta p(H_2)$ ~7.5 bar. The hydrogen equilibrium pressure for Mg/MgH$_2$ at 355 °C is $p_{eq}(H_2)$ = 6.4 bar [30].

Powder X-ray diffraction was conducted to characterize the nanoconfined MgH$_2$ samples before and after five desorption/absorption cycles. This was done by using a SmartLab diffractometer (Cu Kα_1 source, λ = 1.5406 Å, Rigaku, Ettlingen, Germany). The samples were mounted in 0.5 mm-diameter Lindemann glass capillaries, and the diffraction patterns were collected with an angular step of 3° per minute. The Rietveld analysis was performed in Topas (Bruker, Cambridge, UK) along with crystallite size refinement using fundamental parameters after an instrument calibration using LaB$_6$. The crystallite size was calculated using the LVol-IB method (volume averaged column height calculated from the integral breadth), which provides a measure of the volume-weighted crystallite size.

The distribution of MgH$_2$ in the samples before and after five desorption/absorption cycles was studied using an Talos F200X (S)TEM-microscope (FEI, Copenhagen, Denmark) equipped with an advanced energy dispersive X-ray spectroscopy (EDS) system operated at 200 kV. Samples were dispersed on a copper grid coated in a holey carbon film after suspension in (dry) cyclohexane. Sample grids were attached to the TEM sample holder in ambient conditions, i.e., exposing the sample to air for several minutes.

4. Conclusions

MgH$_2$ was infiltrated into four different carbon aerogel scaffolds using a comprehensive activation process. Multiple infiltrations showed a limited increase in the amount of MgH$_2$ (18.2 cm^3/mol) due to the large molar volume of MgBu$_2$ (188.2 cm^3/mol). The volumetric loading of MgH$_2$ after three loading steps was 17–20 vol % in the various scaffolds. Despite the vigilant infiltration and activation procedure, hydrogen cycling resulted in the production of butane from the conversion of residual MgBu$_2$ in the scaffold. It appears as though batch-wise hydrogenation of MgBu$_2$ is inefficient in fully converting it to MgH$_2$, and future studies may benefit from high pressure flow-through hydrogenation to decrease the MgBu$_2$ content. The nanoconfined MgH$_2$ samples also displayed significant hydrogen capacity loss after cycling that appears to be due to the formation of large quantities of MgO from interactions between MgH$_2$ and the carbon aerogel scaffold. Carbon aerogel scaffolds are not pure carbon, and can contain C–O and C=O groups that could be reduced by Mg at high temperature. Overall, we observe hydrogen release of 1.3 to 3.1 wt % in the first cycle, which for some samples is higher than previously reported ref. [11–13], and 1.0 to 1.9 in the fifth cycle, which may be slightly lower. Further work must be directed towards further purifying carbon aerogel scaffolds or finding alternative, less reactive scaffolds. Hydrogen kinetics was also found to decrease due to Mg/MgH$_2$ growth after cycling at high temperature. It is likely that Mg is able to migrate out of the pore network under vacuum (or low pressure) at high temperature. Other nanoconfinement studies should focus on unreactive scaffold design, improved flow-through MgH$_2$ activation procedures, and work towards understanding the migration of active metal hydride material within the scaffold at high temperature.

Supplementary Materials: The following are available online at www.mdpi.com/2304-6740/5/3/57/s1, Table S1: The mass of scaffold, initial pore volume, and gain in mass of each infiltration of MgBu$_2$ for the four nanoconfined samples, Figure S1: Sievert's measurements of nanoconfined samples under *condition* 2, Figure S2: Sievert's measurements of the first 3 h of nanoconfined samples under *condition* 1, Figure S3: Mass spectroscopic analysis of the gas release from the as-prepared Mg_X1 at 348 °C, Figure S4: Rietveld refinement and difference plots of as-prepared and cycled Mg_CX1.

Acknowledgments: This research project received funding from the People Program (Marie Curie Actions) of the European Union's Seventh Framework Program FP7/2007–2013/ under REA grants agreement No. 607040 (Marie Curie ITN ECOSTORE). Furthermore, the work was supported by the Danish National Research Foundation, Center for Materials Crystallography (DNRF93), The Innovation Fund Denmark (project HyFill-Fast), and by the Danish Research Council for Nature and Universe (Danscatt). We are grateful to the Carlsberg Foundation.

Author Contributions: Priscilla Huen was involved in all stages of the work, including planning, conducting experiments, and analyzing the data; Mark Paskevicius conducted part of the Sievert's measurements and the Rietveld refinement of diffraction patterns; Bo Richter performed the TEM-EDS experiments; Dorthe B. Ravnsbæk acted as co-supervisor and was involved in the discussion of the results and work planning; Torben R. Jensen acted as the main supervisor and helped with the data analysis and work planning; Priscilla Huen, Mark Paskevicius, and Torben R. Jensen wrote the paper; and all the authors contributed to the revision of the paper.

Conflicts of Interest: The authors declare no conflict of interest.

References

1. Mazloomi, K.; Gomes, C. Hydrogen as an energy carrier: Prospects and challenges. *Renew. Sustain. Energy Rev.* **2012**, *16*, 3024–3033. [CrossRef]
2. Holladay, J.D.; Hu, J.; King, D.L.; Wang, Y. An overview of hydrogen production technologies. *Catal. Today* **2009**, *139*, 244–260. [CrossRef]
3. Ley, M.B.; Jepsen, L.H.; Lee, Y.-S.; Cho, Y.W.; Bellosta von Colbe, J.M.; Dornheim, M.; Rokni, M.; Jensen, J.O.; Sloth, M.; Filinchuk, Y.; et al. Complex hydrides for hydrogen storage—New perspectives. *Mater. Today* **2014**, *17*, 122–128. [CrossRef]
4. Møller, K.T.; Jensen, T.R.; Akiba, E.; Li, H. Hydrogen—A sustainable energy carrier. *Prog. Nat. Sci. Mater. Int.* **2017**, *27*, 34–40. [CrossRef]
5. Haynes, W.M. *CRC Handbook of Chemistry and Physics*, 95th ed.; CRC Press: Boca Raton, FL, USA, 2014; ISBN 9781482208689.
6. Lai, Q.; Paskevicius, M.; Sheppard, D.A.; Buckley, C.E.; Thornton, A.W.; Hill, M.R.; Gu, Q.; Mao, J.; Huang, Z.; Liu, H.K.; et al. Hydrogen Storage Materials for Mobile and Stationary Applications: Current State of the Art. *ChemSusChem* **2015**, *8*, 2789–2825. [CrossRef] [PubMed]
7. Paskevicius, M.; Jepsen, L.H.; Schouwink, P.; Černý, R.; Ravnsbæk, D.B.; Filinchuk, Y.; Dornheim, M.; Besenbacher, F.; Jensen, T.R. Metal borohydrides and derivatives—Synthesis, structure and properties. *Chem. Soc. Rev.* **2017**, *46*, 1565–1634. [CrossRef] [PubMed]
8. Webb, C.J. A review of catalyst-enhanced magnesium hydride as a hydrogen storage material. *J. Phys. Chem. Solids* **2015**, *84*, 96–106. [CrossRef]
9. Crivello, J.-C.; Denys, R.V.; Dornheim, M.; Felderhoff, M.; Grant, D.M.; Huot, J.; Jensen, T.R.; de Jongh, P.; Latroche, M.; Walker, G.S.; et al. Mg-based compounds for hydrogen and energy storage. *Appl. Phys. A* **2016**, *122*, 85. [CrossRef]
10. Crivello, J.-C.; Dam, B.; Denys, R.V.; Dornheim, M.; Grant, D.M.; Huot, J.; Jensen, T.R.; de Jongh, P.; Latroche, M.; Milanese, C.; et al. Review of magnesium hydride-based materials: Development and optimisation. *Appl. Phys. A* **2016**, *122*, 97. [CrossRef]
11. Nielsen, T.K.; Manickam, K.; Hirscher, M.; Besenbacher, F.; Jensen, T.R. Confinement of MgH$_2$ Nanoclusters within Nanoporous Aerogel Scaffold Materials. *ACS Nano* **2009**, *3*, 3521–3528. [CrossRef] [PubMed]
12. Zhang, S.; Gross, A.F.; Van Atta, S.L.; Lopez, M.; Liu, P.; Ahn, C.C.; Vajo, J.J.; Jensen, C.M. The synthesis and hydrogen storage properties of a MgH$_2$ incorporated carbon aerogel scaffold. *Nanotechnology* **2009**, *20*, 204027. [CrossRef] [PubMed]
13. Gross, A.F.; Ahn, C.C.; Van Atta, S.L.; Liu, P.; Vajo, J.J. Fabrication and hydrogen sorption behaviour of nanoparticulate MgH$_2$ incorporated in a porous carbon host. *Nanotechnology* **2009**, *20*, 204005. [CrossRef] [PubMed]

14. Jia, Y.; Sun, C.; Cheng, L.; Abdul Wahab, M.; Cui, J.; Zou, J.; Zhu, M.; Yao, X. Destabilization of Mg–H bonding through nano-interfacial confinement by unsaturated carbon for hydrogen desorption from MgH₂. *Phys. Chem. Chem. Phys.* **2013**, *15*, 5814. [CrossRef] [PubMed]

15. Nielsen, T.K.; Javadian, P.; Polanski, M.; Besenbacher, F.; Bystrzycki, J.; Jensen, T.R. Nanoconfined NaAlH₄: Determination of Distinct Prolific Effects from Pore Size, Crystallite Size, and Surface Interactions. *J. Phys. Chem. C* **2012**, *116*, 21046–21051. [CrossRef]

16. De Jongh, P.E.; Adelhelm, P. Nanosizing and Nanoconfinement: New Strategies Towards Meeting Hydrogen Storage Goals. *ChemSusChem* **2010**, *3*, 1332–1348. [CrossRef] [PubMed]

17. Gosalawit-Utke, R.; Thiangviriya, S.; Javadian, P.; Laipple, D.; Pistidda, C.; Bergemann, N.; Horstmann, C.; Jensen, T.R.; Klassen, T.; Dornheim, M. Effective nanoconfinement of 2LiBH₄–MgH₂ via simply MgH₂ premilling for reversible hydrogen storages. *Int. J. Hydrogen Energy* **2014**, *39*, 15614–15626. [CrossRef]

18. Bérubé, V.; Radtke, G.; Dresselhaus, M.; Chen, G. Size effects on the hydrogen storage properties of nanostructured metal hydrides: A review. *Int. J. Energy Res.* **2007**, *31*, 637–663. [CrossRef]

19. Nielsen, T.K.; Besenbacher, F.; Jensen, T.R. Nanoconfined hydrides for energy storage. *Nanoscale* **2011**, *3*, 2086–2098. [CrossRef] [PubMed]

20. Gutowska, A.; Li, L.; Shin, Y.; Wang, C.M.; Li, X.S.; Linehan, J.C.; Smith, R.S.; Kay, B.D.; Schmid, B.; Shaw, W.; et al. Nanoscaffold Mediates Hydrogen Release and the Reactivity of Ammonia Borane. *Angew. Chem.* **2005**, *117*, 3644–3648. [CrossRef]

21. Ngene, P.; van Zwienen, M.; de Jongh, P.E. Reversibility of the hydrogen desorption from LiBH₄: A synergetic effect of nanoconfinement and Ni addition. *Chem. Commun.* **2010**, *46*, 8201. [CrossRef] [PubMed]

22. Paskevicius, M.; Filsø, U.; Karimi, F.; Puszkiel, J.; Pranzas, P.K.; Pistidda, C.; Hoell, A.; Welter, E.; Schreyer, A.; Klassen, T.; et al. Cyclic stability and structure of nanoconfined Ti-doped NaAlH₄. *Int. J. Hydrogen Energy* **2016**, *41*, 4159–4167. [CrossRef]

23. Zhao-Karger, Z.; Hu, J.; Roth, A.; Wang, D.; Kübel, C.; Lohstroh, W.; Fichtner, M. Altered thermodynamic and kinetic properties of MgH₂ infiltrated in microporous scaffold. *Chem. Commun.* **2010**, *46*, 8353. [CrossRef] [PubMed]

24. De Jongh, P.E.; Wagemans, R.W.P.; Eggenhuisen, T.M.; Dauvillier, B.S.; Radstake, P.B.; Meeldijk, J.D.; Geus, J.W.; de Jong, K.P. The Preparation of Carbon-Supported Magnesium Nanoparticles using Melt Infiltration. *Chem. Mater.* **2007**, *19*, 6052–6057. [CrossRef]

25. Utke, R.; Thiangviriya, S.; Javadian, P.; Jensen, T.R.; Milanese, C.; Klassen, T.; Dornheim, M. 2LiBH₄–MgH₂ nanoconfined into carbon aerogel scaffold impregnated with ZrCl₄ for reversible hydrogen storage. *Mater. Chem. Phys.* **2016**, *169*, 136–141. [CrossRef]

26. Nielsen, T.K.; Javadian, P.; Polanski, M.; Besenbacher, F.; Bystrzycki, J.; Skibsted, J.; Jensen, T.R. Nanoconfined NaAlH₄: Prolific effects from increased surface area and pore volume. *Nanoscale* **2014**, *6*, 599–607. [CrossRef] [PubMed]

27. Fichtner, M. Nanoconfinement effects in energy storage materials. *Phys. Chem. Chem. Phys.* **2011**, *13*, 21186. [CrossRef] [PubMed]

28. Roedern, E.; Hansen, B.R.S.; Ley, M.B.; Jensen, T.R. Effect of Eutectic Melting, Reactive Hydride Composites, and Nanoconfinement on Decomposition and Reversibility of LiBH₄–KBH₄. *J. Phys. Chem. C* **2015**, *119*, 25818–25825. [CrossRef]

29. Bramwell, P.L.; Ngene, P.; de Jongh, P.E. Carbon supported lithium hydride nanoparticles: Impact of preparation conditions on particle size and hydrogen sorption. *Int. J. Hydrogen Energy* **2017**, *42*, 5188–5198. [CrossRef]

30. Paskevicius, M.; Sheppard, D.A.; Buckley, C.E. Thermodynamic Changes in Mechanochemically Synthesized Magnesium Hydride Nanoparticles. *J. Am. Chem. Soc.* **2010**, *132*, 5077–5083. [CrossRef] [PubMed]

31. Cuevas, F.; Korablov, D.; Latroche, M. Synthesis, structural and hydrogenation properties of Mg-rich MgH₂–TiH₂ nanocomposites prepared by reactive ball milling under hydrogen gas. *Phys. Chem. Chem. Phys.* **2012**, *14*, 1200–1211. [CrossRef] [PubMed]

32. Pasquini, L.; Boscherini, F.; Callini, E.; Maurizio, C.; Pasquali, L.; Montecchi, M.; Bonetti, E. Local structure at interfaces between hydride-forming metals: A case study of Mg-Pd nanoparticles by X-ray spectroscopy. *Phys. Rev. B* **2011**, *83*, 184111. [CrossRef]

33. Andreasen, A.; Sørensen, M.B.; Burkarl, R.; Møller, B.; Molenbroek, A.M.; Pedersen, A.S.; Vegge, T.; Jensen, T.R. Dehydrogenation kinetics of air-exposed MgH_2/Mg_2Cu and $MgH_2/MgCu_2$ studied with in situ X-ray powder diffraction. *Appl. Phys. A* **2006**, *82*, 515–521. [CrossRef]

34. Andreasen, A.; Sørensen, M.B.; Burkarl, R.; Møller, B.; Molenbroek, A.M.; Pedersen, A.S.; Andreasen, J.W.; Nielsen, M.M.; Jensen, T.R. Interaction of hydrogen with an Mg–Al alloy. *J. Alloys Compd.* **2005**, *404–406*, 323–326. [CrossRef]

35. Jensen, T.; Andreasen, A.; Vegge, T.; Andreasen, J.; Stahl, K.; Pedersen, A.; Nielsen, M.; Molenbroek, A.; Besenbacher, F. Dehydrogenation kinetics of pure and nickel-doped magnesium hydride investigated by in situ time-resolved powder X-ray diffraction. *Int. J. Hydrogen Energy* **2006**, *31*, 2052–2062. [CrossRef]

36. Dornheim, M.; Eigen, N.; Barkhordarian, G.; Klassen, T.; Bormann, R. Tailoring Hydrogen Storage Materials Towards Application. *Adv. Eng. Mater.* **2006**, *8*, 377–385. [CrossRef]

37. Barkhordarian, G.; Klassen, T.; Bormann, R. Catalytic Mechanism of Transition-Metal Compounds on Mg Hydrogen Sorption Reaction. *J. Phys. Chem. B* **2006**, *110*, 11020–11024. [CrossRef] [PubMed]

38. Nielsen, T.K.; Jensen, T.R. MgH_2–Nb_2O_5 investigated by in situ synchrotron X-ray diffraction. *Int. J. Hydrogen Energy* **2012**, *37*, 13409–13416. [CrossRef]

39. Alegre, C.; Sebastián, D.; Baquedano, E.; Gálvez, M.E.; Moliner, R.; Lázaro, M. Tailoring Synthesis Conditions of Carbon Xerogels towards Their Utilization as Pt-Catalyst Supports for Oxygen Reduction Reaction (ORR). *Catalysts* **2012**, *2*, 466–489. [CrossRef]

40. Chumphongphan, S.; Filsø, U.; Paskevicius, M.; Sheppard, D.A.; Jensen, T.R.; Buckley, C.E. Nanoconfinement degradation in NaAlH4/CMK-1. *Int. J. Hydrogen Energy* **2014**, *39*, 11103–11109. [CrossRef]

41. Li, W.-C.; Lu, A.-H.; Weidenthaler, C.; Schüth, F. Hard-Templating Pathway to Create Mesoporous Magnesium Oxide. *Chem. Mater.* **2004**, *16*, 5676–5681. [CrossRef]

42. Lin, C.; Ritter, J.A. Carbonization and activation of sol-gel derived carbon xerogels. *Carbon* **2000**, *38*, 849–861. [CrossRef]

43. Brunauer, S.; Emmett, P.H.; Teller, E. Adsorption of Gases in Multimolecular Layers. *J. Am. Chem. Soc.* **1938**, *60*, 309–319. [CrossRef]

44. Deboer, J. Studies on pore systems in catalysts VII. Description of the pore dimensions of carbon blacks by the t method. *J. Catal.* **1965**, *4*, 649–653. [CrossRef]

45. Barrett, E.P.; Joyner, L.G.; Halenda, P.P. The Determination of Pore Volume and Area Distributions in Porous Substances. I. Computations from Nitrogen Isotherms. *J. Am. Chem. Soc.* **1951**, *73*, 373–380. [CrossRef]

inorganics

MDPI

Article

Interface Enthalpy-Entropy Competition in Nanoscale Metal Hydrides

Nicola Patelli [1,*], Marco Calizzi [2] and Luca Pasquini [1,*]

[1]   Department of Physics and Astronomy, Alma Mater Studiorum Università di Bologna,
      Viale C. Berti-Pichat 6/2, 40127 Bologna, Italy
[2]   Laboratory of Materials for Renewable Energy, Institute of Chemical Sciences and Engineering,
      Ecole Polytechnique Fédérale de Lausanne, Valais/Wallis, Rue de l'Industrie 17, 440 1951 Sion, Switzerland;
      marco.calizzi@epfl.ch
*   Correspondence: nicola.patelli@unibo.it (N.P.); luca.pasquini@unibo.it (L.P.)

Received: 2 November 2017; Accepted: 8 January 2018; Published: 12 January 2018

Abstract: We analyzed the effect of the interfacial free energy on the thermodynamics of hydrogen sorption in nano-scaled materials. When the enthalpy and entropy terms are the same for all interfaces, as in an isotropic bi-phasic system, one obtains a compensation temperature, which does not depend on the system size nor on the relative phase abundance. The situation is different and more complex in a system with three or more phases, where the interfaces have different enthalpy and entropy. We also consider the possible effect of elastic strains on the stability of the hydride phase and on hysteresis. We compare a simple model with experimental data obtained on two different systems: (1) bi-phasic nanocomposites where ultrafine TiH_2 crystallite are dispersed within a Mg nanoparticle and (2) Mg nanodots encapsulated by different phases.

Keywords: MgH_2; TiH_2; interface; entropy; enthalpy; compensation; destabilization; thermodynamics; nanoparticles; nanodots

---

1. Introduction

Nanoparticles are halfway between atomic world and bulk world. A whole new and non-trivial variety of effects arise whenever a system is pushed towards the atomic scale. Material's properties change significantly when system size becomes comparable to a characteristic length scale, such as the mean free path of particles or excitations that carry charges, energy and momentum. The high density of interfaces confers to nano-systems unique physical and chemical properties, and promotes their chemical reactivity. The thermodynamics of nano-scaled systems is strongly altered with respect to bulk material and size-dependent effects [1,2], becoming more and more complicated when dealing with compounds and heterogeneous ones [3]. Far from putting a limit to the exploitation of nanoparticles, this fascinating behavior allows for the tailoring of material properties by designing unique functional devices at nanoscale. This work aims to discuss hydride formation in composite nanomaterials, relating peculiar H-sorption properties to confinement effects [4] and to the physical properties of interfaces [5].

Before going into details with the effects of nanosizing, we recall some general concepts about the thermochemical equilibrium of hydride formation. The Van't Hoff equation relates the equilibrium pressure $p_{H_2}^{eq}$ and temperature T to the reaction's free energy ΔG^0 at standard conditions:

$$\ln\left(\frac{p_{H_2}^{eq}}{p^0}\right) = \frac{\Delta G^0}{RT} = \frac{\Delta H^0}{RT} - \frac{\Delta S^0}{R} \tag{1}$$

where ΔH^0 and ΔS^0 are, respectively, the enthalpy and the entropy at standard conditions. The workhorse for the characterization of the thermodynamics of hydride formation are the pressure-composition isotherms (PCI). All real systems exhibit some hysteresis, i.e., a pressure shift between the absorption and desorption branches of a PCI cycle, due to coherency strains [6,7]. This is called intrinsic hysteresis and is represented in Figure 1a. In presence of hysteresis, the equilibrium pressure $p_{H_2}^{eq}$ is calculated as the geometric average of the absorption (p_{abs}) and desorption (p_{des}) plateau pressures. The hysteresis is usually quantified by the ratio:

$$\Delta G_{its}^{hyst} = RT \ln\left(\frac{p_{abs}}{p_{des}}\right)_{bulk} \tag{2}$$

Additional sources of strain, such as the presence of defects or interfaces in the material, can lead to a wider extrinsic hysteresis, as shown in Figure 1b, where the equilibrium pressure is not altered because of the symmetrical shift of both absorption and desorption plateau. A simple relation between ΔG_{ext}^{hyst} and plateau pressures holds:

$$\Delta G_{ext}^{hyst} = RT \ln\left(\frac{p_{abs}}{p_{des}}\right) - \Delta G_{its}^{hyst} \tag{3}$$

A true thermodynamical bias, i.e., a change $\delta\left(\Delta G^0\right)$ in the free energy, corresponds to a rigid shift of both PCI branches as schematized in Figure 1c.

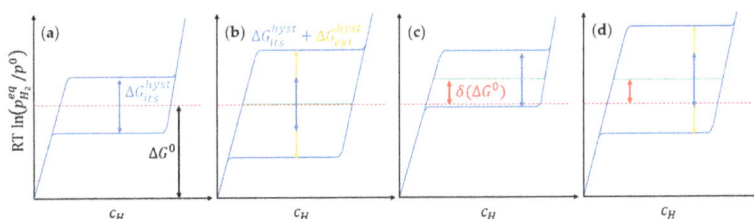

Figure 1. A sketch to represent hysteresis and thermodynamical bias effects on hydrogen sorption isotherms. From the left to the right: (**a**) Reference bulk material showing an intrinsic hysteresis; (**b**) The effect of additional extrinsic hysteresis that increases the separation between absorption and desorption curves with no bias induction; (**c**) a real thermodynamical bias induced, e.g., by interface effects; (**d**) a more realistic case where both extrinsic hysteresis and thermodynamical bias are present.

Many efforts have been made to design and grow nano-sized materials with more favorable hydrogen sorption thermodynamics than their bulk counterparts by inducing a thermodynamical bias in the right direction. For instance, if the material is too stable, an upward shift of the PCI branches, i.e., a positive thermodynamical bias, is desired. There are two main contributions to the thermodynamical bias when a material is refined to the nanoscale. The first one originates with the interface free energy [5,7] according to the following equation:

$$\delta\left(\Delta G^0\right)_{int} = \Delta G_{nano}^0 - \Delta G_{bulk}^0 = \frac{V_M}{V_M}\left(\sum_i A_{MH|i}^{int}\gamma_{MH|i}^{int} - \sum_j A_{M|j}^{int}\gamma_{M|j}^{int}\right) \tag{4}$$

where $\gamma_{MH|i}^{int}$ and $\gamma_{M|j}^{int}$ are the interface free energy for unit area for metal hydride and metal, respectively, while $A_{MH|i}^{int}$ and $A_{M|j}^{int}$ are the corresponding interface areas.

This contribution clearly scales with the volume fraction occupied by interfaces (or surfaces), and is roughly inversely proportional to the material's length scale. In this sense, the effect of the interfaces is short-ranged, i.e., it vanishes rapidly as the spatial separation between interfaces increases above a few nanometers.

The second contribution can arise in case of elastic confinement, i.e., if the material is prevented to freely expand upon hydride formation [4]. The hydride formation enthalpy of core-shell nanoparticles

(NPs) subjected to elastic constraints, ΔH^0_{constr}, is less negative with respect to ΔH^0_{free} of the corresponding free NPs [6]. The thermodynamical bias in this case is proportional to the volume strain ϵ_V in the constrained hydride NPs:

$$\delta\left(\Delta G^0\right)_{el} = \delta\left(\Delta H^0\right)_{el} = \Delta H^0_{constr} - \Delta H^0_{free} = -BV_H\epsilon_V \tag{5}$$

where B is the bulk modulus and ϵ_V is the volume strain. In Equation (5), we have assumed that $\Delta S^0_{constr} = \Delta S^0_{free}$. Elastic strain engineering has the potential to induce a significant bias up to length scales of a few tens of nanometers. The main problem is that the onset of plastic deformation strongly suppresses the effect and induces an unwanted extrinsic hysteresis [5].

Comparing Equations (1), (4) and (5), one obtains:

$$RT\ln\left(\frac{p^{eq}_{nano}}{p^{eq}_{bulk}}\right) = \delta\left(\Delta G^0\right)_{int} + \delta\left(\Delta G^0\right)_{el} \tag{6}$$

and the total thermodynamical bias $\delta\left(\Delta G^0\right)$ can be estimated from the ratio between the equilibrium pressures in the nanomaterial and in the bulk material.

In a real nano-system there may be a combination of extrinsic hysteresis and thermodynamical bias, leading to an asymmetric shift of PCI branches as sketched in Figure 1d. The comprehension of these separate contributions to the difference in thermodynamics and kinetics of hydrogen sorption is of paramount importance in the perspective of specifically designed multifunctional materials. We will here analyze and compare our results in the framework of these guidelines.

2. Results

The materials analyzed in this work are composite Mg–Ti–H nanoparticles (NPs) synthetized by gas phase condensation, as previously reported [8]. A broader description of the NPs growth technique is given in the Methods and Material section of this work. The combination of scanning transmission electron microscopy and X-ray diffraction suggested that the composite NPs are MgH_2 single crystals, in which ultrafine TiH_2 crystallites are dispersed. A schematic representation of the composite NPs is shown in Figure 2. They will be compared to the results obtained on multilayered, nanoconfined Mg/Ti/Pd nanodots (NDs) [5] sketched in Figure 3.

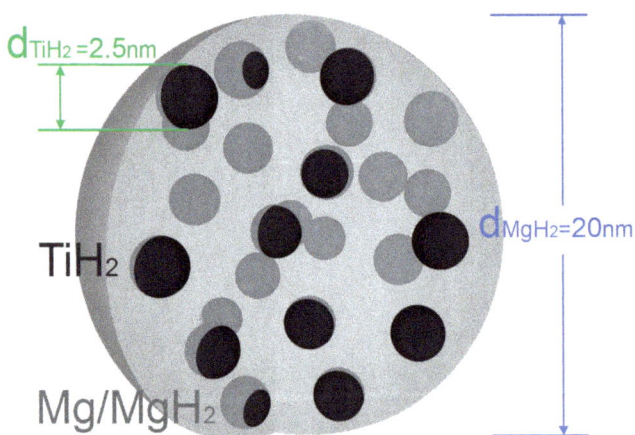

Figure 2. 3D model for MgH_2–TiH_2 composite nanoparticles (NPs); d_{TiH_2} and d_{MgH_2} are the average TiH_2 and MgH_2 crystallite size.

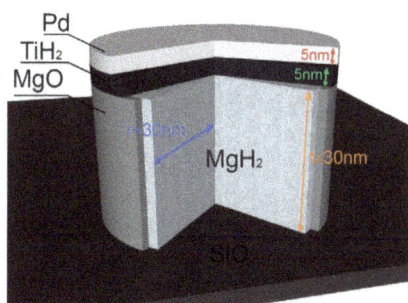

Figure 3. 3D Model for MgH$_2$–TiH$_2$ nanodots (NDs); r = 30 nm is the radius and t = 30 nm is the thickness of MgH$_2$. MgH$_2$ is covered with TiH$_2$ layer (5 nm) and a top Pd layer (5 nm).

H-sorption thermodynamics and kinetics were measured in-situ in an ultra-high-vacuum (UHV) chamber directly connected to the NPs deposition chamber. The hydrogen sorption properties were investigated in the low temperature regime (340 K < T < 425 K) to determine p_{abs} and p_{des}. Figure 4 shows sorption pseudo-kinetics obtained at 375 K. The hydrogen absorption and desorption rates were calculated from the initial slope of the kinetics, resulting in 0.28 wt % H$_2$/min in absorption and 0.018 wt % H$_2$/min. These values are roughly one order of magnitude lower than observed at 423 K [8], but are remarkable for MgH$_2$ at such low temperature and mild pressure, especially for hydrogen desorption. Figure 4 also suggests that the sorption rates rapidly decrease when $p(H_2)/p_{eq}$ approaches unity because the thermodynamic driving force tends to zero in this limit. Even if the chamber volume is large, the pressure approaches the low equilibrium values after a relatively small mass loss (<0.5%) for a Mg-based material. Therefore, several steps are needed in order to complete the desorption, as reported in the inset of Figure 4. These steps altogether make it possible to construct a PCI curve and to determine the plateau pressure p_{des} for desorption. A similar argument applies to the determination of the absorption plateau pressure p_{abs}.

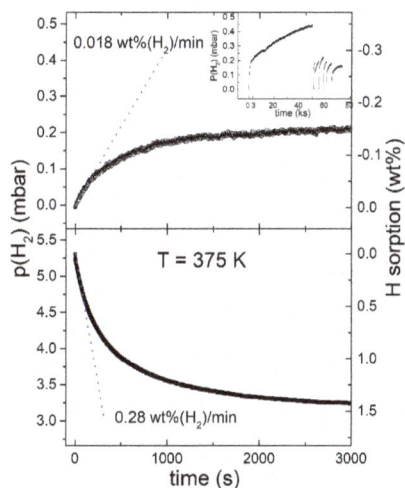

Figure 4. Hydrogen desorption (**top**) and absorption (**bottom**) for Ti 15 atom % NPs at 375 K. Sorption rates obtained by linear fits (dotted blue lines) of the initial data are shown. Several consecutive desorption steps are shown in the inset, to demonstrate that the slowdown in the kinetics is to be ascribed to p_{des} approaching.

The fast kinetics at low temperature make it possible to measure absorption and desorption plateau pressures and to calculate the equilibrium values down to 355 K. In Figure 5, the p_{abs} and p_{des} values for a collection of NPs with varying Ti content ($X_{Ti} = Ti/(Mg + Ti) = 6, 15, 30$ atom %) are shown along with those measured on NDs. The corresponding enthalpy and entropy values are collected in Table 1. The discrepancy between absorption and desorption is due to hysteresis. In general, the determination of thermodynamic parameters from only absorption (or desorption) pressures is prone to large errors. This is particularly evident for the NDs that, due to strong hysteretic behavior, show remarkably different slopes and intercepts in absorption and desorption, resulting in unrealistic enthalpy/entropy values (see Table 1). The correct results are retrieved by fitting the equilibrium pressure. Looking at the equilibrium data in Table 1, we notice that the formation enthalpy for both NPs and NDs is slightly less negative than for bulk Mg (although the error on the NDs is quite large). For the NPs, the entropy is also less negative (by about 10%) compared to the bulk, whereas for the NDs, the difference is smaller (about 3%) and well within the uncertainty. In Figure 5, we also notice that the equilibrium pressure of the NDs (dotted black line) is larger than that of the NPs (solid black line) and of bulk Mg [8] by a factor of two. The NDs, therefore, realize the picture outlined Figure 1d, showing both an upward shift of the equilibrium pressure (thermodynamical bias) and a large extrinsic hysteresis.

Table 1. Hydride formation enthalpy and entropy for a collection of Mg–Ti–H NPs with varying Ti content (6, 15, 30 atom %) and for Mg NDs (radius = 30 nm) corresponding to the linear fits displayed in Figure 5. ΔH^0 and ΔS^0 are determined by the linear fit of the equilibrium pressures. The data of reference bulk Mg are also listed [9]. The values obtained from the separate fits of absorption and desorption pressures reveal a discrepancy arising from hysteretic effects.

	ΔH^0 (kJ/mol H_2)	ΔS^0 (J/mol H_2 K)	ΔH_{abs} (kJ/mol H_2)	ΔS_{abs} (J/mol H_2 K)	ΔH_{des} (kJ/mol H_2)	ΔS_{des} (J/mol H_2 K)
NPs	-68.1 ± 0.9	-119.0 ± 2	-64.0	-112	71.1	123
NDs	-70.0 ± 3.5	-129 ± 9	-47	-86	86	154
Ref. Mg	-74.06 ± 0.42	-133.4 ± 0.7	-	-	-	-

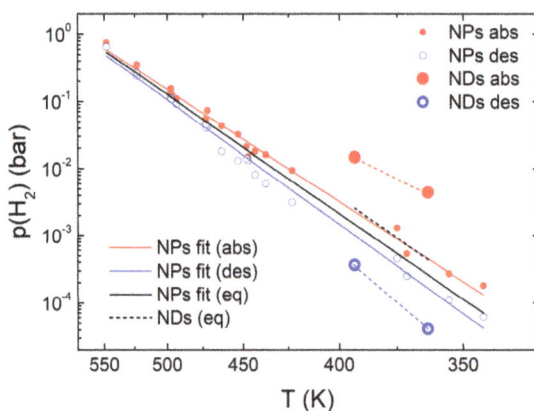

Figure 5. Van't Hoff plot of sorption (red) and desorption (blue) equilibrium pressures (log scale) versus T (reciprocal scale) measured for NPs samples [8] and NDs ($t = r = 30$ nm) [5]. The black solid and dashed lines are the best linear fit on equilibrium pressure values for NPs and NDs respectively. Red and blue solid lines are the best linear fit on absorption and desorption data for NPs.

The combination of these two effects does not result in a true destabilization of the NDs, that can only be claimed if p_{des} is higher than the bulk equilibrium value [10].

The pressure hysteresis values measured on NPs and NDs are collected in Figure 6. The NPs, which are relatively free to expand and contract upon hydrogen sorption, exhibit a small hysteresis, from ~3 at low temperature to ~1.5 at high temperature. The NDs show a huge hysteresis of ~100 at low temperature. The observation of a high hysteresis is quite common in constrained systems such as thin films clamped on a rigid substrate [10–12]. Figure 6 suggests, qualitatively, that the hysteresis rises with increasing dimensionality of the constraint (1D for thin films and 3D for NDs) and with decreasing confinement length (decreasing ND diameter).

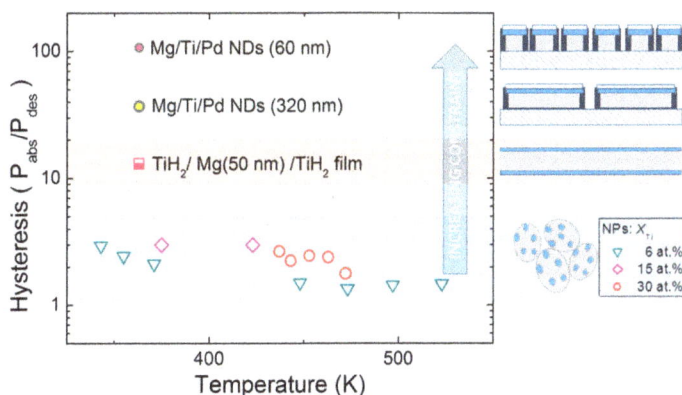

Figure 6. Hysteresis value p_{abs}/p_{des} vs. temperature for a collection of Mg–Ti–H NPs with varying Ti content (X_{Ti} = 6, 15, 30 atom %) [8], confronted with hysteresis value for NDs of different diameter (60 nm and 320 nm) [5] and for a quasi-free 50 nm thick Mg film [7].

A brief comment to the temperature extent of Van't Hoff plot is due. Mg NPs are subjected to severe coarsening when the temperature rises above 475 K. As shown in our precedent work [8], NPs remain stable upon cycling in the 340 K < *T* < 425 K range, while the nanostructure coarsens dramatically upon cycling at higher temperatures, where Mg sublimation from the clean Mg surface occurs and diffusion is faster. This establishes an upper temperature limit. On the other side, the exploration of significantly lower temperatures is hindered by the rapidly decreasing sorption rates and by the fact that the equilibrium pressure enters the high vacuum regime. Nevertheless, to the best of our knowledge, these composite NPs are the only system, with the exception of thin films coated by a Pd catalyst layer, in which the Mg to MgH$_2$ reversible transformation can be observed at such low temperature.

3. Discussion

Starting from the model presented in Figure 2, if we assume that all the contributions to the thermodynamical bias come from an interface region of thickness δ^{int} between *M* (metal) or *MH* (hydride) and a second phase (here indicated with α) dispersed as spherical nanocrystals inside the *M*-matrix, Equation (4) turns into

$$\delta\left(\Delta G^0\right)_{NPs} = \frac{V_M}{V_M}\frac{6V_\alpha}{d_\alpha}\Delta\gamma_\alpha^{int}F\left(\frac{V_\alpha}{V_M}\right) \tag{7}$$

where $\Delta\gamma_\alpha^{int} = \gamma_{MH|\alpha}^{int} - \gamma_{M|\alpha}^{int}$, and the function $F \le 1$ takes into account a possible decrease of the interface area due to coalescence of the α-phase crystallites. $F \approx 1$ when the α-phase crystallites are separated, a condition reasonably satisfied in the limit of low α-phase content. For high V_α/V_M, e.g., close to the percolative threshold, the α-phase crystallites will start merging, leading to a collapse of the interface area. Equation (7) is completely general in the limit $A_{MH|\alpha}^{int} = A_{M|\alpha'}^{int}$ i.e., when the $M \leftrightarrow MH$ transformation is complete and α-phase morphology is unchanged upon phase transformation.

Assuming $F \approx 1$ in the NPs, where the α-phase is TiH_2 and $M = Mg$, Equation (7) becomes:

$$\delta\left(\Delta G^0\right)_{NPs} = \frac{V_{Mg}}{V_{Mg}}\frac{6V_{TiH_2}}{d_{TiH_2}}\Delta\gamma_{TiH_2}^{int} \tag{8}$$

where V_{Mg} is the molar volume of Mg, V_{TiH_2} and V_{Mg} are the volumes occupied by TiH_2 and Mg, respectively, and d_{TiH_2} is the average TiH_2 grain size inside the Mg matrix. Exploiting the enthalpic end entropic contribution to the interface free energy, i.e.,:

$$\Delta\gamma_{TiH_2}^{int} = \Delta h_{TiH_2}^{int} - T\Delta s_{TiH_2}^{int} \tag{9}$$

we can separately evaluate these contributions to the thermodynamical bias since they are simply multiplied by a geometrical term and Equation (8) becomes

$$\delta\left(\Delta G^0\right)_{NPs} = \frac{V_{Mg}}{V_{Mg}}\frac{6V_{TiH_2}}{d_{TiH_2}}\left(\Delta h_{TiH_2}^{int} - T\Delta s_{TiH_2}^{int}\right) = \delta\left(\Delta H^0\right)_{NPs} - T\delta\left(\Delta S^0\right)_{NPs} \tag{10}$$

Hao et al. reported density functional theory calculations of the Mg | TiH_2 and MgH_2 | TiH_2 interface energy, yielding $\Delta h_{TiH_2}^{int}$ = 0.59 ÷ 0.69 J/m^2 [13].

Unfortunately, it is more difficult either to calculate the interfacial entropy or to extract it from experiments. The entropy has both a configurational and a vibrational term. The configurational term is due to the fraction of atom in the small (about 1 nm thick) interfacial region that accommodates the crystalline mismatch between different phases. According to literature [14], a rather generous estimation for this value is the one associated with the glass transition entropy, $\Delta s_{at}^{conf} \approx 1k_B/H_{at}^{int}$. The vibrational entropy term is even more challenging to assess and depends critically on the nature of the interfaces. For nanocrystalline materials, an increased vibrational entropy at interfaces $\Delta s_{at}^{vib} \approx 0.2k_B/H_{at}^{int}$ seems a realistic estimate [14]. It therefore appears reasonable to consider an overall entropy per interface H atom of $\Delta s_{at} = \Delta s_{at}^{conf} + \Delta s_{at}^{vib} \approx 1k_B/H_{at}^{int}$.

The interface-induced entropy variation per unit area, Δs^{int}, can then be calculated with the approximation that the volume of H atoms at the interface is the average $\langle\overline{V_H}\rangle$ = 1.7 cm^3/mol of the corresponding volumes in MgH_2 (2.08 cm^3/mol) and in TiH_2 (1.32 cm^3/mol):

$$\Delta s_{TiH_2}^{int} = \frac{1}{A^{int}}\frac{V^{int}}{\langle\overline{V_H}\rangle}\Delta s_{at} = \frac{\delta^{int}\cdot\Delta s_{at}}{\langle\overline{V_H}\rangle} \tag{11}$$

where V^{int} is the interface volume. Assuming δ^{int} = 1 nm, we obtain $\Delta s_{TiH_2}^{int} \approx 0.9 \times 10^{-3}$ J/K·m^2.

Now we can further manipulate Equation (10) to make explicit the dependence on the parameter X_{Ti} (atomic fraction Ti/(Mg + Ti)), yielding:

$$\delta\left(\Delta G^0\right)_{NPs} = \frac{X_{Ti}}{1-X_{Ti}}\frac{6V_{TiH_2}}{d_{TiH_2}}\left(\Delta h_{TiH_2}^{int} - T\Delta s_{TiH_2}^{int}\right) = \delta\left(\Delta H^0\right)_{NPs} - T\delta\left(\Delta S^0\right)_{NPs} \tag{12}$$

Values calculated with Equation (10), where no percolation effects are considered, are reported in Figure 7 for the composite NPs as a function of X_{Ti} and for different TiH_2 mean crystallite sizes. We assumed $\Delta h_{TiH_2}^{int}$ = 0.64 J/m^2 $\Delta s_{TiH_2}^{int}$ = 0.9 × 10^{-3} J/K·m^2 and used V_{TiH_2} = 13.26 cm^3/mol. Experimental values for NPs are also displayed with error bands in the explored range of compositions. The experimental range shown is not relative to nominal X_{Ti} = 6, 15, 30 atom %, but refers to X_{Ti} = 11, 20, 43 atom % values calculated only considering MgH_2 and TiH_2 crystallites and neglecting the MgO phase that is not participating to H_2 sorption process [8]. The experimentally determined

$\delta\left(\Delta H^{0}\right) \approx 6\,\text{kJ/mol}\,(\text{H}_2)$ is compatible with $X_{Ti} = 20$ and 43 atom % for $d_{TiH_2} = 2.5$ and 6 nm, respectively. However, for $X_{Ti} = 11$ atom %, the calculations only partially explain the enthalpy change, unless the existence of very small TiH_2 domains (about 1 nm diameter) is invoked.

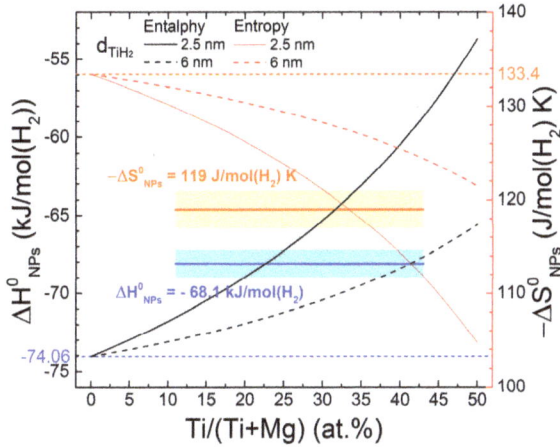

Figure 7. MgH_2 formation enthalpy ΔH^{0}_{NPs} (black lines) and the opposite of formation entropy ΔS^{0}_{NPs} (red lines) in composite MgH_2–TiH_2 NPs as a function of Ti atomic percentage as predicted by Equation (12). Calculations are reported for two different TiH_2 grain sizes: 2 nm (solid lines) and 6 nm (dashed lines) using $\Delta h^{int}_{TiH_2} = 0.64\,\text{J/m}^2$ and $\Delta s^{int}_{TiH_2} = 0.9 \times 10^{-3}\,\text{J/m}^2 \cdot \text{K}$. As a reference [9], the blue dashed line and the orange dashed line are the bulk MgH_2 formation enthalpy and entropy, respectively. Experimental values for NPs, as in [8], are also displayed with error bands. The explored range of compositions is the one determined considering only MgH_2 and TiH_2 phases, $X_{Ti} = 11, 20, 43$ atom % instead of the correspondent nominal values ($X_{Ti} = 6, 15, 30$ atom %).

The calculated entropy critically depends on the chosen Δs_{at}, for which we have just done a simple guess. A more detailed analysis of vibrational and configurational entropy per interface atom would help in refining the parameters for this model. The entropy curves in Figure 7 account only for a fraction of the experimental bias $\delta\left(\Delta S^{0}\right) \approx 14\,\text{J/mol}\,(\text{H}_2)$. Therefore, the experiments suggest that the interface entropy in the composite NPs may be significantly larger than our guessed value $\approx 1 k_B / \text{H}^{int}_{at}$. However, the experimental error could also be larger than the statistical error obtained from the best fitting procedure, due to coarsening of Mg and TiH_2 upon high temperature cycling.

Equations (8)–(10) have interesting implications on the possible presence of a unique enthalpy-entropy compensation temperature in nanomaterials that does not depend on microstructural parameters. This is a non-trivial issue which has stimulated discussions and controversies, since in many cases, the apparent enthalpy-entropy correlation is a phantom statistical phenomenon [10,15]. Starting from (10) and enthalpy and entropy being proportional to the same geometrical factor, as evident in the shapes of the curves in Figure 7, the temperature T^{comp}, at which entropy and enthalpy compensate each other, depends neither on Ti content nor on TiH_2 crystallite size and is simply given by

$$\delta\left(\Delta H^{0}\right)_{NPs} - T^{comp}\delta\left(\Delta S^{0}\right)_{NPs} = 0 \iff \Delta h^{int}_{TiH_2} - T^{comp}\Delta s^{int}_{TiH_2} = 0 \iff T^{comp} = \frac{\Delta h^{int}_{TiH_2}}{\Delta s^{int}_{TiH_2}} \quad (13)$$

We stress again that this result is valid only if the average $\Delta h^{int}_{TiH_2}$ and $\Delta s^{int}_{TiH_2}$ are independent or weakly dependent on system size. This assumption is corroborated by measurements on MgH_2 thin films of varying thickness and on MgH_2 nanoclusters embedded in TiH_2 [7,16].

A slightly different expression holds for Mg-based NDs where one has also to take into account the presence of the Mg | MgO lateral interface and of the SiO_2 substrate. With the assumption of a cylindrical ND, it is straightforward from Equation (4) that

$$\delta\left(\Delta G^0\right)_{NDs} = V_{Mg}\left(\frac{1}{t}\Delta\gamma^{int}_{TiH_2} + \frac{2}{r}\Delta\gamma^{int}_{MgO} + \frac{1}{t}\Delta\gamma^{int}_{SiO_2}\right) = \delta\left(\Delta H^0\right)_{NDs} - T\delta\left(\Delta S^0\right)_{NDs} \qquad (14)$$

Comparing this result with the corresponding Equation (10) one can see that the geometry of the system does not enter simply as a multiplicative term. Therefore, by changing thickness or diameter independently, one can make a determined specific free energy more relevant. Then, in principle, the possibility to tailor the geometry of the system makes it possible to tune the formation enthalpy and entropy and to change T^{comp}.

To the best of our knowledge, there are no literature data on $\Delta\gamma^{int}_{MgO}$ and $\Delta\gamma^{int}_{SiO_2}$. A rather crude simplification consists in taking the same interfacial free energy for every interface, turning Equation (14) into the simple:

$$\delta\left(\Delta G^0\right)_{NDs} = 2V_{Mg}\left(\frac{1}{t} + \frac{1}{r}\right)\Delta\gamma^{int}_{TiH_2} \qquad (15)$$

Here, similarly to NPs, one has a free energy term multiplied by the molar volume and by a geometrical factor, which is the surface-to-volume ratio of the nanosized object. Therefore, for a ND embedded in a homogeneous matrix the compensation temperature is independent on the number and extension of interfaces and can again be calculated using Equation (13).

The application of the simple Equation (15) to the NDs sketched in Figure 3 yields a thermodynamical bias of ~0.6 kJ/mol H_2. This small value is consistent with the low fraction of interfaces in a ND, which is 30 nm thick and 60 nm in diameter. Despite this, the equilibrium pressure of the NDs lies distinctly above those measured in NPs, suggesting that the elastic strain plays a role in the thermodynamics of NDs. The clamping provided by the surrounding phases, which do not expand upon hydrogen absorption, results in a compressive strain that makes the hydride slightly less stable. We believe that this argument does not apply to the NPs, which are relatively free to expand. Even though the effect of elastic strain seems detectable in NDs, it does not provide a significant destabilization of MgH_2, because the pressure change is too small for practical applications. Even worse, the high extrinsic hysteresis in the constrained ND system counterbalances the destabilizing elastic strains during desorption, resulting in an overall reduction of the desorption pressure. We can argue that the hysteresis could be reduced and a true destabilization achieved if it were possible to avoid the onset of plastic deformation. However, the volume changes induced by hydrogen sorption in Mg are so high (~32%) that the resulting stresses exceed by far the yield stress of almost all involved phases.

Finally, we remark that Equation (15) well explains the thermodynamical bias observed in ultra-small Mg clusters ($r \approx 1$ nm) immersed in a TiH_2 matrix, for which the interface free energy becomes relevant [16,17].

4. Materials and Methods

Mg–Ti–H composite NPs were grown by gas phase condensation in an ultra-high vacuum (UHV) chamber (Thermoionics, Port Townsend, WA, USA). Metallic precursor, Mg ingots (Alpha Aesar, Heysham, UK −99.95% purity) and Ti powders (Alpha Aesar, Heysham, UK −99.9%, 150 mesh) were co-evaporated using two tungsten crucibles under a equimolar He/H_2 atmosphere (99.9999% purity) at a pressure of 2.6 mbar. More details can be found in our previous work where a full morphological and structural characterization is also reported [8]. H-sorption measurements were performed in a secondary UHV chamber (Thermoionics, Port Townsend, WA, USA) of the synthesis apparatus, in order to minimize contaminations by oxygen or water vapors. The secondary chamber, the calibrated volume of which is 4.24 L, was used as a Sievert apparatus in the low temperature regime (340 K < *T* < 425 K) at pressures lower than 200 mbar. For measurements at higher temperatures, the NPs were extracted under Ar atmosphere and transferred without air exposure to an external dedicated Sievert apparatus for higher temperature characterization. The nanodots were prepared by template-assisted molecular beam epitaxy, using SiO_2 substrates coated by ultra-thin porous alumina nano-masks, as described in [5].

5. Conclusions

Interface free energy and elastic confinement have the potential to induce a thermodynamical bias in nanoscale hydrides. The independent evaluation of interface enthalpy and entropy is important, in order to assess the temperature dependence and magnitude of the bias. In general, the enthalpy-entropy compensation temperature depends on various microstructural and compositional parameters. However, it can be shown that the compensation temperature is unchanged if the following conditions hold: (i) the specific interface enthalpy and entropy (or their average) is the same for all the interfaces, and (ii) the interface free energy is the sole source of a bias (i.e., absence of elastic constraints).

We have applied these arguments to interpret experimental results obtained on a biphasic nanocomposite, in which ultrafine TiH_2 crystallites are dispersed in a Mg (or MgH_2) NP [8]. The calculations of the enthalpy change, based on DFT data for the interface enthalpy, are in reasonable agreement with the experiments, although they tend to underestimate the enthalpy change at low TiH_2 content. The experimental calculations comparison is much more difficult in the case of entropy, because there are no useful literature data on the interface entropy. An improved understanding and calculation of vibrational and configurational interface entropy appears to be of great importance in order to estimate the entropy change in nano-hydrides, which may be the key to tailoring their thermodynamics.

In the case of constrained systems, such as Mg NDs surrounded by rigid interfaces, an additional positive (i.e., destabilizing) contribution to the bias arises from elastic strains. However, the concomitant development of plastic deformation upon hydrogen sorption reduces the bias and brings about a large hysteresis, which results in a lower desorption pressure compared to the bulk counterparts.

Author Contributions: All the authors equally contributed to this work.

Conflicts of Interest: The authors declare no conflict of interest.

References

1. Kim, K.C.; Dai, B.; Johnson, J.K.; Sholl, D.S. Assessing nanoparticle size effects on metal hydride thermodynamics using the Wulff construction. *Nanotechnology* **2009**, *20*, 204001. [CrossRef] [PubMed]
2. Berube, V.; Chen, G.; Dresselhaus, M.S. Impact of nanostructuring on the enthalpy of formation of metal hydrides. *Int. J. Hydrogen Energy* **2008**, *33*, 4122–4131. [CrossRef]
3. Calvo, F. Thermodynamics of nanoalloys. *Phys. Chem. Chem. Phys.* **2015**, *17*, 27922–27939. [CrossRef] [PubMed]
4. Baldi, A.; Gonzalez-Silveira, M.; Palmisano, V.; Dam, B.; Griessen, R. Destabilization of the Mg-H System through Elastic Constraints. *Phys. Rev. Lett.* **2009**, *102*, 226102. [CrossRef] [PubMed]
5. Molinari, A.; D'Amico, F.; Calizzi, M.; Zheng, Y.; Boelsma, C.; Mooij, L.P.A.; Lei, Y.; Hahn, H.; Dam, B.; Pasquini, L. Interface and strain effects on the H-sorption thermodynamics of size-selected Mg nanodots. *Int. J. Hydrogen Energy* **2016**, *41*, 9841–9851. [CrossRef]
6. Pasquini, L.; Sacchi, M.; Brighi, M.; Boelsma, C.; Bals, S.; Perkisas, T.; Dam, B. Hydride destabilization in core-shell nanoparticles. *Int. J. Energy Res.* **2014**, *39*, 2115–2123. [CrossRef]
7. Mooij, L.P.A.; Baldi, A.; Boelsma, C.; Shen, K.; Wagemaker, M.; Pivak, Y.; Schreuders, H.; Griessen, R.; Dam, B. Interface Energy Controlled Thermodynamics of Nanoscale Metal Hydrides. *Adv. Energy Mater.* **2011**, *1*, 754–758. [CrossRef]

8.   Patelli, N.; Calizzi, M.; Migliori, A.; Morandi, V.; Pasquini, L. Hydrogen Desorption below 150 °C in MgH_2–TiH_2 Composite Nanoparticles: Equilibrium and Kinetic Properties. *J. Phys. Chem. C* **2017**, *121*, 11166–11177. [CrossRef]

9.   Paskevicius, M.; Sheppard, D.A.; Buckley, C.E. Thermodynamic Changes in Mechanochemically Synthesized Magnesium Hydride Nanoparticles. *J. Am. Chem. Soc.* **2010**, *132*, 5077–5083. [CrossRef] [PubMed]

10. Griessen, R.; Strohfeldt, N.; Giessen, H. Thermodynamics of the hybrid interaction of hydrogen with palladium nanoparticles. *Nat. Mater.* **2016**, *15*, 311–317. [CrossRef] [PubMed]

11. Pivak, Y.; Schreuders, H.; Dam, B. Thermodynamic Properties, Hysteresis Behavior and Stress-Strain Analysis of MgH_2 Thin Films, Studied over a Wide Temperature Range. *Crystals* **2012**, *2*, 710–729. [CrossRef]

12. Pivak, Y.; Schreuders, H.; Slaman, M.; Griessen, R.; Dam, B. Thermodynamics, stress release and hysteresis behavior in highly adhesive Pd–H films. *Int. J. Hydrogen Energy* **2011**, *36*, 4056–4067. [CrossRef]

13. Hao, S.; Sholl, D.S. Effect of TiH_2—Induced Strain on Thermodynamics of Hydrogen Release from MgH_2. *J. Phys. Chem. C* **2012**, *116*, 2045–2050. [CrossRef]

14. Fultz, B. Vibrational thermodynamics of materials. *Prog. Mater. Sci.* **2010**, *55*, 247–352. [CrossRef]

15. Cornish-Bowden, A. Enthalpy-entropy compensation: A phantom phenomenon. *J. Biosci.* **2002**, *27*, 121–126. [CrossRef] [PubMed]

16. Asano, K.; Westerwaal, R.J.; Anastasopol, A.; Mooij, L.P.A.; Boelsma, C.; Ngene, P.; Schreuders, H.; Eijt, S.W.H.; Dam, B. Destabilization of Mg Hydride by Self-Organized Nanoclusters in the Immiscible Mg–Ti System. *J. Phys. Chem. C* **2015**, *119*, 12157–12164. [CrossRef]

17. Asano, K.; Westerwaal, R.J.; Schreuders, H.; Dam, B. Enhancement of Destabilization and Reactivity of Mg Hydride Embedded in Immiscible Ti Matrix by Addition of Cr: Pd-Free Destabilized Mg Hydride. *J. Phys. Chem. C* **2017**, *121*, 12631–12635. [CrossRef]

inorganics

Article

Synthesis of LiAlH₄ Nanoparticles Leading to a Single Hydrogen Release Step upon Ti Coating

Lei Wang and Kondo-Francois Aguey-Zinsou *

Merlin group, School of Chemical Engineering, The University of New South Wales, Sydney 2052, Australia; lei.wang@unsw.edu.au
* Correspondence: f.aguey@unsw.edu.au; Tel.: +61-293-857-970

Academic Editor: Torben R. Jensen
Received: 20 May 2017; Accepted: 3 June 2017; Published: 7 June 2017

Abstract: Lithium aluminum hydride (LiAlH₄) is an interesting high capacity hydrogen storage material with fast hydrogen release kinetics when mechanically activated with additives. Herein, we report on a novel approach to produce nanoscale LiAlH₄ via a bottom-up synthesis. Upon further coating of these nanoparticles with Ti, the composite nanomaterial was found to decompose at 120 °C in one single and extremely sharp exothermic event with instant hydrogen release. This finding implies a significant thermodynamic alteration of the hydrogen properties of LiAlH₄ induced by the synergetic effects of the Ti catalytic coating and nanosizing effects. Ultimately, the decomposition path of LiAlH₄ was changed to LiAlH₄ → Al + LiH + 3/2H₂.

Keywords: alanate; lithium aluminum hydride; nanosizing; core–shell; hydrogen storage

1. Introduction

Hydrogen holds enormous potential to be the ultimate energy carrier of the future [1–5]. Yet the lack of an effective method for storing hydrogen with high density currently restrains its widespread utilization. Solid state hydride materials offer promising possibilities to deliver high capacity, safe, and compact hydrogen storage systems [5,6]. However, so far no material satisfies all the requirements for practical mobile application [3]. For hydrogen storage purposes, LiAlH₄ is one of the most promising and interesting candidates owing to its competitive total hydrogen capacity as well as its low temperature for hydrogen release with fast kinetics [3,6]. When doped with catalysts, about 75% of its stored hydrogen can be released below 100 °C in about 1.5 h, while 30 min is required when at 150 °C [7–11]. A particularity of LiAlH₄ is the exothermic nature of its initial hydrogen release following the widely accepted Steps 1–3 [12–14].

Step 1a	150–175 °C	
LiAlH₄(solid) → LiAlH₄(liquid)	endothermic	
Step 1b	150–200 °C	
LiAlH₄(liquid) → 1/3 Li₃AlH₆(solid) + 2/3 Al + H₂↑	exothermic	5.3 mass % H₂
Step 2	200–270 °C	
Li₃AlH₆(solid) → 3 LiH + Al + 3/2 H₂↑	endothermic	2.6 mass % H₂
Step 3	400–440 °C	
LiH + Al → LiAl + 1/2 H₂↑	endothermic	2.6 mass % H₂

Step 1a corresponds to the melting of LiAlH₄ quickly followed by the exothermic Step 1b with a reported ΔH of decomposition of −10 kJ·mol⁻¹ H₂ [1,2]. Hence, the reversibility of this step is generally believed to be thermodynamically impossible under practical conditions of temperature and pressure [3,4]. Indeed, theoretical studies have suggested the need for extreme hydrogen pressures

>100 MPa at room temperature for the direct regeneration of $LiAlH_4$ from Li_3AlH_6 [2,3,5]. In Step 2, Li_3AlH_6 decomposes via an endothermic reaction with a ΔH of 25 $kJ \cdot mol^{-1}$ H_2 [6]. Hence the regeneration of Li_3AlH_6 from LiH and Al can be considered thermodynamically feasible. The last decomposition occurring at temperatures higher than 400 °C precludes this step from practical hydrogen storage application, reducing the total hydrogen capacity to 7.9 mass % H_2 (below 300 °C).

Accordingly, current investigations have mainly focused at the destabilization of $LiAlH_4$ via catalyst doping and/or mechanical milling with additives to further improve its dehydrogenation kinetics and potentially achieve some degree of reversibility. Hence, upon mechanically milling $LiAlH_4$ with $TiCl_3$, $ZrCl_4$, or nanosized Ni, lower hydrogen desorption temperatures have been achieved [7–11]; however, with no successful rehydrogenation. Indirect rehydrogenation routes have also been demonstrated through the use of wet synthesis approaches involving the formation of solvent adducts of $LiAlH_4$—i.e., THF, Et_2O, diglyme, and Me_2O—and in most cases activated/catalyzed Al [12–16]. Such an indirect route not only needs to be performed off-board while requiring the additional process of desolvation, but also can compromise the integrity of activated doped $LiAlH_4$. For example, for the multi-step regeneration method developed by Graetz et al., and involving the use of THF, successive doping and redoping is required to maintain the same hydrogen storage performance before and after each desolvation of THF [13]. Chen et al. achieved a partial direct reversibility of titanium-catalyzed Li_3AlH_6, i.e., to reverse Step 2, without the need of solvent desolvation under a hydrogen pressure of 40 MPa only [17]. However, the evidence of regenerated Li_3AlH_6 was a few non-exclusive XRD peaks without structure determination of the rehydrided Li_3AlH_6 and despite this promising result no follow-up study was reported.

Previous work from our group revealed the possibility of direct rehydrogenation of $LiAlH_4$ via nanoconfinement in mesoporous carbon and this could be achieved under mild conditions of pressures and temperatures following the assumed route of $3LiH + Al + 3/2H_2 \rightarrow Li_3AlH_6 + 2Al + 3H_2 \rightarrow 3LiAlH_4$ [18]. Furthermore, positive alternations of hydrogen storage properties were observed with a core–shell method for the individual stabilization of $NaBH_4$ or LiH nanoparticles avoiding some drawbacks of porous host scaffolds [19,20]. This impetus led to the current work which investigated a simple route for the synthesis of isolated $LiAlH_4$ nanoparticles and their stabilization for hydrogen cycling through a core–shell approach, with Ti acting as a shell facilitating the retention of molten $LiAlH_4$ upon dehydrogenation as well as a "gateway" for hydrogen absorption/desorption. Ti based additives have been widely studied via top–down ball milling for $LiAlH_4$, as well as with other alanate systems e.g., $NaAlH_4$ and $Mg(AlH_4)_2$ with promising results, i.e., much improved hydrogen storage properties [21–26]. Indeed, Kojima et al. determined the catalytic effects of various metal chlorides on $LiAlH_4$, and revealed their positive effects in terms of hydrogen release in the order of $TiCl_3 > ZrCl_4 > VCl_3 > NiCl_2 > ZnCl_2$ [27]. Amama et al. also investigated the properties of ball milled $LiAlH_4$ with various dopants, and concluded that Ti-based additives lead to the most enhancement in terms of dehydrogenation kinetics and reduction of the decomposition temperature [9]. Based on these previous findings and the potential of a nanosizied approach [28,29], it is thus interesting to investigate the effect of Ti doping/coating on nanosized $LiAlH_4$.

2. Results

In principle, the nanosynthesis of $LiAlH_4$ can be achieved via common solvent evaporation methods [30]. In this procedure, a solution of the nanoparticles' precursor is evaporated under high vacuum [31,32] and this results in the rapid nucleation of the solute materials because of the supersaturated nature of the solution. The growth of the nuclei is then limited via steric repulsion of suitable stabilizers. Currently, the syntheses of most nanoparticles through solvent evaporation are performed in an aqueous solution [33,34]. However, for hydride materials which are highly oxidized by water, the synthesis should be performed under non-aqueous environments. To achieve the preparation of nanosized $LiAlH_4$ particles, we thus used a THF solution of 2M $LiAlH_4$ with 1-dodecanethiol as a stabilizer and weak capping ligand.

As shown in Figure 1, at the end of the evaporation process the XRD pattern of the material confirmed the preparation of LiAlH$_4$ with broader diffraction peaks indicating its nanostructure. The nanosized nature of the material was confirmed by TEM analysis showing isolated spherical nanoparticles with a particle size ranging from 2 to 16 nm (Figure 2).

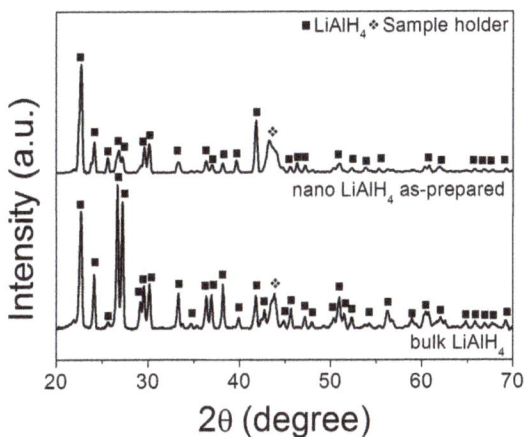

Figure 1. XRD of as-prepared nanosized LiAlH$_4$ via solvent evaporation and commercial bulk LiAlH$_4$.

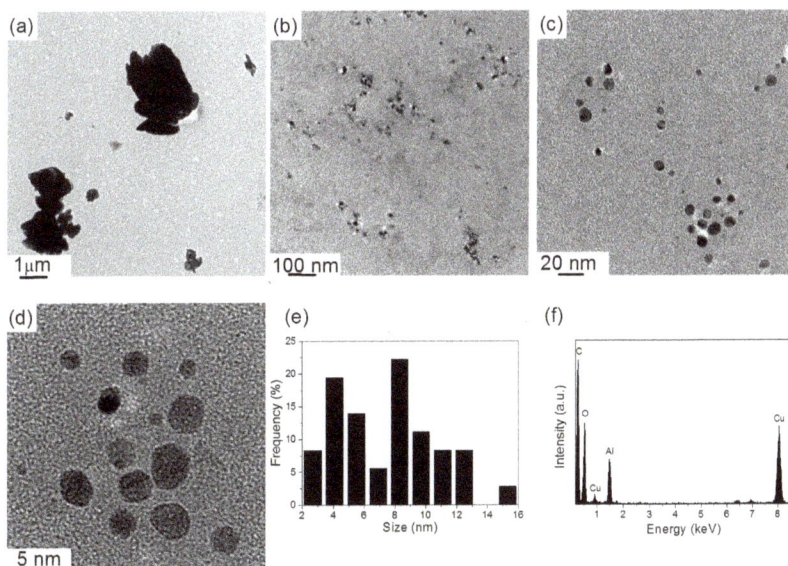

Figure 2. TEM images of (**a**) commercial bulk LiAlH$_4$ and (**b–d**) as-prepared nanosized LiAlH$_4$ via solvent evaporation; (**e**) corresponding particle size distribution and (**f**) EDS spectrum.

EDS analysis further confirmed the Al content of these nanoparticles that were thus believed to correspond to LiAlH$_4$ (Figure 2f). In comparison, bulk LiAlH$_4$ was imaged as large particles with no defined morphology (Figure 2a), hence the synthesis of nanosized LiAlH$_4$ was successful.

The hydrogen desorption properties of these nanoparticles were determined by TGA/DSC/MS. As shown in Figure 3, the as-prepared LiAlH$_4$ nanoparticles desorbed hydrogen in a similar way to bulk LiAlH$_4$ with the release of hydrogen occurring in two steps before 300 °C. The small additional exothermic peak at 145 °C was assigned to the reaction of the thiol head of the surfactant with LiAlH$_4$ at the surface of the nanoparticles. The similarities of the decompositions behavior is inherent to the inability of the surfactant to contain the melted LiAlH$_4$ and thus a loss of the nanosize feature in the favor of larger molten agglomerates behaving like bulk LiAlH$_4$. The decomposition of the dodecanthiol as evidenced by mass spectrometry near 220 °C [35] further supports this hypothesis (Figure 3d).

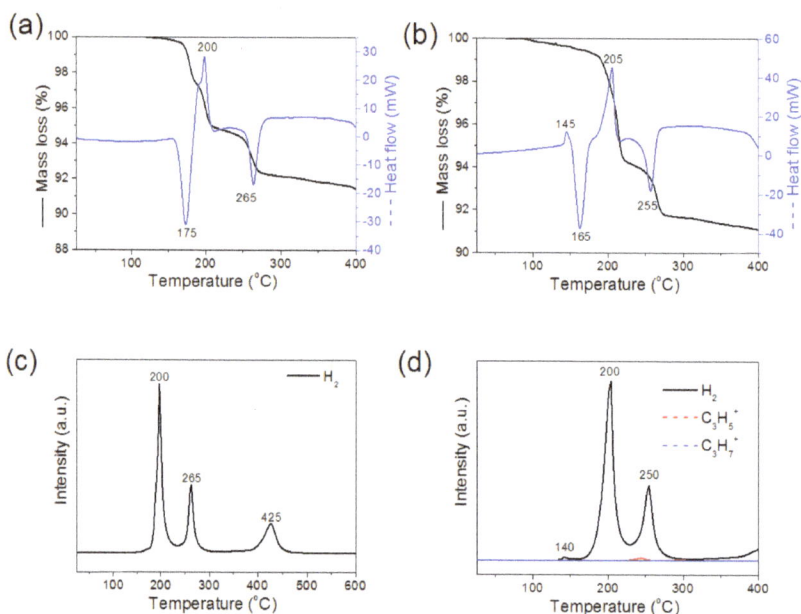

Figure 3. TGA/DSC for (**a**) commercial bulk LiAlH$_4$; (**b**) as-prepared nanosized LiAlH$_4$ via solvent evaporation, and (**c**,**d**) respective hydrogen desorption profile as followed by mass spectrometry. The fragments C$_3$H$_5$$^+$ and C$_3$H$_7$$^+$ correspond to the decomposition of the 1-dodecanethiol surfactant [36]. The decreasing heat flow peaks correspond to an endothermic process.

In order to better stabilize these nanoparticles and contain the melt, we aimed at a core–shell strategy [20]. whereby the core LiAlH$_4$ is contained within a metallic shell. To this aim, Ti was thought to be the best option because it is a well-known catalyst for other alanate systems including NaAlH$_4$ and Mg(AlH$_4$)$_2$ [21,26,37]. In particular, Ti-based additives have shown benefit for enhancing dehydrogenation kinetics and reduction of the decomposition temperature of alanates [9]. Ti also forms a hydride and has good hydrogen diffusion properties [38,39]. In addition, TiCl$_3$ can be readily reduced by LiAlH$_4$, and thus could form through the transmetalation method [19,20,31]. A core–shell structure assuming that the reduction rate of the titanium precursor at the surface of the LiAlH$_4$ nanoparticles can be controlled to form a continuous shell [40].

To coat the nanosized LiAlH$_4$ particles with Ti, LiAlH$_4$ nanoparticles were suspended in pentane where they are insoluble and a solution of TiCl$_3$ in pentane was added dropwise. In this process, TiCl$_3$ is reduced at the surface of LiAlH$_4$ nanoparticles following reaction (1).

$$3LiAlH_4 + TiCl_3 \rightarrow 3LiCl + Ti + 3Al + 6H_2 \tag{1}$$

After ageing overnight, washing, and drying, the resulting black suspension was characterized. As shown by XRD, the diffraction pattern of the material mainly contained peaks with a weaker intensity related to the monoclinic phase of $LiAlH_4$ and additional peaks assigned to tetragonal TiH and residual $TiCl_3$ (Figure 4). No metallic Ti was observed and this was consistent with the ability of $LiAlH_4$ to reduce metal halides with the formation of their corresponding hydride [41]. Non-stoichiometric TiH_{2-x} has previously been reported to form when heating metallic Ti under hydrogen pressure at temperatures >300 °C [39,42]. However, under the current experimental conditions, it is possible that the hydrogen release upon reaction (1) directly interacts with the newly formed Ti nuclei to generate the corresponding hydride. The remaining $TiCl_3$ observed by XRD was unexpected owing to the large molar ratio (16:1) of $LiAlH_4$ compared to $TiCl_3$, and this indicated that the reduction of $TiCl_3$ by $LiAlH_4$ stopped during the process. This could be the result of the formation of a TiH layer at the surface of the $LiAlH_4$ nanoparticles significantly decreasing the contact area of $LiAlH_4$ and $TiCl_3$, and thus precluding any further reduction.

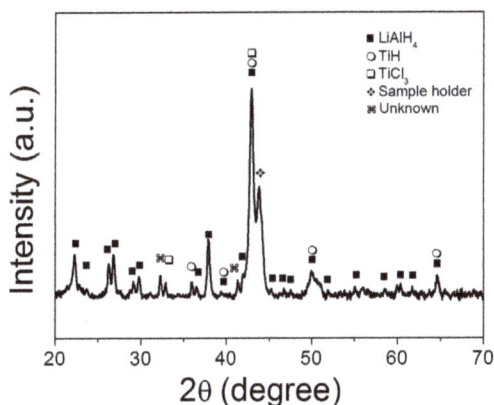

Figure 4. XRD of as-prepared Ti coated nanosized $LiAlH_4$.

Additional characterization by TEM revealed largely spherical particles with a broad particle size distribution ranging from 5 to 50 nm (Figure 5a,b,e). This corresponds to a significant increase in particle size as compared to the uncoated $LiAlH_4$ nanoparticles (Figure 2) and thus indicates the deposition of a TiH layer. Indeed, EDS analysis and elemental mapping revealed the Ti and Al content of the nanoparticles imaged and the even distribution of Ti on the $LiAlH_4$ nanoparticles (Figure 5c,d). Hence, it was believed that Ti was successfully deposited at the surface of the $LiAH_4$ nanoparticles. The weaker intensity of the XRD peaks related to $LiAlH_4$ in the Ti coated nanosized $LiAlH_4$ material also corroborates this conclusion (Figure 4), because upon coating the diffraction peak intensity of the core has been reported to significantly decrease [31].

Analysis of the hydrogen desorption properties of these Ti modified nanoparticles revealed a unique behavior (Figure 6). Nanosized $LiAlH_4$ modified with Ti almost fully decomposed through a single and extremely exothermic event recorded by DSC at 120 °C (Figure 6a,b). In comparison, bulk $LiAlH_4$ ball milled with $TiCl_3$ (material denoted bulk-$LiAlH_4$/Ti) decomposed following the known decomposition path of $LiAlH_4$—i.e., via Steps 1 and 2 (Figure 6c,d)—in agreement with previous reports [9,43,44]. This was confirmed by the XRD of bulk $LiAlH_4$/Ti after heating at 150, 200 and 300 °C.

As shown in Figure 6b, at 150 °C, bulk-$LiAlH_4$/Ti started partial decomposition to Li_3AlH_6 according to Step 1. Then at 180 °C, the $LiAlH_4$ phase totally converted to Li_3AlH_6 and at 300 °C, Li_3AlH_6 fully decomposed to LiH and Al following Step 2. In comparison, the diffraction pattern of nanosized $LiAlH_4$ coated with Ti heated at 150 °C after the single exothermic event showed that no

LiAlH$_4$ remained, while only a very small amount of Li$_3$AlH$_6$ was detected (Figure 7a). The remaining Li$_3$AlH$_6$ could be the result of unevenly or partially coated LiAlH$_4$ nanoparticles evolving to bulk size materials upon melting. However, this could not be confirmed since no apparent melting was detected by DSC peak (Figure 6a).

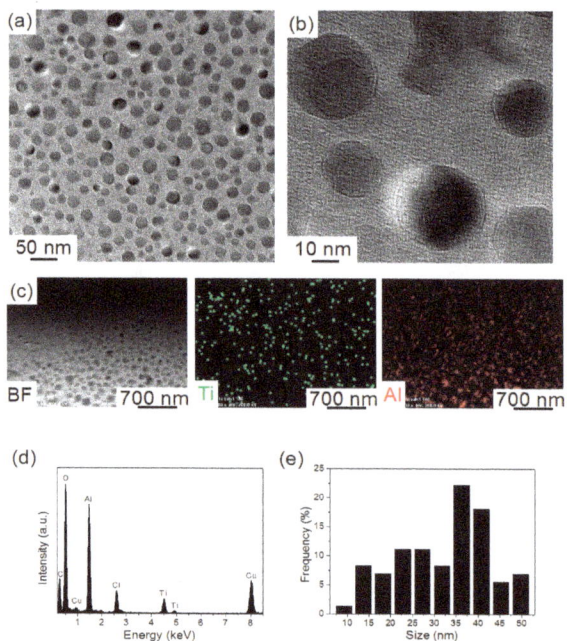

Figure 5. TEM images of (**a,b**) as-prepared Ti coated nanosized LiAlH$_4$ and corresponding (**c**) elemental mapping analysis; (**d**) EDS spectrum and (**e**) particle size distribution.

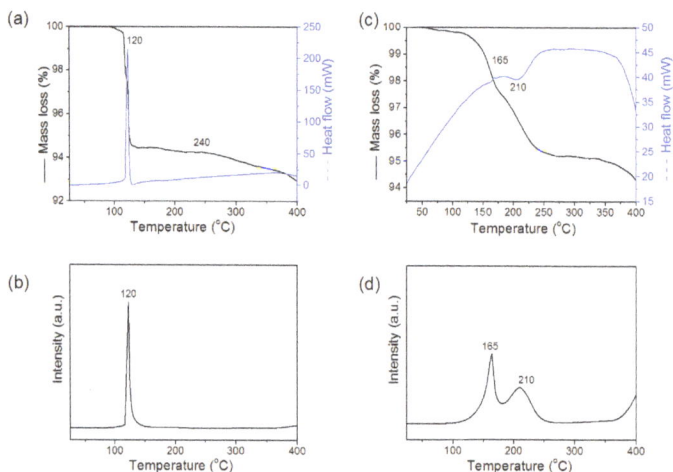

Figure 6. TGA/DSC for (**a**) as-prepared Ti coated nanosized LiAlH$_4$, (**c**) commercial bulk LiAlH$_4$ ball milled with TiCl$_3$, (**b,d**) respective hydrogen desorption profile as followed by mass spectrometry. No other gases than hydrogen were detected.

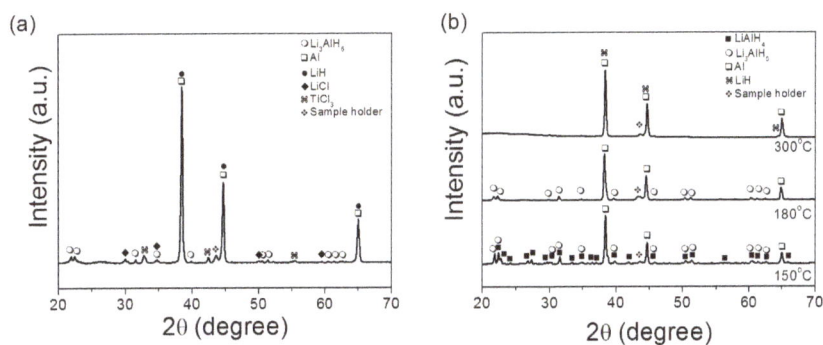

Figure 7. XRD of (**a**) as-prepared Ti coated nanosized LiAlH$_4$ after hydrogen release at 150 °C and (**b**) commercial bulk LiAlH$_4$ ball milled with TiCl$_3$ after thermal decomposition at 150, 180 and 300 °C.

Another possibility for the partial decomposition of Li$_3$AlH$_6$ is that the intended core–hell structure disintegrated under the sharp exothermic event as proven by TEM analysis of the dehydrogenated material (Figure 8), and consequently sprayed around undecomposed Li$_3$AlH$_6$. The observed suppression of melting as per bulk LiAlH$_4$ is also an indicator that LiAlH$_4$ was confined within a TiH shell.

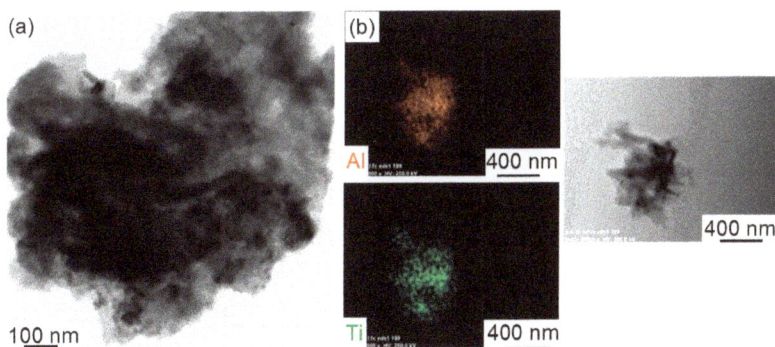

Figure 8. TEM images of (**a**) as-prepared Ti coated nanosized LiAlH$_4$ after hydrogen release at 150 °C and (**b**) corresponding elemental mapping analysis.

The possible core–shell structure disintegration could also be the reason of an appearance of LiCl and the disappearance of diffraction peaks related to TiH (Figure 7a). LiCl was not present in the as-prepared Ti coated nanosized LiAlH$_4$ (Figure 4), and this potentially indicates that more LiAlH$_4$ was exposed to react with TiCl$_3$ to form LiCl following reaction (1) during the exothermic event. The lack of TiH phase after the exothermic release of hydrogen also indicates a deterioration of the coating owing to the heat generated during the exothermic release of hydrogen. TiH has been reported to decompose at temperatures above 470 °C [45].

Furthermore, upon hydrogen release, nanosized LiAlH$_4$ coated with Ti lost about 5.7 mass % almost instantly near 120 °C. Taking into account the weight of TiCl$_3$ added to nanosized LiAlH$_4$, the theoretical hydrogen release for Step 1 should be reduced from 5.3 to 4.2 mass %, and from 2.6 to 2.1 mass % for both Steps 2 and 3. This corresponds to a theoretical hydrogen content of 6.3 mass % up to Step 2. The 5.7 mass % loss of the Ti coated nanosized LiAlH$_4$ thus corresponds to a close to full hydrogen release. Possible hydrogen release from TiH was unlikely to be noticeable in the hydrogen desorption profile. Assuming that nanoscale TiH was decomposed, this would correspond to an additional amount of hydrogen of less than 0.1 mass %. Accordingly, this result implies a significant thermodynamic alternation induced by the synergetic effects of the Ti catalytic coating with $LiAlH_4$ nanosizing potentially involving the decompositions of both $LiAlH_4$ and Li_3AlH_6 in a rapid manner following the direct reaction (2).

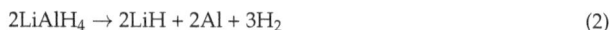

$$2LiAlH_4 \rightarrow 2LiH + 2Al + 3H_2 \tag{2}$$

Although, the currently modified properties of Ti modified nanosized $LiAHl_4$ are unsuitable for hydrogen storage purposes, this finding implies that the strategy of catalytic coating on hydride nanoparticles offers a more versatile avenue to modifying the hydrogen storage properties of a hydride to a larger extent when compared to mechanically activated $LiAlH_4$ with Ti.

3. Materials and Methods

All experiments were performed under an inert atmosphere in an argon filled glove box (O_2 and $H_2O < 1$ ppm) from LC Technology (Salisbury, MD, USA) $LiAlH_4$ 2.0 M in THF, titanium (III) chloride, 1-dodecanethiol, and anhydrous pentane were purchased from Sigma Aldrich (Sydney, Australia) and used as received. Lithium aluminum hydride ($LiAlH_4$, 95%) was purchased from Sigma Aldrich (Sydney, Australia), and purified by dissolving it in a large amount of diethyl ether and recrystallizing it from the filtrate solution.

3.1. Synthesis of $LiAlH_4$ Nanoparticles

In a typical synthesis, 5 mL of a commercial $LiAlH_4$ solution (2.0 M in THF) was placed in a vial and stirred gently with 10 μL (2 mg) of 1-dodecanethiol. The vial was then closed tightly and the solution was left to age overnight. The solvent was then evaporated under a moderate vacuum until a viscous paste was formed. The later was fully dried at 30 °C under dynamic vacuum. The yield was around 410 mg, which is slightly higher than the 382 mg theoretical yield owing to remaining THF bonded to $LiAlH_4$.

3.2. Coating of the $LiAlH_4$ Nanoparticles with Ti

The as-synthesized $LiAlH_4$ nanoparticles (100 mg) were suspended in 20 mL of pentane, under magnetic stirring at 600 rpm to form a milky suspension. 25 mg of $TiCl_3$ ($8.1 \cdot 10^{-3}$ mol·L^{-1}) was then suspended in a 10 mL solution of pentane and then added dropwise to the suspension of the $LiAlH_4$ nanoparticles at a rate of 50 μL·min^{-1}. After a few minutes, the suspension turned black and was allowed to age overnight before separation by centrifugation and drying under dynamic vacuum overnight.

3.3. Preparation of Reference Material Ball Milled $LiAlH_4$ with $TiCl_3$

As a reference, commercial bulk $LiAlH_4$ was ball milled with $TiCl_3$ (5 mol %) using a Retsch Mixer Mill MM 400 (Haan, Germany). The mixture was milled for 10 min at 20 Hz three times with a ball to powder weight ratio of 135:1.

3.4. Characterization

Transmission Electron Microscopy (TEM) and Energy Dispersive X-ray Spectroscopy (EDS) were performed with a Philips CM200 (Sydney, Australia) operated at 200 kV. The materials were dispersed in pentane, dropped onto a carbon coated copper grid, and dried in an argon filled glovebox before transfer to the microscope in a quick manner as to minimize air exposure. X-ray Diffraction (XRD) was performed by using a PANalytical X'pert Multipurpose XRD system (Sydney, Australia) operated at 40 mA and 45 kV with a monochromated Cu Kα radiation ($\lambda = 1.541$ Å); step size = 0.01, 0.02 or 0.05; time per step = 10 or 20 s/step. The materials were protected against oxidation from air by a Kapton foil covering a stainless steel sample holder. Hydrogen desorption profiles were acquired

Inorganics **2017**, *5*, 38

by Thermogravimetric Analysis (TGA)/Differential Scanning Calorimetry (DSC) coupled with Mass Spectrometry (MS) using a Mettler Toledo TGA/DSC (Sydney, Australia) 1 coupled with an Omnistar MS. Measurements were conducted with alumina crucibles at 10 °C·min^{-1} under an argon flow of 25 mL·min^{-1}. Masses between m/z = 2 and 100 were recorded. Reversibility was checked by manually performing hydrogen absorption/desorption cycles with a homemade Sievert apparatus and a stainless steel sample holder operated at 300 °C, with 7 MPa hydrogen pressure for absorption and 0.01 MPa for desorption. No significant reversibility was observed.

4. Conclusions

An effective downsizing and coating approach was developed and demonstrated that it is possible to synthesize isolated LiAlH$_4$ nanoparticles to a certain extent and ultimately modify the hydrogen storage properties of LiAlH$_4$. Once coated with Ti, nanosized LiAlH$_4$ was found to decompose at 120 °C in a single exothermic event with a 5.7 mass % and instant hydrogen release. This finding implies a significant thermodynamic alternation induced by the synergetic effects of the Ti catalytic coating with LiAlH$_4$ nanosizing that ultimately changed its decomposition path to LiAlH$_4 \rightarrow$ Al + LiH + 3/2H$_2$. This finding also demonstrates that the hydrogen storage properties of LiAlH$_4$ should be size dependent and thus provides a new avenue to modify the properties of LiAlH$_4$ through particle size effects. Due to the violent exothermic nature of its decomposition, this material is considered not viable for practical hydrogen storage purposes. However, Ti coated nanosized LiAlH$_4$ offers a unique alternative as compared to hydrides which only supply hydrogen via an endothermic reaction. It enables applications where larger amounts of hydrogen and heat are both needed at the same time and rapidly, e.g., for compact explosives and fire starters.

Acknowledgments: Financial support by UNSW Internal Research Grant program is gratefully acknowledged. We appreciate the use of instruments in the Mark Wainwright Analytical Centre at UNSW.

Author Contributions: Lei Wang carried out all the experimental work which was conceived and designed with Kondo-Francois Aguey-Zinsou.

Conflicts of Interest: The authors declare no conflict of interest.

References

1. Dornheim, M. *Thermodynamics of Metal Hydrides: Tailoring Reaction Enthalpies of Hydrogen Storage Materials*; In Tech: Rijeka, Croatia, 2011.
2. Jang, J.W.; Shim, J.H.; Cho, Y.W.; Lee, B.J. Thermodynamic calculation of LiH $\leftrightarrow$ Li$_3$AlH$_6$ $\leftrightarrow$ LiAlH$_4$ reactions. *J. Alloys Compd.* **2006**, *420*, 286–290. [CrossRef]
3. Varin, R.A.; Czujko, T.; Wronski, Z.S. *Nanomaterials for Solid State Hydrogen Storage*; Springer: Berlin, Germany, 2009.
4. Lai, Q.; Paskevicius, M.; Sheppard, D.A.; Buckley, C.E.; Thornton, A.W.; Hill, M.R.; Gu, Q.; Mao, J.; Huang, Z.; Liu, H.K.; et al. Hydrogen storage materials for mobile and stationary applications: Current state of the art. *ChemSusChem* **2015**, *8*, 2789–2825. [CrossRef] [PubMed]
5. Lacina, D.; Yang, L.; Chopra, I.; Muckerman, J.; Chabal, Y.; Graetz, J. Investigation of LiAlH$_4$–THF formation by direct hydrogenation of catalyzed Al and LiH. *Phys. Chem. Chem. Phys.* **2012**, *14*, 6569–6576. [CrossRef] [PubMed]
6. Orimo, S.I.; Nakamori, Y.; Eliseo, J.R.; Züttel, A.; Jensen, C.M. Complex hydrides for hydrogen storage. *Chem. Rev.* **2007**, *107*, 4111–4132. [CrossRef] [PubMed]
7. Varin, R.A.; Zbroniec, L. The effects of nanometric nickel (n-Ni) catalyst on the dehydrogenation and rehydrogenation behavior of ball milled lithium alanate (LiAlH$_4$). *J. Alloy. Compd.* **2010**, *506*, 928–939. [CrossRef]
8. Li, Z.; Zhai, F.; Wan, Q.; Liu, Z.; Shan, J.; Li, P.; Volinsky, A.A.; Qu, X. Enhanced hydrogen storage properties of LiAlH$_4$ catalyzed by CoFe$_2$O$_4$ nanoparticles. *RSC Adv.* **2014**, *4*, 18989–18997. [CrossRef]
9. Amama, P.B.; Grant, J.T.; Shamberger, P.J.; Voevodin, A.A.; Fisher, T.S. Improved Dehydrogenation Properties of Ti-Doped LiAlH$_4$: Role of Ti Precursors. *J. Phys. Chem.* **2012**, *116*, 21886–21894. [CrossRef]

10. Easton, D.S.; Schneibel, J.H.; Speakman, S.A. Factors affecting hydrogen release from lithium alanate (LiAlH$_4$). *J. Alloy Compd.* **2005**, *398*, 245–248. [CrossRef]

11. Fu, J.; Tegel, M.; Kieback, B.; Röntzsch, L. Dehydrogenation properties of doped LiAlH$_4$ compacts for hydrogen generator applications. *Int. J. Hydrog. Energy* **2014**, *39*, 16362–16371. [CrossRef]

12. Liu, X.; McGrady, G.S.; Langmi, H.W.; Jensen, C.M. Facile cycling of Ti-doped LiAlH$_4$ for high performance hydrogen storage. *J. Am. Chem. Soc.* **2009**, *131*, 5032–5033. [CrossRef] [PubMed]

13. Graetz, J.; Wegrzyn, J.; Reilly, J.J. Regeneration of lithium aluminum hydride. *Chem. Inf.* **2008**, *130*, 17790–17794. [CrossRef] [PubMed]

14. Ashby, E.C.; Brendel, G.J.; Redman, H.E. Direct Synthesis of Complex Metal Hydrides. *Inorg. Chem.* **1963**, *2*, 499–504. [CrossRef]

15. Wang, J.; Ebner, A.D.; Ritter, J.A. Physiochemical Pathway for Cyclic Dehydrogenation and Rehydrogenation of LiAlH$_4$. *J. Am. Chem. Soc.* **2006**, *128*, 5949–5954. [CrossRef] [PubMed]

16. Wang, J.; Ebner, A.D.; Ritter, J.A. Synthesis of metal complex hydrides for hydrogen storage. *J. Phys. Chem.* **2007**, *111*, 14917–14924. [CrossRef]

17. Chen, J.; Kuriyama, N.; Xu, Q.; Takeshita, H.T.; Sakai, T. Reversible hydrogen storage via titanium-catalyzed LiAlH$_4$ and Li$_3$AlH$_6$. *J. Phys. Chem.* **2001**, *105*, 11214–11220. [CrossRef]

18. Wang, L.; Rawal, A.; Quadir, M.Z.; Aguey-Zinsou, K.F. Nanoconfined lithium aluminium hydride (LiAlH$_4$) and hydrogen reversibility. *Int. J Hydrog. Energy* **2017**, *42*, 14144–14153. [CrossRef]

19. Wang, L.; Quadir, M.Z.; Aguey-Zinsou, K.F. Ni coated LiH nanoparticles for reversible hydrogen storage. *Int. J Hydrog. Energy* **2016**, *41*, 6376–6386. [CrossRef]

20. Christian, M.L.; Aguey-Zinsou, K.F. Core–shell strategy leading to high reversible hydrogen storage capacity for NaBH$_4$. *ACS Nano* **2012**, *6*, 7739–7751. [CrossRef] [PubMed]

21. Bogdanović, B.; Schwickardi, M. Ti-doped NaAlH$_4$ as a hydrogen-storage material—Preparation by Ti-catalyzed hydrogenation of aluminum powder in conjunction with sodium hydride. *Appl. Phys.* **2001**, *72*, 221–223. [CrossRef]

22. Blanchard, D.; Brinks, H.W.; Hauback, B.C.; Norby, P. Desorption of LiAlH$_4$ with Ti- and V-based additives. *Mat. Sci. Eng. Solid* **2004**, *108*, 54–59. [CrossRef]

23. Kojima, Y.; Kawai, Y.; Haga, T.; Matsumoto, M.; Koiwai, A. Direct formation of LiAlH$_4$ by a mechanochemical reaction. *J. Alloy. Compd.* **2007**, *441*, 189–191. [CrossRef]

24. Liu, X.; Langmi, H.W.; Beattie, S.D.; Azenwi, F.F.; McGrady, G.S.; Jensen, C.M. High-yield direct synthesis of LiAlH$_4$ from LiH and Al in the Presence of TiCl$_3$ and Me$_2$O. *J. Am. Chem. Soc.* **2011**. [CrossRef]

25. Zhou, C.; Fang, Z.Z.; Ren, C.; Li, J.; Lu, J. Effect of Ti intermetallic catalysts on hydrogen storage properties of magnesium hydride. *J. Phys. Chem.* **2013**, *117*, 12973–12980. [CrossRef]

26. Walker, G. *Solid-State Hydrogen Storage: Materials and Chemistry*; Elsevier: Amsterdam, The Netherlands, 2008.

27. Kojima, Y.; Kawai, Y.; Matsumoto, M.; Haga, T. Hydrogen release of catalyzed lithium aluminum hydride by a mechanochemical reaction. *J. Alloy Compd.* **2008**, *462*, 275–278. [CrossRef]

28. Sun, Y.; Shen, C.; Lai, Q.; Liu, W.; Wang, D.W.; Aguey-Zinsou, K.F. Tailoring magnesium based materials for hydrogen storage through synthesis: Current state of the art. *Energy Storage Mater.* **2017**. [CrossRef]

29. De Jongh, P.E.; Adelhelm, P. Nanosizing and nanoconfinement: New strategies towards meeting hydrogen storage goals. *ChemSusChem* **2010**, *3*, 1332–1348. [CrossRef] [PubMed]

30. Wan, X.; Shaw, L.L. Novel dehydrogenation properties derived from nanoscale LiBH$_4$. *Acta Mater.* **2011**, *59*, 4606–4615. [CrossRef]

31. Ghosh Chaudhuri, R.; Paria, S. Core/shell nanoparticles: Classes, properties, synthesis mechanisms, characterization, and applications. *Chem. Rev.* **2011**, *112*, 2373–2433. [CrossRef] [PubMed]

32. Cushing, B.L.; Kolesnichenko, V.L.; O'Connor, C.J. Recent advances in the liquid-phase syntheses of inorganic nanoparticles. *Chem. Rev.* **2004**, *104*, 3893–3946. [CrossRef] [PubMed]

33. Bensebaa, F. Chapter 2—Wet production methods. In *Interface Science and Technology*; Farid, B., Ed.; Elsevier: Amsterdam, The Netherlands, 2013; pp. 85–146.

34. Wang, L.S.; Hong, R.Y. Synthesis, surface modification and characterisation of nanoparticles. *Adv. Nanocompos.* **2011**, *18*, 7533–7548.

35. Kühnle, A.; Vollmer, S.; Linderoth, T.R.; Witte, G.; Besenbacher, F. Adsorption of dodecanethiol on Cu (110): Structural ordering upon thiolate formation. *Langmuir* **2002**, *18*, 5558–5565. [CrossRef]

36. Stein, S.E. *"Mass Spectra" in NIST Chemistry WebBook. In NIST Standard Reference Database Number 69*; NIST Mass Spec Data Center: Gaithersburg, MD, USA, 2016.

37. Chaudhuri, S.; Muckerman, J.T. First-principles study of Ti-catalyzed hydrogen chemisorption on an al surface: A critical first step for reversible hydrogen storage in NaAlH$_4$. *J. Phys. Chem.* **2005**, *109*, 6952–6957. [CrossRef] [PubMed]

38. Wipf, H.; Kappesser, B.; Werner, R. Hydrogen diffusion in titanium and zirconium hydrides. *J. Alloy Compd.* **2000**, *310*, 190–195. [CrossRef]

39. Livanov, V.A.; Bukhanova, A.A.; Kolachev, B. *Hydrogen in Titanium*; Israel Program for Scientific Translations: Jerusalem, Israel, 1965.

40. Gu, J.; Zhang, Y.-W.; Tao, F. Shape control of bimetallic nanocatalysts through well-designed colloidal chemistry approaches. *Chem. Soc. Rev.* **2012**, *41*, 8050–8065. [CrossRef] [PubMed]

41. Patnaik, P. *Handbook of Inorganic Chemicals*; McGraw-Hill: New York, NY, USA, 2003.

42. Rittmeyer, P.; Wietelmann, U. *Hydrides. Ullmann's Encyclopedia of Industrial Chemistry*; Wiley-VCH Verlag GmbH & Co: Berlin, Germany, 2000.

43. Liu, X.; Beattie, S.D.; Langmi, H.W.; McGrady, G.S.; Jensen, C.M. Ti-doped LiAlH$_4$ for hydrogen storage: Rehydrogenation process, reaction conditions and microstructure evolution during cycling. *Int. J. Hydrog. Energy* **2012**, *37*, 10215–10221. [CrossRef]

44. Resan, M.; Hampton, M.D.; Lomness, J.K.; Slattery, D.K. Effects of various catalysts on hydrogen release and uptake characteristics of LiAlH. *Int. J. Hydrog. Energy* **2005**, *30*, 1413–1416. [CrossRef]

45. Bhosle, V.; Baburaj, E.G.; Miranova, M.; Salama, K. Dehydrogenation of TiH2. *Mat. Sci. Eng. A Struct.* **2003**, *356*, 190–199. [CrossRef]

![inorganics logo] *inorganics*

MDPI

Article

Microstructure and Hydrogen Storage Properties of $Ti_1V_{0.9}Cr_{1.1}$ Alloy with Addition of x wt % Zr (x = 0, 2, 4, 8, and 12)

Salma Sleiman and Jacques Huot *

Hydrogen Research Institute, Université du Québec à Trois-Rivières, 3351 des Forges, Trois-Rivières, QC G9A 5H7, Canada; salma.sleiman@uqtr.ca
* Correspondence: jacques.huot@uqtr.ca; Tel.: +1-819-376-5011 (ext. 3576); Fax: +1-819-376-5164

Received: 4 October 2017; Accepted: 28 November 2017; Published: 3 December 2017

Abstract: The effect of adding Zr on microstructure and hydrogen storage properties of BCC $Ti_1V_{0.9}Cr_{1.1}$ synthesized by arc melting was studied. The microstructures of samples with Zr were multiphase with a main BCC phase and secondary Laves phases C15 and C14. The abundance of secondary phases increased with increasing amount of zirconium. We found that addition of Zr greatly enhanced the first hydrogenation kinetics. The addition of 4 wt % of Zr produced fast kinetics and high hydrogen storage capacity. Addition of higher amount of Zr had for effect of decreasing the hydrogen capacity. The reduction in hydrogen capacity might be due to the increased secondary phase abundance. The effect of air exposure was also studied. It was found that, for the sample with 12 wt % of Zr, exposure to the air resulted in appearance of an incubation time in the first hydrogenation and a slight reduction of hydrogen capacity.

Keywords: hydrogen storage; bcc alloys; kinetics; air exposure

1. Introduction

Hydrogen is considered a possible alternative energy carrier in the future. One of the most challenging barriers of using hydrogen is to establish a safe, reliable, compact, and cost-effective means of storing hydrogen [1–4]. Because of their high volumetric hydrogen capacity, metal hydrides could potentially replace the conventional storage in gaseous or liquid state. Moreover, metal hydrides are, in many respects, safe and cost effective [5,6]. Among the many different metal hydride systems, the Ti-based body-centred cubic (BCC) solid solution alloys are considered to be good candidates for hydrogen storage tanks because of their tunable pressure plateau and safety [7–10]. However, these alloys suffer from slow and difficult first hydrogenation, the so-called activation step [11]. Different approaches have been taken to overcome this drawback. These include heat treatment [12], addition of Zr_7Ni_{10} [12–16], or Zr [17,18] and by element substitution [19,20]. Another approach was made by Edalati K. et al. [21] who used high-pressure torsion to induce microstructural modification and so enhancing hydrogen storage properties of Ti–V BCC alloys. With respect to Zr_7Ni_{10} addition, the ease of activation may be due to the two-phase microstructure consisting of a main BCC phase and a secondary phase. Recently, Banerjee et al. [22] found that with increasing secondary phase concentration, faster kinetics was obtained but the hydrogen storage capacity was reduced. Moreover, Shashikala et al. [19] noticed that Zr is the element responsible for producing the secondary phase that ensures a rapid reaction with hydrogen and observed that hydrogen absorption capacity was decreased as the Zr content increased due to a greater fraction of Laves phase. In addition, Martinez and Santos found that $TiV_{0.9}Cr_{1.1}$ alloy with addition of 4 wt % of Zr_7Ni_{10} showed fast hydrogenation kinetics and high capacity (3.6 wt %) [14]. Lately, Bellon et al. showed that by substituting Zr for V in the alloy TiCrV resulted in a two-phase structure made up of a main BCC structure and a less abundant cubic-type Laves phase C15. They concluded that the Laves phase acts as a catalyst for the hydrogenation [17]. In this paper, we report our investigation on the effect of addition of Zr on the microstructure and hydrogen storage properties of $Ti_1V_{0.9}Cr_{1.1}$. Contrary to the work of Bellon et al., in the present investigation zirconium was not substituting for vanadium but instead added to the $TiV_{0.9}Cr_{1.1}$ composition. By varying the Zr addition from 2 to 12 wt % we aimed to see how the amount of zirconium affect the relative abundance of BCC and Laves phases as well as their composition and the effect on hydrogen storage.

2. Results and Discussion

2.1. Microstructure Analysis

Figure 1 shows the backscattered electron micrographs of $Ti_1V_{0.9}Cr_{1.1} + x$ wt % Zr alloys where $x = 0, 2, 4, 8,$ and 12. It is clear that the microstructure changes with x content. Pure $Ti_1V_{0.9}Cr_{1.1}$ (Figure 1a) shows a network of black areas structure. From EDX analysis these black areas were found to be titanium precipitates. Even with a slight doping of Zr ($x = 2$), a bright secondary phase appeared. All the modified alloys were found to be multi-phase: a matrix phase and a bright phase. Dendrites appeared from $x = 8$ and were clearly observed for $x = 12$. It is seen that more Zr leads to higher secondary phase abundance.

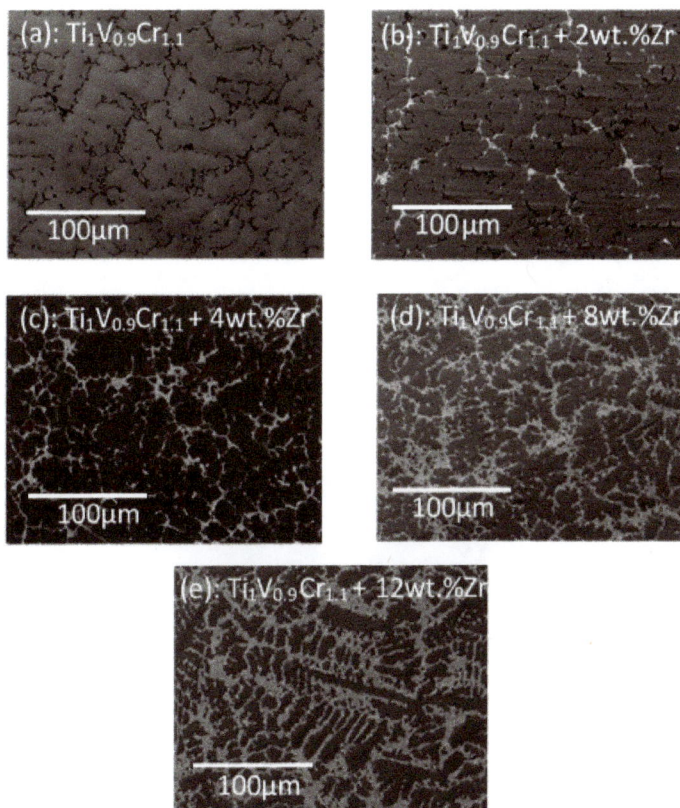

Figure 1. Backscattered electrons (BSE) micrographs of: $Ti_1V_{0.9}Cr_{1.1} + x$ wt % Zr with $x = 0$ (**a**); 2 (**b**); 4 (**c**); 8 (**d**); and 12 (**e**).

Using Image J software, the bright phase surface areas were measured. The percentage of the bright areas are 0, 4, 12, 26, and 35 respectively for $x = 0, 2, 4, 8,$ and 12. It shows that, as the amount of doped Zr increases, the bright phase surface area is getting bigger. To determine the chemical composition, EDX measurement was performed on all alloys. Table 1 shows the bulk measured atomic abundance compared to the nominal composition for all alloys studied. We see that the bulk measured composition agrees with the nominal one in all cases. This was expected as all these elements have similar melting points.

Table 1. Bulk atomic abundance: nominal and as measured by EDX of: $Ti_1 V_{0.9} Cr_{1.1} + x$ wt % Zr alloys for $x = 0, 2, 4, 8,$ and 12. Error on the last significant digit is indicated in parentheses.

Sample		Ti (at %)	V (at %)	Cr (at %)	Zr (at %)
$x = 0$	Nominal composition	33	30	37	0
	Measurement	34.0(2)	29.8 (3)	36.2 (3)	0.0
$x = 2$	Nominal composition	32.6	29.7	36.6	1.1
	Measurement	32.8 (2)	29.4 (2)	36.4 (3)	1.4 (1)
$x = 4$	Nominal composition	32.3	29.3	36.2	2.2
	Measurement	34.0 (3)	28.2 (1)	35.2 (3)	2.6 (1)
$x = 8$	Nominal composition	31.6	28.7	35.4	4.3
	Measurement	32.3 (2)	27.7 (2)	34.5(3)	5.5 (1)
$x = 12$	Nominal composition	30.9	28.1	34.7	6.3
	Measurement	31.7 (2)	26.2 (2)	33.9 (2)	8.2 (1)

Using EDX, we also measured the atomic composition of each phase for all alloys. In Figure 2a representative of elemental mapping is shown for the compound $Ti_1 V_{0.9} Cr_{1.1} + 4$ wt % Zr alloy. Similar features were seen for the other alloys.

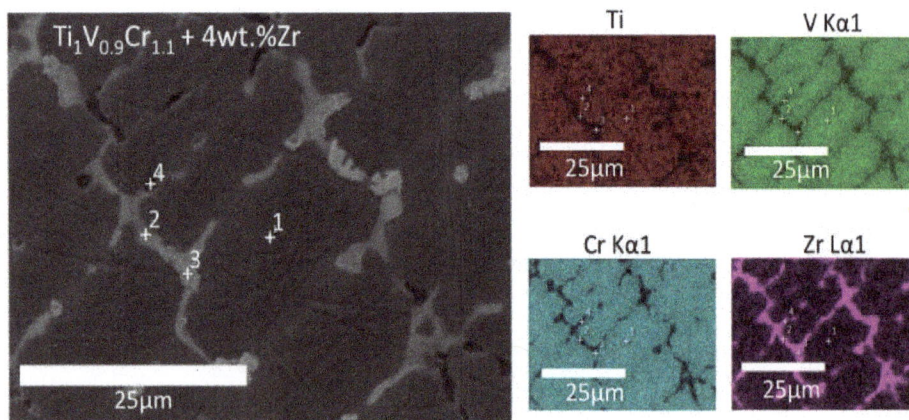

Figure 2. Backscattered electron micrograph of: $Ti_1 V_{0.9} Cr_{1.1} + 4$ wt % Zr alloy with elements mapping.

In these micrographs, specific locations are indicated by numbers. The matrix phase is numbered 1. Close inspection has shown that the bright secondary phase is in fact composed of two phases having different shades of white. Thus, two secondary phases were indexed. They are indicated by point 2 and point 3. Finally, the dark area is identified as point 4. The quantitative analysis at these specific locations are listed in the Tables 2–5. Table 2 presents the composition of the matrix (point 1) for all alloys.

Table 2. EDX analysis showing the elemental composition of the matrix (point 1) of: $Ti_1V_{0.9}Cr_{1.1} + x$ wt % Zr alloys for $x = 0, 2, 4, 8$, and 12.

Sample	Ti (at %)	V (at %)	Cr (at %)	Zr (at %)
$x = 0$	27.8	33.8	38.4	0.0
$x = 2$	27.4	33.4	38.9	0.3
$x = 4$	25.8	35.0	38.9	0.3
$x = 8$	29.3	31.6	38.2	0.9
$x = 12$	28.5	31.7	38.9	0.9

From Table 2, we see that the matrix is a Ti–V–Cr solid solution with inclusion of only a very small amount of Zr (<1 at %). The matrix composition does not significantly change when more zirconium is added. On average, the atomic composition of the matrix is $Ti_{0.84}V_1Cr_{1.2}Zr_{0.03}$ which is quite different than the nominal value. The main difference is that, compared to the nominal composition, the titanium proportion is reduced, the vanadium amount increased, and the chromium abundance stays almost the same. The zirconium abundance in the matrix is always less than 1 at % which is far from the nominal composition. As the amount of zirconium was very small in the matrix, we expect that zirconium will be mainly confined in the secondary phases. This could be seen in Tables 3 and 4 which show the atomic composition of the two secondary phases. Secondary phase one (point 2) is rich in Ti and Zr while secondary phase two (point 3) has a relatively high content of the four elements.

Table 3. EDX analysis showing the elemental composition of the secondary phase 1 (point 2) of $Ti_1V_{0.9}Cr_{1.1} + x$ wt % Zr alloys for $x = 0, 2, 4, 8$, and 12.

Sample	Ti (at %)	V (at %)	Cr (at %)	Zr (at %)
$x = 2$	64.0	7.0	6.2	22.8
$x = 4$	64.0	6.7	6.2	23.1
$x = 8$	58.3	6.8	4.6	30.3
$x = 12$	55.8	7.2	5.1	32.0

From Table 3, we see that the secondary phase 1 (point 2) is close to composition $Ti_{1.9}V_{0.2}Cr_{0.2}Zr_{0.7}$ for $x = 2$ or 4 and to composition $Ti_{1.7}V_{0.2}Cr_{0.2}Zr_{0.9}$ for $x = 8$ and 12. To our knowledge, there are no quaternary alloys that have stoichiometry close to these. Therefore, this secondary phase may be metastable. It has to be mentioned that, when x is higher than 8 there is a higher proportion of zirconium and lower proportion of titanium while the abundances of the two other elements are constant.

Table 4 shows the composition of secondary phase 2 (point 3). This phase has an almost constant composition $Ti_{1.0}V_{0.4}Cr_{1.1}Zr_{0.5}$ and also has a high zirconium content but here chromium is the most abundant element. Contrary to secondary phase 1, the relative atomic abundance of this phase does not vary with increasing of x.

Table 4. EDX analysis showing the elemental composition of secondary phase 2 (point 3) of: $Ti_1V_{0.9}Cr_{1.1}$ + x wt % Zr alloys for $x = 0, 2, 4, 8$, and 12.

Sample	Ti (at %)	V (at %)	Cr (at %)	Zr (at %)
$x = 2$	33.7	12.3	38.0	16.0
$x = 4$	32.8	12.3	37.9	17.0
$x = 8$	32.8	15.5	33.6	18.1
$x = 12$	28.5	14.8	36.5	20.2

Close inspection of Figure 2 also indicates that secondary phase 2 is usually on the edge of the bright surfaces while secondary phase 1 is more in the centre of the bright areas. Thus, it seems that secondary phase 2 is bridging secondary phase 1 with the matrix. Close inspection of Tables 3 and 4 demonstrates that the chemical composition of the secondary phases slightly changes when x increases. The main characteristics of these phases are however constant. When x increases, secondary phase 1 remains rich in titanium and zirconium with the proportion of titanium slightly decreasing and zirconium increasing. Regarding secondary phase 2, this phase is rich in chromium but also the proportions of zirconium and titanium relatively important. When x increase zirconium content increases, titanium abundance decreases and chromium only slightly decreases. Chemical composition of the black phase (point 4) was measured and the results are reported in Table 5.

Table 5. EDX analysis showing the elemental composition at point 4 of: $Ti_1V_{0.9}Cr_{1.1} + x$ wt % Zr alloys for $x = 0, 2,$ and 4.

Sample	Ti (at %)	V (at %)	Cr (at %)	Zr (at %)
$x = 0$	87.7	7.0	5.3	0.0
$x = 2$	88.2	3.3	2.0	6.5
$x = 4$	79.6	6.8	5.6	8.0

It is evident that the black areas are titanium precipitates. The amount of precipitation tends to decrease with increasing Zr content and for 8 and 12 wt % of Zr there are no Ti-precipitates.

2.2. Structural Characterization

Figure 3 shows the XRD pattern of all the studied samples in the as-cast state. The major diffraction peaks of all as cast alloys can be identified as BCC phase (space group *Im-3m*). Small peaks appeared as the zirconium doping increases. These peaks are surely related to the secondary phases. Due to the smallness of the peaks and their relatively small numbers, indexing them is not obvious. However, as demonstrated by Akiba and Iba, Laves phases could be closely related to BCC alloys [7]. Laves phases are well known intermetallic phases that crystallize in three different structures: a cubic $MgCu_2$-type (Space group Fd-3m), a hexagonal $MgZn_2$-type (space group $P6_3/mmc$), or a hexagonal $MgNi_2$-type (Space group $P6_3/mmc$). Following the *Strukturbericht* nomenclature, these phases will be thereafter respectively designated C15, C14, and C36. In the specific case of Ti–V–Cr BCC alloys, it was also shown by Bellon et al. [17] that when zirconium is substituted for vanadium in these alloys there is the appearance of a C15 Laves phase.

We analyzed our diffraction patterns by Rietveld's refinement. The obtained crystal parameters and the abundance of each phase in all samples are shown in Table 6. In these refinements, we tried to index the secondary phases' peaks with Laves phases. We found that the peaks could be indexed by using two Laves phases, namely the C15 and C14 phases. Tentatively, the C14 phase was assigned to the composition $Ti_{1.9}V_{0.2}Cr_{0.2}Zr_{0.7}$ (secondary phase 1) and the C15 phase to the composition $Ti_{1.0}V_{0.4}Cr_{1.1}Zr_{0.5}$ (secondary phase 2). However, as it is difficult to distinguish between Ti, V, and Cr by X-ray diffraction the exact assignment could be only done by using neutron diffraction. Also, the problem with these compositions is that the ratio A/B is not 2 as it should be. For these reasons, we should stress that the indexing of C14 and C15 is tentative and more experiments are needed in order to have the definitive crystal structure of the secondary phases. We are now undertaking neutron experiments and these will be reported in a future paper. In the present case, we could conclude that using two Laves phases for fitting the secondary phase peaks confirm the finding of electron microscopy where two different secondary phases were found. We should also point out that, on Figure 3 only the main peaks of C14 and C15 have been indexed. As these two phases have small abundances, only the main peaks are showing in the diffraction patterns and indexing all possible peaks will have confused the reader, especially for overlapping peaks that are so small that they are undistinguishable from the background. Referring to Table 6 at which the phase abundances are presented for all studied alloys, we see that the total abundance of C15 + C14 phases roughly matches the abundance measured from image J.

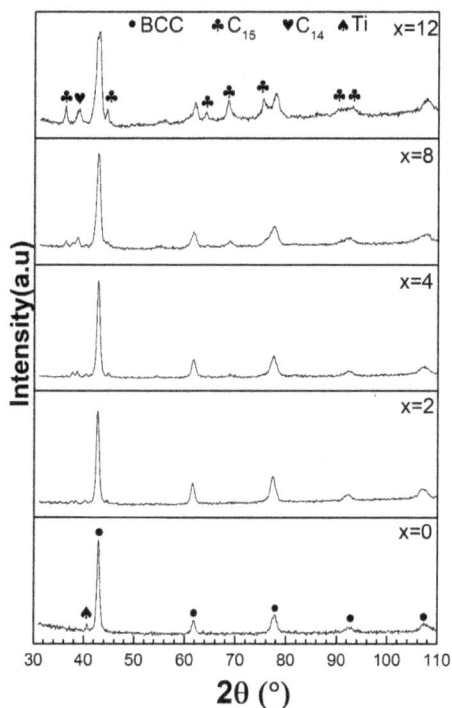

Figure 3. X-ray diffraction patterns of as cast: $Ti_1V_{0.9}Cr_{1.1} + x$ wt % Zr alloys with $x = 0, 2, 4, 8,$ and 12.

Table 6. Crystal parameters and abundance of each phase in as cast: $Ti_1V_{0.9}Cr_{1.1} + x$ wt % Zr alloys for $x = 0, 2, 4, 8,$ and 12. Error on the last significant digit is indicated in parentheses.

Sample	Phase	Lattice Parameter (Å)	Crystallite Size (nm)	Phase Abundance (%)	Bright Area Abundance (%)
$x = 0$	BCC Ti	3.0379 (9) $a = 2.961$ (4) $c = 4.773$ (1)	24 (2) 13 (2)	97 (3) 3 (3)	0
$x = 2$	BCC C15 C14	3.0479 (5) 7.172 (7) $a = 5.903$ (9) $c = 7.26$ (2)	22 (1) 14 (7) 8 (2)	92 (2) 2 (1) 6 (2)	4
$x = 4$	BCC C15 C14	3.0470 (7) 7.153 (5) $a = 5.935$ (1) $c = 8.09$ (4)	23 (2) 20 (8) 5 (1)	89 (3) 3 (1) 8 (3)	12
$x = 8$	BCC C15 C14	3.0452 (1) 7.180 (5) $a = 5.849$ (7) $c = 8.109$ (2)	38 (9) 6 (1) 6 (1)	68 (5) 21(4) 11 (2)	26
$x = 12$	BCC C15 C14	3.0310 (1) 6.099 (7) $a = 5.815$ (2) $c = 8.01$ (1)	8 (3) 10 (1) 3 (1)	66 (4) 22 (2) 12 (2)	35

2.3. Activation Process

The first hydrogenation (activation) of as-cast $Ti_1V_{0.9}Cr_{1.1} + x$ wt % Zr alloys was performed at room temperature under a hydrogen pressure 2MPa after the alloys were crushed in argon. Results are presented in Figure 4. The first hydrogenation for $x = 0$ is practically impossible. This sample did not absorb hydrogen even after 900 min of hydrogen exposure. Adding only 2 wt % of Zr to the alloy resulted in a good hydrogen uptake, but the full hydrogenation still takes about 10 h. Increasing x value to 4 wt % had the effect of greatly enhancing the kinetics and full hydrogenation was achieved after 150 min. Further increase of zirconium content slightly improved the kinetics but the total capacity decreased. Among all doped samples, the one with 12 wt % Zr has the fastest activation kinetics, reaching maximum capacity within 3 min. For hydrogen storage purposes, the optimum amount of zirconium seems to be 4 wt %. However, even if the capacity slightly decreases with addition of more zirconium, the fact is that even when the amount of secondary phases is high, the decrease in capacity is not very important. For example, in the alloy with 12 wt % of Zr the secondary phases comprise between 35 wt % (as measured from SEM images) and 34 wt % (as determined from Rietveld refinement). Taking the low estimation of 34 wt %, if these secondary phases do not absorb hydrogen, and assuming that pure BCC absorbs about 4 wt % of hydrogen, we should expect a capacity of about 2.7 wt % which is clearly not the case. The measured capacity could not be explained by the presence of the BCC alloy itself. Thus, we have evidence that, at least one of the secondary phases absorbs hydrogen. Nonetheless, this then raises the question of the kinetic curves. Except for the 2 wt % Zr alloy, all activation curves do not show any kinks or slope change. This is the signature of a single-phase alloy which is evidently not the case here. However, it has been shown by Akiba and Iba [7] that the BCC Laves phase related alloys could actually be formed by a BCC phase and a Laves phase and still display a single phase behaviour upon hydrogen absorption. It seems to be the case in the present system.

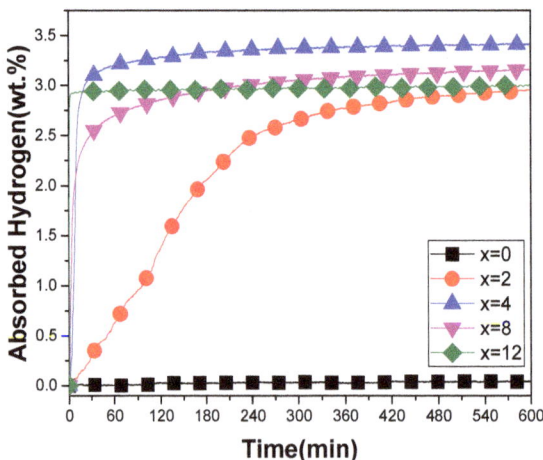

Figure 4. Activation curves of: $Ti_1V_{0.9}Cr_{1.1} + x$ wt % Zr for $x = 0, 2, 4, 8,$ and 12 at room temperature under 2MPa of hydrogen.

Interphase boundaries probably play a role in the activation. Similar investigation in other systems seems to point that way. Nanocrystallinity is also well known to enhance hydrogenation/dehydrogenation kinetics [6]. However, in the present case it is not the main factor as evidenced by Table 6. From this table, we see that the samples with $x = 2$ and $x = 4$ have almost the same crystallite size but their activation curve is drastically different. Moreover, the sample with $x = 8$ has a larger crystallite size but the hydrogenation is faster. Therefore, nanocrystallinity alone could not explain the behaviour of this system. Dislocations and defects could lead to a faster first hydrogenation in metal hydrides and was shown for cold rolled alloys [23,24].

Another explanation for the quick activation of alloys with zirconium addition may be stability of the FCC hydride. As the matrix of the Zr-containing alloys has a stoichiometry different than the nominal value and also, taking into account the fact that the BCC alloy contains some zirconium, it may well result in some change of the formation enthalpy of the FCC hydride phase. A more stable hydride may result in an easier activation but to have a definitive answer, the pressure-composition isotherms have to be registered. Such an investigation is planned for the near future.

2.4. XRD Patterns after Hydrogenation

In this set of experiments, in order to study the crystal structure of the hydride phase, the absorption experiment was stopped after reaching full hydrogenation and thereafter the pressure just lowered to one bar of hydrogen. No vacuum was applied to the sample to be sure that no desorption occurred during removal of the sample holder. In separated tests, we exposed the samples to a pressure of 10 kPa and no desorption was noticed.

Figure 5 shows X-ray diffraction patterns of all alloys after first exposure to hydrogen. We see that for $x = 0$, the structure is still BCC which confirms what was seen on Figure 4. For $x = 2$, the crystal structure is essentially FCC which is usually the crystal structure adopted by a fully hydrided BCC alloy. The FCC phase is also seen in the patterns for $x = 4, 8$, and 12 but there are clearly other phases present. Indexation of the supplementary peaks is helped by doing a Rietveld refinement. Even if, in a Rietveld refinement prior knowledge of the phase present is mandatory, we could use this technique to first fit the patterns by using the known phase such as FCC in the present case. The residue of the fit then shows the unindexed peaks more clearly and it is easier to figure out which crystal structure is possible for the remaining peaks. This is the procedure we performed for all patterns and by doing it we found that the supplementary peaks could be associated with a C14 and a C15 phases.

The crystal parameters and abundance of each phase in all hydrogenated samples as determined by Rietveld's analysis are presented in Table 7. For $x = 0$, the sample did not absorb hydrogen and the crystal structure is still BCC. There is practically no change in the lattice parameter and crystallite size, which means that there was no reaction with hydrogen. In the case of $x = 2$, the FCC phase abundance in the hydrogenated sample is less than the BCC abundance in the as-cast alloy (80% vs. 92%) while both C14 and C15 phases are more abundant in the hydrogenated sample. The reason for this behaviour is unclear but a premature conclusion should be avoided because the abundance values are associated with relatively high experimental errors. For $x = 4$, there is a presence of both BCC and FCC phases in the hydrogenated alloy. This is surprising, considering that the hydrogen capacity is slightly higher than for $x = 2$. It is well known that the monohydride of the BCC phase is a BCT (Body Centred Tetragonal) but, as the lattices parameters are very close to each other it is practically impossible to distinguish between the BCC and BCT phases. However, the lattice parameter of the BCC phase in the hydrogenated state is much bigger than in the as-cast state. From the difference in volume of the unit cells and assuming that, on average, a hydrogen atom occupies a volume of 2.5 Å^3, we could estimate that the BCC phase in the hydrogenated state has a ratio of H/M of about 0.56. When $x = 8$, we see that the hydrogenated alloy has 72 wt % of FCC phase which corresponds closely to the abundance of BCC phase in the as-cast alloy (68 wt %). Only the C15 phase is present in the hydrogenated sample and the abundance is almost the total of the C14 and C15 phases in the as-cast alloy. Thus, it seems that, upon hydrogenation, the C14 phase turns into a C15 phase. Finally, for $x = 12$, the FCC phase has an abundance of 56 wt % which is lower than the abundance of the BCC phase in the as-cast alloy (66 wt %). As for $x = 8$, there is no C14 phase in the hydrogenated pattern.

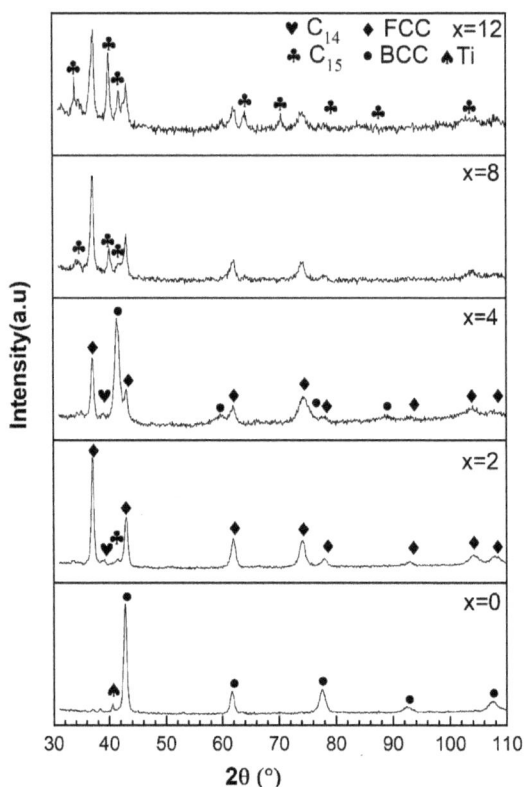

Figure 5. X-ray diffraction patterns of: $Ti_1V_{0.9}Cr_{1.1} + x$ wt % Zr alloys in the hydrogenated state for $x = 0, 2, 4, 8,$ and 12.

Table 7. Crystal parameters and abundance of each phase in hydrogenated: $Ti_1V_{0.9}Cr_{1.1} + x$ wt % Zr alloys for $x = 0, 2, 4, 8,$ and 12. Error on the last significant digit is indicated in parentheses.

Sample	Phase	Lattice Parameter (Å)	Crystallite Size (nm)	Phase Abundance (%)
$x = 0$	BCC Ti	3.0375 (4) $a = 2.9622$ (2) $c = 4.773$ (6)	20 (5) 11 (3)	93 (8) 7 (3)
$x = 2$	FCC C15 C14	4.2856 (6) 7.413 (1) $a = 6.141$ (7) $c = 7.262$ (2)	34 (3) 4 (2) 6 (3)	80 (2) 10 (2) 10 (1)
$x = 4$	BCC FCC C14	3.1443 (1) 4.2912 (2) $a = 5.809$ (2) $c = 8.03$ (4)	25 (5) 42 (2) 11 (3)	56 (4) 41 (4) 3 (1)
$x = 8$	FCC C15	4.2875(2) 7.605(7)	16 (2) 8 (3)	72 (6) 28 (3)
$x = 12$	FCC C15	4.289 (2) 7.646 (4)	12 (2) 12 (1)	56 (6) 44 (6)

For high addition of zirconium $x > 8$, the C14 phase was not detected by X-ray diffraction. This may be due to the fact that, upon hydrogenation, the crystallite size is reduced thus, broadening the peaks of this phase. As C14 phase is already present in relatively small amount (less than 12 wt %) this makes that phase practically undetectable by X-ray diffraction as the peaks are merged into the background.

2.5. Air Exposure Effect

For industrial production, it could be beneficial to be able to handle the cast alloys in air. This motivated us to investigate the air resistance of the alloy with 12 wt % of zirconium. Figure 6 shows the activation curves for this alloy crushed in air, in argon and after two days of air exposure. The sample crushed in air has a very short incubation time of 1.76 min while the sample after two days of air exposure presents a longer incubation time of 3.75 h. However, after incubation time the intrinsic kinetic is as fast for the sample exposed to the air two days as for the samples only crushed in air or in argon. The longer incubation time shown by the sample stored in air for two days is most probably due to the presence of surface oxide.

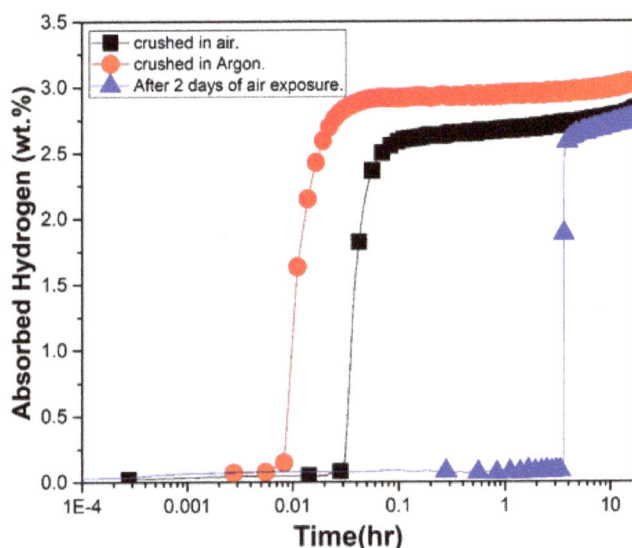

Figure 6. First hydrogenation curves of $Ti_1V_{0.9}Cr_{1.1}$ + 12 wt % Zr crushed under argon and under air and after two days of air exposure at room temperature and under 2MPa of hydrogen.

3. Materials and Methods

Synthesis of $Ti_1V_{0.9}Cr_{1.1}$ + x wt % Zr alloys were performed using an arc melting apparatus. The raw materials were purchased from Alfa-Aesar (Ward Hill, MA, USA) and had the following purity: Ti sponge (99.95%), V pieces (99%), Cr pieces (99%), and Zr sponge (99.95%). They were mixed at the desired stoichiometry and melted together under argon atmosphere. To ensure homogeneity, each pellet was melted, turned over, and remelted three times. Each pellet was then hand crushed using a steel mortar and pestle under argon atmosphere. Crushing in air was performed only for the air exposed sample. The material was filled into a sample holder and kept under vacuum for one hour at room temperature before exposing it to hydrogen. The hydrogen sorption properties were measured using a homemade Sievert's type apparatus. The measurements were done at room temperature under a hydrogen pressure of 20 MPa for absorption. The crystal structure was determined by X-ray powder diffraction on a Bruker D8 focus apparatus with Cu Kα radiation (Bruker AXS INC, Madison, WI, USA). Lattice parameters were evaluated by Rietveld's refinement of the X-ray patterns using Topas software [25]. Microstructure and chemical analysis were performed using a JEOL-JSM 5500 scanning electron microscopy (Ebay, Carson, CA, USA) equipped with an EDX (energy dispersive X-ray) apparatus from Oxford Instruments. The percentage of different phases was analyzed by image J software [26,27].

4. Conclusions

The first hydrogenation kinetics of $Ti_1V_{0.9}Cr_{1.1}$ + x wt % Zr (x = 2, 4, 8, and 12) alloys was found to be increased with the amount of Zr. Doping with Zr generated a multiphase microstructure made of a main BCC phase and secondary phases. The secondary phases abundance increased with Zr content. The two secondary phases were found to be Laves phases: a hexagonal C14 phase and a cubic C15 phase. Upon hydrogenation, most of the BCC phase transforms to the dihydride FCC phase but a small fraction of the BCC phase also seems to transform to a C15 phase with bigger lattice parameter, indicating that this phase also absorbed hydrogen. For high zirconium loading (x > 8), only C15 phase is present in the hydride pattern beside the FCC phase. Therefore, hydrogenation seems to promote the transformation of C14 and some fraction of BCC phase into C15 phase. The faster first hydrogenation of the alloys containing zirconium could be explained by the presence of the C14 and C15 secondary phases. This made the first hydrogenation at room temperature much faster. Addition of 4 wt % of Zr produced fast kinetics and the highest hydrogen storage capacity while higher abundance of zirconium (x > 4) leads to decrease in hydrogen storage capacity. We also found that air exposure has a relatively small impact on the activation of $Ti_1V_{0.9}Cr_{1.1}$ + 12 wt % Zr alloy.

Acknowledgments: This investigation was supported by a NSERC (Natural Sciences and Engineering Research Council of Canada) discovery grant which also covered the cost of publishing in open access. We would like to thank A. Lejeune for electron microscopy experiments.

Author Contributions: All experiments, except electron microscopy, were performed by Salma Sleiman under the supervision of Jacques Huot.

Conflicts of Interest: The authors declare no conflict of interest.

References

1. Jena, P. Materials for hydrogen storage: Past, present, and future. *J. Phys. Chem. Lett.* **2011**, *2*, 206–211. [CrossRef]
2. Schlapbach, L.; Züttel, A. Hydrogen-storage materials for mobile applications. *Nature* **2001**, *414*, 353–358. [CrossRef] [PubMed]
3. Züttel, A. Materials for hydrogen storage. *Mater. Today* **2003**, *6*, 24–33. [CrossRef]
4. Züttel, A. Hydrogen storage and distribution systems. *Mitig. Adapt. Strateg. Glob. Chang.* **2007**, *12*, 343–365. [CrossRef]
5. Gambini, M.; Manno, M.; Vellini, M. Numerical analysis and performance assessment of metal hydride-based hydrogen storage systems. *Int. J. Hydrog. Energy* **2008**, *33*, 6178–6187. [CrossRef]
6. Sakintuna, B.; Lamari-Darkrim, F.; Hirscher, M. Metal hydride materials for solid hydrogen storage: A review. *Int. J. Hydrog. Energy* **2007**, *32*, 1121–1140. [CrossRef]
7. Akiba, E.; Iba, H. Hydrogen absorption by laves phase related BCC solid solution. *Intermetallics* **1998**, *6*, 461–470. [CrossRef]
8. Kuriiwa, T.; Tamura, T.; Amemiya, T.; Fuda, T.; Kamegawa, A.; Takamura, H.; Okada, M. New V-based alloys with high protium absorption and desorption capacity. *J. Alloys Compd.* **1999**, *293*, 433–436. [CrossRef]
9. Okada, M.; Kuriiwa, T.; Tamura, T.; Takamura, H.; Kamegawa, A. Ti–V–Cr BCC Alloys with high protium content. *J. Alloys Compd.* **2002**, *330*, 511–516. [CrossRef]
10. Tamura, T.; Kazumi, T.; Kamegawa, A.; Takamura, H.; Okada, M. Protium absorption properties and protide formations of Ti–Cr–V alloys. *J. Alloys Compd.* **2003**, *356*, 505–509. [CrossRef]

11. Itoh, H.; Arashima, H.; Kubo, K.; Kabutomori, T. The influence of microstructure on hydrogen absorption properties of Ti–Cr–V alloys. *J. Alloys Compd.* **2002**, *330*, 287–291. [CrossRef]

12. Fruchart, D.; De Rango, P.; Charbonnier, J.; Miraglia, S.; Rivoirard, S.; Skryabina, N.; Jehan, M. Nanocrystalline Composite for Storage of Hydrogen. U.S. Patent No. 8,012,452, 6 September 2011.

13. Bibienne, T.; Bobet, J.-L.; Huot, J. Crystal structure and hydrogen storage properties of body centered cubic 52Ti–12V–36Cr alloy doped with Zr_7Ni_{10}. *J. Alloys Compd.* **2014**, *607*, 251–257. [CrossRef]

14. Martínez, A.; dos Santos, D.S. Hydrogen absorption/desorption properties in the TiCrV based alloys. *Mater. Res.* **2012**, *15*, 809–812. [CrossRef]

15. Miraglia, S.; de Rango, P.; Rivoirard, S.; Fruchart, D.; Charbonnier, J.; Skryabina, N. Hydrogen sorption properties of compounds based on BCC $Ti_{1-x}V_{1-y}Cr_{1+x+y}$ alloys. *J. Alloys Compd.* **2012**, *536*, 1–6. [CrossRef]

16. Skryabina, N.; Fruchart, D.; Medvedeva, N.A.; de Rango, P.; Mironova, A.A. Correlation between the hydrogen absorption properties and the vanadium concentration of Ti–V–Cr based alloys. *Solid State Phenom.* **2017**, *257*, 165–172. [CrossRef]

17. Bellon, D.; Martinez, A.; Barreneche, D.; dos Santos, D. A structural study of the hydrogen absorption properties by replacing vanadium with zirconium in metal alloys. *J. Phys. Conf. Ser.* **2016**, *687*, 012057. [CrossRef]

18. Kamble, A.; Sharma, P.; Huot, J. Effect of doping and particle size on hydrogen absorption properties of BCC solid solution 52Ti-12V-36Cr. *Int. J. Hydrog. Energy* **2017**, *42*, 11523–11527. [CrossRef]

19. Shashikala, K.; Banerjee, S.; Kumar, A.; Pai, M.; Pillai, C. Improvement of hydrogen storage properties of TiCrV alloy by Zr substitution for Ti. *Int. J. Hydrog. Energy* **2009**, *34*, 6684–6689. [CrossRef]

20. Yoo, J.-H.; Shim, G.; Yoon, J.-S.; Cho, S.-W. Effects of substituting Al for Cr in the $Ti_{0.32}Cr_{0.43}V_{0.25}$ alloy on its microstructure and hydrogen storage properties. *Int. J. Hydrog. Energy* **2009**, *34*, 1463–1467. [CrossRef]

21. Edalati, K.; Shao, H.; Emami, H.; Iwaoka, H.; Akiba, E.; Horita, Z. Activation of titanium–vanadium alloy for hydrogen storage by introduction of nanograins and edge dislocations using high-pressure torsion. *Int. J. Hydrog. Energy* **2016**, *41*, 8917–8924. [CrossRef]

22. Banerjee, S.; Kumar, A.; Ruz, P.; Sengupta, P. Influence of laves phase on microstructure and hydrogen storage properties of Ti–Cr–V based alloy. *Int. J. Hydrog. Energy* **2016**, *41*, 18130–18140. [CrossRef]

23. Huot, J.; Tousignant, M. Hydrogen sorption enhancement in cold-rolled and ball-milled $CaNi_5$. *J. Mater. Sci.* **2017**, *52*, 11911–11918. [CrossRef]

24. Tousignant, M.; Huot, J. Hydrogen sorption enhancement in cold rolled $LaNi_5$. *J. Alloys Compd.* **2014**, *595*, 22–27. [CrossRef]

25. Bruker, A.X.S. TOPAS V3: General profile and structure analysis software for powder diffraction data. In *User's Manual*; Bruker AXS: Karlsruhe, Germany, 2005.

26. Abramoff, M.D.; Magalhaes, P.J.; Ram, S.J. Image processing with imageJ. *Biophotonics Int.* **2004**, *11*, 36–42.

27. Collins, T.J. ImageJ for microscopy. *Biotechniques* **2007**, *43*, 25–30. [CrossRef] [PubMed]

inorganics

Article

Improvement in the Electrochemical Lithium Storage Performance of MgH$_2$

Shuo Yang [1], Hui Wang [1,2,*], Liuzhang Ouyang [1], Jiangwen Liu [2] and Min Zhu [1,2]

[1] School of Materials Science and Engineering and Key Laboratory of Advanced Energy Storage Materials of Guangdong Province, South China University of Technology, Guangzhou 510641, China; 201520115810@mail.scut.edu.cn (S.Y.); meouyang@scut.edu.cn (L.O.); memzhu@scut.edu.cn (M.Z.)

[2] China-Australia Joint Laboratory for Energy & Environmental Material, South China University of Technology, Guangdong 510641, China; mejwliu@scut.edu.cn

* Correspondence: mehwang@scut.edu.cn

Received: 15 October 2017; Accepted: 11 December 2017; Published: 26 December 2017

Abstract: Magnesium hydride (MgH$_2$) exhibits great potential for hydrogen and lithium storage. In this work, MgH$_2$-based composites with expanded graphite (EG) and TiO$_2$ were prepared by a plasma-assisted milling process to improve the electrochemical performance of MgH$_2$. The resulting MgH$_2$–TiO$_2$–EG composites showed a remarkable increase in the initial discharge capacity and cycling capacity compared with a pure MgH$_2$ electrode and MgH$_2$–EG composite electrodes with different preparation processes. A stable discharge capacity of 305.5 mAh·g^{-1} could be achieved after 100 cycles for the 20 h-milled MgH$_2$–TiO$_2$–EG-20 h composite electrode and the reversibility of the conversion reaction of MgH$_2$ could be greatly enhanced. This improvement in cyclic performance is attributed mainly to the composite microstructure by the specific plasma-assisted milling process, and the additives TiO$_2$ and graphite that could effectively ease the volume change during the de-/lithiation process as well as inhibit the particle agglomeration.

Keywords: hydrogen storage materials; MgH$_2$; anode material; electrochemical performance

1. Introduction

Magnesium hydride (MgH$_2$) has been intensively investigated as a hydrogen and heat energy storage medium [1–4] because of its high hydrogen storage capacity (7.6 wt % 110 kg/m^3), low cost, and abundance. MgH$_2$ is an intrinsically ionic compound. The strong Mg–H bond determines the high hydrogen desorption enthalpy (~74 kJ/mol H$_2$) and the high decomposition temperature (>350 °C) of MgH$_2$. Additionally, MgH$_2$ suffers from sluggish hydrogen sorption kinetics due to the slow hydrogen diffusion in the MgH$_2$ lattice and poor hydrogen dissociation on the surface of Mg [5–8]. To overcome the thermodynamic and kinetic problems of MgH$_2$, many effective methods including nanostructuring, alloying, catalyzing, and compositing have been developed by ball milling, film deposition, and chemical synthetic processes [9–14].

In 2008, MgH$_2$ was first reported as a potential anode material for lithium-ion batteries (LIBs) by Y. Oumellal et al. [15] since it exhibited rather high lithium storage capacity (MgH$_2$ + Li ↔ Mg + LiH, ~2038 mAh·g^{-1}), low work potential (~0.5 V versus Li$^+$/Li), and low voltage hysteresis (<0.2 V). The last point is especially superior over other kinds of conversion reaction materials, such as metal oxides, nitrides, fluorides etc. However, the large volume expansion (>85%) during the conversion reaction of magnesium hydride with lithium resulted in fast capacity fading, which is even more severe than other conversion anode materials as the metal hydride has low conductivity and high activity in the liquid electrolyte. Since then, many efforts have been devoted to improving the electrochemical performance of MgH$_2$ [16–21]. As reported by S. Brutti et al. [22], the ball-milled MgH$_2$ sample with the addition of Super P shows a relatively high discharge capacity (~1600 mAh·g^{-1}) and coulombic efficiency (~60%), which are mainly attributed to the reduction in crystallite size and enhancement of the electronic conductivity of MgH$_2$. Furthermore, the addition of metal oxides is also helpful in the conversion reaction of MgH$_2$ with lithium. For example, Kojima et al. [23,24] added the Nb$_2$O$_5$/Al$_2$O$_3$ into the active material MgH$_2$, thus enhancing the coulombic efficiency of all solid-state lithium-ion batteries. However, the improved effect in the coulombic efficiency, the reversibility of the conversion reaction, and the cycling performance of a MgH$_2$ anode is rather limited [25–27].

In this work, the expanded graphite (EG) and TiO$_2$ were milled with Mg/MgH$_2$ by dielectric barrier discharge plasma-assisted vibratory milling (P-milling), which is especially advantageous for the preparation of composites containing few-layered graphite [28]. The electrochemical performances of the obtained composites, MgH$_2$–EG and MgH$_2$–TiO$_2$–EG, were compared. Additionally, the new electrode preparation method was also developed to improve the performances of hydride anodes. It was demonstrated that the reversibility of the conversion reaction and the cyclic stability of a MgH$_2$ anode could be greatly enhanced.

2. Results and Discussion

The MgH$_2$–EG composite was prepared by the hydrogogenation treatment of the milled mixture (denoted as Mg–EG) of Mg powder and expandable graphite. Figure 1a shows the XRD patterns of the as-milled and hydrogenated Mg–EG composites. After milling for 10 h, the hexagonal Mg remains the major phase of the composite, and the weak MgO peak is due to the slight oxidation during sample transfer. It is stated that the weak MgF$_2$ diffractions indicate the reaction of Mg with the electrode bar composed of polytetrafluoroethylene, which was also reported in previous work [20]. After hydrogenation, all the Mg diffraction peaks disappear, indicating complete hydrogenation of the Mg powder. The rutile-type α-MgH$_2$ phase is indexed by strong and sharp diffraction peaks, implying grain growth of MgH$_2$ due to the hydrogenation treatment.

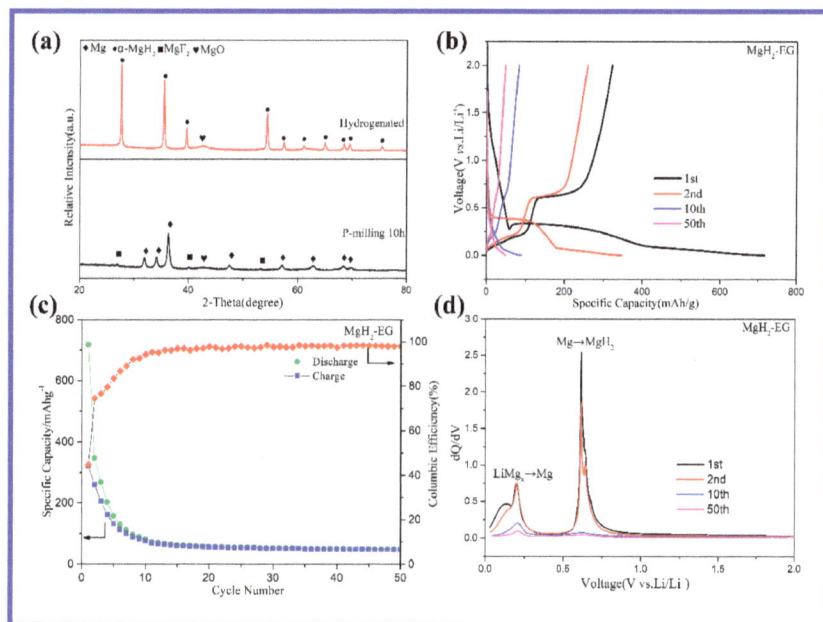

Figure 1. (**a**) XRD patterns of the as-milled and hydrogenated Mg–EG composites; (**b**) discharge/charge curves of MgH$_2$–EG at a current rate of 100 mA·g^{-1}; (**c**) cycling performance of MgH$_2$–EG electrode; (**d**) differential capacity plots (dQ/dV) of the MgH$_2$–EG electrode at different cycles.

Figure 1b shows the galvanostatic discharge/charge curves of the MgH_2–EG electrode at different cycles. In the first discharge profile, the potential drops rapidly from the initial open circuit potential (OCP) to 0.27 V, and then increases to 0.33 V, followed by a well-defined potential plateau assigning to the conversion reaction of MgH_2 with lithium. The slight polarization of the MgH_2–EG composite is owing to the kinetic limitation caused by the poor electronic conductivity of MgH_2 and the weak electronic contact between the active material and the nickel foam [29]. After that, the potential gradually drops further to 0.10 V with another plateau, which is attributed to the alloying of Mg with Li. Upon charging, two potential plateaus ranging from 0.10 V to 0.21 V and from 0.21 V to 0.60 V are assigned to the de-alloying reaction and the reverse conversion reaction of Mg/LiH, respectively. The first discharge capacity, amounting to 717.4 mAh·g^{-1}, is much less than its theoretical capacity (1704.8 mAh·g^{-1} = 2038 mAh·g^{-1} × 80 wt % (MgH_2) + 372 mAh·g^{-1} × 20 wt % (graphite)). This result reflects the loss of active Mg during milling, as well as the kinetic limitation of coarsening the MgH_2 electrode. In addition, in the initial charge process, the MgH_2–EG electrode shows a total charge capacity of 320.6 mAh·g^{-1}, corresponding to an initial coulombic efficiency (ICE) of 44.7%. This low ICE value also indicates the incomplete reversible formation of MgH_2 in the delithiation process.

The cycling performance of MgH_2–EG electrode is shown in Figure 1c. The rapid capacity fading in the initial several cycles may be attributed to the pulverization of active material leading to the loss of electronic contact between the active material and the nickel foam. After 50 cycles, a capacity of only 48.1 mAh·g^{-1} is maintained in the cell; this value is even less than the capacity (~70 mAh·g^{-1}) contributed by the graphite component. In addition, according to the discharge profile at the 10th cycle (Figure 1b), the plateau corresponding to the conversion reaction of MgH_2 with lithium is invisible. Further, the differential capacity plots (dQ/dV) of different cycles are compared in Figure 1d. The peak centered at 0.61 V, which is assigned to the reverse conversion reaction of Mg with LiH, disappears after 10 cycles. This result further confirms the poor conversion reversibility and cycling performance of the MgH_2–EG electrode.

To improve the electrochemical performances, especially the cycling stability of a MgH_2 electrode, TiO_2 was added to the MgH_2–EG composite to accommodate the large volume variation. In addition, Mg was replaced by MgH_2 as the starting milling material in order to avoid grain growth during the hydrogenation treatment. The XRD patterns of as-milled MgH_2–TiO_2–EG composite with different milling times are shown in Figure 2a. The MgH_2 peaks for the 20 h-milled composite (denoted as MgH_2–TiO_2–EG-20 h) show a relative broadening effect compared to that of the 10 h-milled composite (denoted as MgH_2–TiO_2–EG-10 h), implying a finer grain size of the MgH_2 by longer milling time. The SEM observation shown in Figure 2c,d also displays smaller particle size (~10 μm) for the MgH_2–TiO_2–EG-20 h composite. Actually, the composite particles consist of nanosized primary particles according to the magnified SEM images (Figure 2c,d). With regard to the graphite after P-milling for 20 h, the graphite peak around 26.6° disappears in Figure 2a, implying the formation of a disordered structure of the graphite. It is believed that the graphite could be effectively exfoliated to few-layer graphene (FLG) nanosheets due to the synergic effect of the plasma heating and the impact stress from the milling balls [28,30,31]. Additionally, the hydrogen desorption kinetic curves (Figure 2b) show that both samples could release ~3.4 wt % H_2 within 15 min, corresponding to the actual MgH_2 content of ~44.3 wt % in the MgH_2–TiO_2–EG composite.

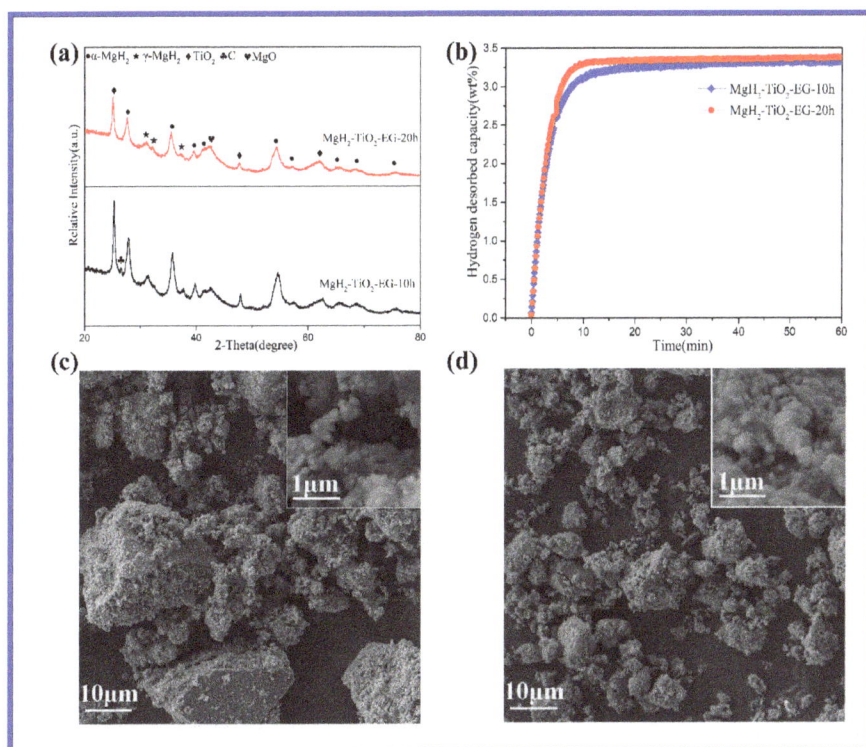

Figure 2. (a) XRD patterns of as-milled MgH$_2$–TiO$_2$–EG-10 h and MgH$_2$–TiO$_2$–EG-20 h composite; (b) desorption kinetic plots of MgH$_2$–TiO$_2$–EG-10 h and MgH$_2$–TiO$_2$–EG-20 h composite measured at 400 °C; and typical SEM images of (c) MgH$_2$–TiO$_2$–EG-10 h composite and (d) MgH$_2$–TiO$_2$–EG-20 h composite.

The galvanostatic charge/discharge curves of the MgH$_2$–TiO$_2$–EG composite with different milling times are compared in Figure 3a,b. In the first discharge profile, the MgH$_2$–TiO$_2$–EG-10 h composite electrode delivers a total discharge capacity of ~1224.6 mAh·g^{-1}, which is very close to its theoretical capacity (1193.9 mAh·g^{-1} = 2038 mAh·g^{-1} × 50 wt % (MgH$_2$) + 335 mAh·g^{-1} × 30 wt % (TiO$_2$) + 372 mAh·g^{-1} × 20 wt % (graphite)), and this result is also much higher than that of the MgH$_2$–EG electrode mentioned above. The ICE for the MgH$_2$–TiO$_2$–EG-10 h composite electrode is 46.4%, which is a little higher than that for the MgH$_2$–EG electrode. As seen in Figure 3c, the MgH$_2$–TiO$_2$–EG-10 h composite electrode exhibits a discharge capacity of 179.1 mAh·g^{-1} at the 100th cycle, with a capacity retention of 33%. Compared with the MgH$_2$–TiO$_2$–EG-10 h electrode, the MgH$_2$–TiO$_2$–EG-20 h electrode shows a similar initial discharge capacity (~1218.6 mAh·g^{-1}) and ICE (48.1%) but possessing a much higher cycling capacity of 305.4 mAh·g^{-1} after 100 cycles and a capacity retention of ~31%. As also shown in Figure 3c, while the MgH$_2$–TiO$_2$–EG-10 h composite electrode experiences rapid capacity fading within the first several cycles, the MgH$_2$–TiO$_2$–EG-20 h electrode delivers more stable capacity within 10 cycles, and it also shows higher coulombic efficiency throughout the cycling. Additionally, as shown in Figure 3d, the distinct anodic peak in the differential capacity plots (dQ/dV) at the 100th cycle clearly demonstrates the reversible formation of MgH$_2$, which indicates that the enhanced cyclic stability of the MgH$_2$–TiO$_2$–EG-20 h composite electrode is due to the enhanced conversion reaction reversibility of MgH$_2$.

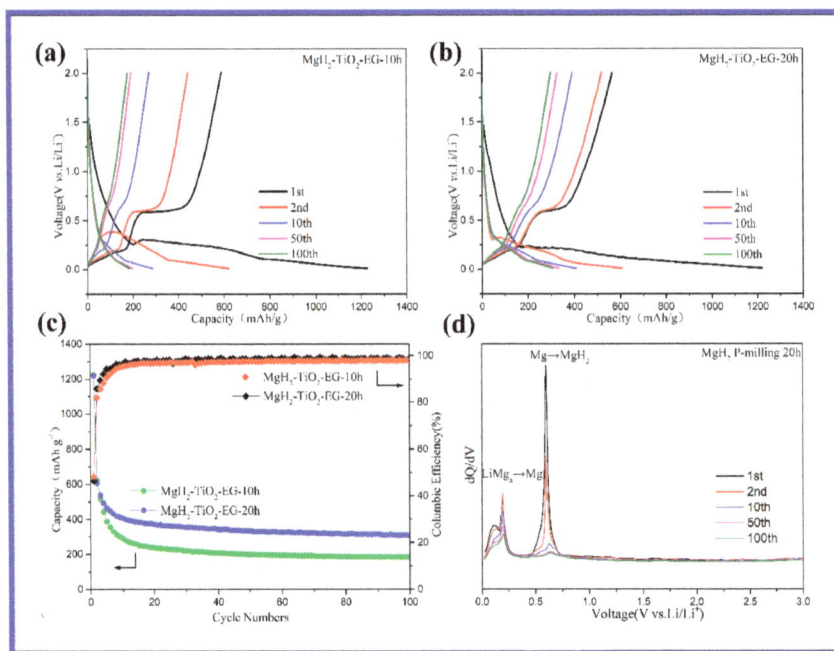

Figure 3. Discharge/charge curves of (**a**) MgH$_2$–TiO$_2$–EG-10 h and (**b**) MgH$_2$–TiO$_2$–EG-20 h electrode at a current rate of 100 mA·g^{-1}; (**c**) cycling performance of MgH$_2$–TiO$_2$–EG electrodes; (**d**) differential capacity plots (dQ/dV) of MgH$_2$–TiO$_2$–EG-20 h electrode at different cycles.

XRD analysis was performed to the change of phase structure of the MgH$_2$–TiO$_2$–EG-20 h electrode after cycling, but the result (not shown here) shows no diffractions and implies poor crystallinity of the active materials. SEM observation was also carried out to investigate the microstructural evolution of the composite electrodes, and the results are shown in Figure 4. Before cycling, the electrode surface of both the MgH$_2$–TiO$_2$–EG-10 h and the MgH$_2$–TiO$_2$–EG-20 h electrodes are composed of irregular particles with sizes less than 20 μm (Figure 4a,c), and there is no remarkable morphological difference between them. After 100 cycles, the surface morphology of both electrodes experiences obvious particle coarsening, which is due to the lithiation and delithiation of the active material, which causes repeated powder pulverization and agglomeration. This result also explains the capacity loss of the electrodes during cycling. Further, it is also shown that the particle coarsening effect for the MgH$_2$–TiO$_2$–EG-10 h electrode is more serious than for the MgH$_2$–TiO$_2$–EG-20 h electrode and the large voids between coarse particles are clearly observed. This microstructural difference indicates that large-volume changes are better accommodated by the TiO$_2$ and graphite additives with finer microstructure and by longer P-milling time, which help to maintain the structural integrity of the electrode.

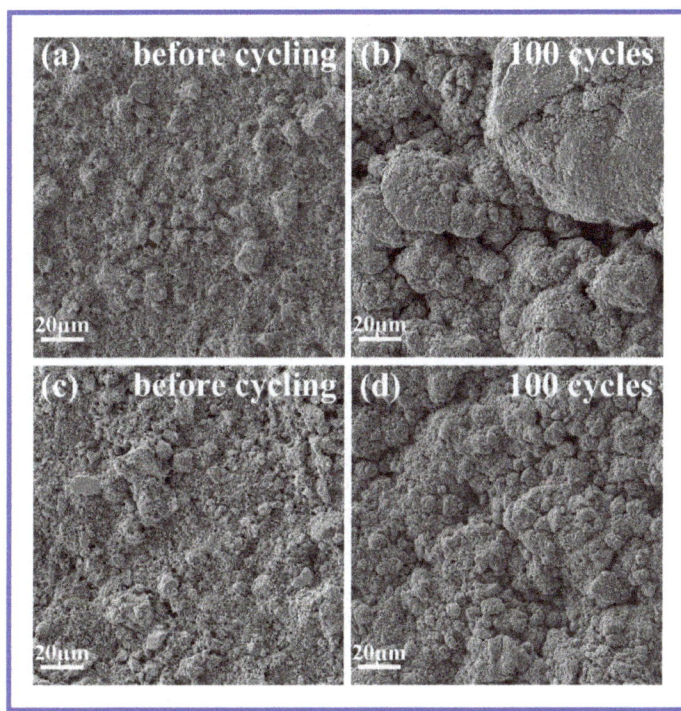

Figure 4. SEM surface morphological evolution of (**a**,**b**) the MgH$_2$–TiO$_2$–EG-10 h electrode and (**c**,**d**) the MgH$_2$–TiO$_2$–EG-20 h electrode.

3. Experimental

3.1. Materials Preparation

To synthesize the MgH$_2$–EG composite, 1.48 g Mg powder (99.9% purity, ~50 μm) and 0.4 g expandable graphite (EG) with a mass ratio of 80:20 were handled in a steel vial. The expandable graphite (99.9% purity, 100 mesh) was preheated at 1000 °C and held for 90 s under air atmosphere to obtain the worm-like expandable graphite. The handling process was operated in an argon-filled glovebox with an O$_2$ and H$_2$O content of less than 1 ppm to minimize the contamination. The milling was carried out on a dielectric barrier discharge plasma-assisted vibratory miller with the ball to powder weight ratio of 50:1; the details of plasma-assisted milling (P-milling) have been described in previous work [30,31]. After ball milling for 10 h, the as-prepared Mg–EG sample was hydrogenated at 450 °C under 6 MPa H$_2$ for 6 h.

To synthesize the MgH$_2$–TiO$_2$–EG composite, 2 g mixture of MgH$_2$ powder (hydrogen-storage grade), TiO$_2$ powder (99.0% purity, ≥325 mesh), and the worm-like EG with a weight ratio of 5:3:2 were handled in a steel vial and milled with the same parameters for 10 h and 20 h and denoted as MgH$_2$–TiO$_2$–EG-10 h and MgH$_2$–TiO$_2$–EG-20 h, respectively.

3.2. Material Characterization

X-ray diffraction (XRD, Empyrean diffractometer, PANAlytical Inc., Almelo, The Netherland) with Cu Kα radiation was used to characterize the phase structure of the samples. The microstructure was observed by using a scanning electron microscope (SEM, Carl Zeiss Supra 40, Oberkochen, Germany).

To determine the hydrogen content of the as-prepared MgH_2–TiO_2–EG sample, the desorption kinetics were measured at 400 °C using a Sievert-type automatic apparatus.

3.3. Electrochemical Measurement

The electrochemical properties of the active materials were measured using coin-type half-cells (CR2016) assembled in an Ar-filled glovebox. For preparation of the MgH_2–EG electrode, the active material was cold pressed directly on the nickel foam with the pressure of 20 MPa. For preparation of the MgH_2–TiO_2–EG electrode, the active material was first mixed with the conductive agent (Super-P) and the binder (polyvinylidene fluoride (PVdF)) in a mass ratio of 8:1:1 and then dissolved in solvent (*N*-methyl-2-pyrrolidinone (NMP)) to make a slurry with the appropriate viscosity. The slurry was then manually spread onto a Cu foil in the glovebox filled with Ar and dried in a vacuum oven at 80 °C for 12 h. The loading of the active material was ~1.0 mg·cm^{-2}. The cell used Li foil as the counter and reference electrode and a Celgrad 2400 membrane as the separator. The electrolyte was 1 M $LiPF_6$ in ethylene carbonate and diethyl carbonate (1:1 by volume) with 10 wt % fluoroethylene carbonate (FEC).

The galvanostatic charge/discharge tests were performed in a voltage range of 0.01 V to 2.0 V (vs. Li/Li$^+$) at the current density of 100 mA·g^{-1} using a Land test system (Wuhan, China) at a constant temperature (30 °C).

4. Conclusions

In summary, the electrochemical lithium storage properties of MgH_2 were greatly improved by compositing with graphite and TiO_2 via the discharge plasma milling process. The resulting MgH_2–TiO_2–EG composites show a remarkable increase in the initial discharge capacity and cycling capacity compared to pure MgH_2 and MgH_2–EG composite electrodes with different preparation processes. The 20 h-milled MgH_2–TiO_2–EG-20 h composite delivered a stable discharge capacity of 305.5 mAh·g^{-1} even after 100 cycles, and the reversible conversion reaction of MgH_2 has been greatly enhanced. This work demonstrates the potential of the MgH_2–TiO_2 graphite composite by plasma-milled milling for electrochemical applications. The next goal is to obtain a higher cyclic capacity by suppressing the fast capacity fading within the initial discharge/charge cycle, and further elevate the reversible conversion reaction of MgH_2. It is also stated that the possible hydrogen release from hydride materials during discharge/charging should be avoided and given more attention.

Acknowledgments: We acknowledge financial support from the National Natural Science Foundation of China (Grant Nos. 51471070, U1601212), the Fund for Innovative Research Groups of the National Natural Science Foundation of China (Grant No. 51621001), and the Natural Science Foundation of Guangdong Province (2016A030312011).

Author Contributions: Shuo Yang: materials preparation, electrode preparation and electrochemical tests. Hui Wang: data analysis and writing of paper. Jiangwen Liu, Liuzhang Ouyang, Min Zhu: discussion on the research plan and experimental results.

Conflicts of Interest: The authors declare no conflict of interest.

References

1. Aguey-Zinsou, K.F.; Ares-Fernandez, J.R. Hydrogen in magnesium: New perspectives toward functional stores. *Energy Environ. Sci.* **2010**, *3*, 526–543. [CrossRef]
2. Wang, H.; Lin, H.J.; Cai, W.T.; Ouyang, L.Z.; Zhu, M. Tuning kinetics and thermodynamics of hydrogen storage in light metal element based systems—A review of recent progress. *J. Alloys Compd.* **2016**, *658*, 280–300. [CrossRef]
3. Mohtadi, R.; Orimo, S.I. The renaissance of hydrides as energy materials. *Nat. Rev. Mater.* **2017**, *2*, 16091–16106. [CrossRef]
4. Muthukumar, P.; Groll, M. Metal hydride based heating and cooling systems: A review. *Int. J. Hydrogen Energy* **2010**, *35*, 8816–8829. [CrossRef]

5.   Lai, Q.; Paskevicius, M.; Sheppard, D.A.; Buckley, C.E.; Thornton, A.W.; Hill, M.R.; Gu, Q.; Mao, J.; Huang, Z.; Liu, H.K.; et al. Hydrogen Storage Materials for Mobile and Stationary Applications: Current State of the Art. *ChemSusChem* **2015**, *8*, 2789–2825. [CrossRef] [PubMed]

6.   Møller, K.T.; Jensen, T.R.; Akiba, E.; Li, H.-W. Hydrogen—A sustainable energy carrier. *Prog. Natl. Sci. Mater. Int.* **2017**, *27*, 34–40. [CrossRef]

7.   Crivello, J.C.; Dam, B.; Denys, R.V.; Dornheim, M.; Grant, D.M.; Huot, J.; Jensen, T.R.; de Jongh, P.; Latroche, M.; Milanese, C.; et al. Review of magnesium hydride-based materials: development and optimization. *Appl. Phys. A* **2016**, *122*, 97. [CrossRef]

8.   He, T.; Pachfule, P.; Wu, H.; Xu, Q.; Chen, P. Hydrogen carriers. *Nat. Rev. Mater.* **2016**, *1*, 16059. [CrossRef]

9.   Wang, Y.; Wang, Y.J. Recent advances in additive-enhanced magnesium hydride for hydrogen storage. *Prog. Natl. Sci. Mater. Int.* **2017**, *27*, 41–49. [CrossRef]

10.   Yu, X.; Tang, Z.; Sun, D.; Ouyang, L.; Zhu, M. Recent advances and remaining challenges of nanostructured materials for hydrogen storage applications. *Prog. Mater. Sci.* **2017**, *88*, 1–48. [CrossRef]

11.   Sadhasivam, T.; Kim, H.-T.; Jung, S.; Roh, S.-H.; Park, J.-H.; Jung, H.-Y. Dimensional effects of nanostructured Mg/MgH_2 for hydrogen storage applications: A review. *Renew. Sustain. Energy Rev.* **2017**, *72*, 523–534. [CrossRef]

12.   Crivello, J.C.; Denys, R.V.; Dornheim, M.; Felderhoff, M.; Grant, D.M.; Huot, J.; Jensen, T.R.; de Jongh, P.; Latroche, M.; Walker, G.S.; et al. Mg-based compounds for hydrogen and energy storage. *Appl. Phys. A* **2016**, *122*, 85. [CrossRef]

13.   Jia, Y.; Sun, C.; Shen, S.; Zou, J.; Mao, S.S.; Yao, X. Combination of nanosizing and interfacial effect: Future perspective for designing Mg-based nanomaterials for hydrogen storage. *Renew. Sustain. Energy Rev.* **2015**, *44*, 289–303. [CrossRef]

14.   Shao, H.; Xin, G.; Zheng, J.; Li, X.; Akiba, E. Nanotechnology in Mg-based materials for hydrogen storage. *Nano Energy* **2012**, *1*, 590–601. [CrossRef]

15.   Oumellal, Y.; Rougier, A.; Nazri, G.A.; Tarascon, J.M.; Aymard, L. Metal hydrides for lithium-ion batteries. *Nat. Mater.* **2008**, *7*, 916–921. [CrossRef] [PubMed]

16.   Zaidi, W.; Oumellal, Y.; Bonnet, J.P.; Zhang, J.; Cuevas, F.; Latroche, M.; Bobet, J.L.; Aymard, L. Carboxymethylcellulose and carboxymethylcellulose-formate as binders in MgH_2–carbon composites negative electrode for lithium-ion batteries. *J. Power Sources* **2011**, *196*, 2854–2857. [CrossRef]

17.   Oumellal, Y.; Zlotea, C.; Bastide, S.; Cachet-Vivier, C.; Leonel, E.; Sengmany, S.; Leroy, E.; Aymard, L.; Bonnet, J.P.; Latroche, M. Bottom-up preparation of MgH_2 nanoparticles with enhanced cycle life stability during electrochemical conversion in Li-ion batteries. *Nanoscale* **2014**, *6*, 14459–14466. [CrossRef] [PubMed]

18.   Huang, L.; Aymard, L.; Bonnet, J.P. MgH_2–TiH_2 mixture as an anode for lithium-ion batteries: Synergic enhancement of the conversion electrode electrochemical performance. *J. Mater. Chem. A* **2015**, *3*, 15091–15096. [CrossRef]

19.   Meggiolaro, D.; Gigli, G.; Paolone, A.; Reale, P.; Doublet, M.L.; Brutti, S. Origin of the Voltage Hysteresis of MgH_2 Electrodes in Lithium Batteries. *J. Phys. Chem. C* **2015**, *119*, 17044–17052. [CrossRef]

20.   Li, D.X.; Zhang, T.R.; Yang, S.Q.; Tao, Z.L.; Chen, J. Ab initio investigation of structures, electronic and thermodynamic properties for Li–Mg–H ternary system. *J. Alloys Compd.* **2011**, *509*, 8228–8234. [CrossRef]

21.   Meggiolaro, D.; Gigli, G.; Paolone, A.; Vitucci, F.; Brutti, S. Incorporation of Lithium by MgH_2: An Ab Initio Study. *J. Phys. Chem. C* **2013**, *117*, 22467–22477. [CrossRef]

22.   Brutti, S.; Mulas, G.; Piciollo, E.; Panero, S.; Reale, P. Magnesium hydride as a high capacity negative electrode for lithium ion batteries. *J. Mater. Chem.* **2012**, *22*, 14531–14537. [CrossRef]

23.   Ikeda, S.; Ichikawa, T.; Kawahito, K.; Hirabayashi, K.; Miyaoka, H.; Kojima, Y. Anode properties of magnesium hydride catalyzed with niobium oxide for an all solid-state lithium-ion battery. *Chem. Commun.* **2013**, *49*, 7174–7176. [CrossRef] [PubMed]

24.   Ikeda, S.; Ichikawa, T.; Goshome, K.; Yamaguchi, S.; Miyaoka, H.; Kojima, Y. Anode properties of Al_2O_3-added MgH_2 for all-solid-state lithium-ion batteries. *J. Solid State Electrochem.* **2015**, *19*, 3639–3644. [CrossRef]

25.   Brutti, S.; Meggiolaro, D.; Paolone, A.; Reale, P. Magnesium hydride as negative electrode active material in lithium cells: A review. *Mater. Today Energy* **2017**, *3*, 53–59. [CrossRef]

26.   Aymard, L.; Oumellal, Y.; Bonnet, J.P. Metal hydrides: An innovative and challenging conversion reaction anode for lithium-ion batteries. *Beilstein J. Nanotechnol.* **2015**, *6*, 1821–1839. [CrossRef] [PubMed]

27. Sartori, S.; Cuevas, F.; Latroche, M. Metal hydrides used as negative electrode materials for Li-ion batteries. *Appl. Phys. Mater.* **2016**, *122*, 135. [CrossRef]

28. Liu, H.; Hu, R.Z.; Zeng, M.Q.; Liu, J.W.; Zhu, M. Enhancing the performance of Sn–C nanocomposite as lithium ion anode by discharge plasma assisted milling. *J. Mater. Chem.* **2012**, *22*, 8022–8028. [CrossRef]

29. Wang, M.; Li, X.; Gao, M.; Pan, H.; Liu, Y. A Novel synthesis of MgS and its application as electrode material for lithium-ion batteries. *J. Alloys Compd.* **2014**, *603*, 158–166. [CrossRef]

30. Wang, Y.K.; Yang, L.C.; Hu, R.Z.; Sun, W.; Liu, J.W.; Ouyang, L.Z.; Yuan, B.; Wang, H.H.; Zhu, M. A stable and high-capacity anode for lithium-ion battery: Fe$_2$O$_3$ wrapped by few layered graphene. *J. Power Sources* **2015**, *288*, 314–319. [CrossRef]

31. Sun, W.; Hu, R.Z.; Liu, H.; Zeng, M.Q.; Yang, L.C.; Wang, H.H.; Zhu, M. Embedding nano-silicon in graphene nanosheets by plasma assisted milling for high capacity anode materials in lithium ion batteries. *J. Power Sources* **2014**, *268*, 610–618. [CrossRef]

![inorganics logo] *inorganics*

MDPI

Article

Investigation of the Reversible Lithiation of an Oxide Free Aluminum Anode by a LiBH$_4$ Solid State Electrolyte

Jason A. Weeks [1], Spencer C. Tinkey [1], Patrick A. Ward [1], Robert Lascola [2], Ragaiy Zidan [1,*] and Joseph A. Teprovich Jr. [3,*]

[1] Secure Energy Manufacturing, Savannah River National Laboratory, Aiken, SC 29808, USA; Jason.Weeks@srnl.doe.gov (J.A.W.); Spencer.Tinkey@srnl.doe.gov (S.C.T.); patrick.ward@srs.gov (P.A.W.)
[2] Analytical Development, Savannah River National Laboratory, Aiken, SC 29808, USA; robert.lascola@srnl.doe.gov
[3] Department of Chemistry and Biochemistry, College of Science and Math, California State University, Northridge, 18111 Nordhoff St., Northridge, CA 91334, USA
* Correspondence: ragaiy.zidan@srnl.doe.gov (R.Z.); joseph.teprovich@csun.edu (J.A.T.J.); Tel.: +01-818-677-4239 (J.A.T.J.)

Received: 19 September 2017; Accepted: 21 November 2017; Published: 23 November 2017

Abstract: In this study, we analyze and compare the physical and electrochemical properties of an all solid-state cell utilizing LiBH$_4$ as the electrolyte and aluminum as the active anode material. The system was characterized by galvanostatic lithiation/delithiation, cyclic voltammetry (CV), X-ray diffraction (XRD), energy dispersive X-ray spectroscopy (EDS), Raman spectroscopy, electrochemical impedance spectroscopy (EIS), and scanning electron microscopy (SEM). Constant current cycling demonstrated that the aluminum anode can be reversibly lithiated over multiple cycles utilizing a solid-state electrolyte. An initial capacity of 895 mAh/g was observed and is close to the theoretical capacity of aluminum. Cyclic voltammetry of the cell was consistent with the constant current cycling data and showed that the reversible lithiation/delithiation of aluminum occurs at 0.32 V and 0.38 V (vs. Li$^+$/Li) respectively. XRD of the aluminum anode in the initial and lithiated state clearly showed the formation of a LiAl (1:1) alloy. SEM-EDS was utilized to examine the morphological changes that occur within the electrode during cycling. This work is the first example of reversible lithiation of aluminum in a solid-state cell and further emphasizes the robust nature of the LiBH$_4$ electrolyte. This demonstrates the possibility of utilizing other high capacity anode materials with a LiBH$_4$ based solid electrolyte in all-solid-state batteries.

Keywords: solid state electrolyte; LiBH$_4$; aluminum anode; lithium ion battery

1. Introduction

Global reliance on the need for portable power has led to the ubiquitous deployment of lithium ion batteries (LIB), prompting manufacturers to push the limits of current technologies through new cell design and packaging approaches. The recent fires and explosions of lithium ion batteries in consumer electronics and vehicles has served as a driving force for the investigation of solid state electrolytes in LIB. Solid electrolytes have the potential to serve as non-flammable alternatives to liquid electrolytes and allow for the use of metallic lithium as the anode and utilization of high capacity/voltage cathodes. Utilizing these high capacity materials as anodes and cathodes, in conjunction with a solid-state electrolyte in the next generation of LIB, will facilitate a significant increase in energy density, safety, and operation time.

Over the years, a number of solid state ionic conductors have been investigated such as lithium phosphorus oxynitride (LiPON) [1], Li$_7$La$_3$Zr$_2$O$_{12}$ (LLZO) [2], and lithium thiophosphates

(LPS) [3]. Recent interest in utilizing $LiBH_4$ as a solid-state electrolyte was established through the work of Orimo [4], who demonstrated that the ionic conductivity of lithium can be greater than 1 mS/cm at temperatures above the orthorhombic to hexagonal phase transition that occurs at 380 K [5]. This work has been expanded to achieve high conductivity in $LiBH_4$ based solid electrolytes through the addition of Li halide salts [6–8], nanoconfinement [9–12], nanoionic destabilization [13,14], ion substitution [15–21], and eutectic formation [22].

It has also been demonstrated that $LiBH_4$ can be utilized as a solid-state electrolyte in a $Li/LiBH_4/LiCoO_2$ configuration [23]. This work utilized a PLD thin film of $LiCoO_2$ that was coated with a Li_3PO_4 protecting layer to mitigate a chemical reaction from occurring at the $LiBH_4/Li_{1-x}CoO_2$ interface. Since this work, researchers have also expanded the utilization of a $LiBH_4$ electrolyte with high capacity electrode materials including sulfur [24–26], silicon [14], TiS_2 [27,28] and MgH_2/TiH_2 [26]. Aluminum anodes have previously been investigated as a potential high capacity anode in lithium ion cells due to its low cost, low lithiation/delithiation potential (0.32 V and 0.38 V respectively), high electrical conductivity, and high theoretical capacity (993 mAh/g) in traditional liquid based electrolytes [29]. However, to the best of our knowledge, no one has demonstrated the reversible lithiation of aluminum in the solid state or with a $LiBH_4$ electrolyte.

An aluminum composite was chosen as the anode for the cell, because aluminum anodes have much higher theoretical capacities than conventional carbon-anodes (372 mAh/g) [30]. Carbon does have a lower plateau potential of lithiation (<0.09 V vs. Li^+/Li) [30] than aluminum (~0.3 V vs. Li^+/Li) [31]. However, this low plateau potential of the carbon anode could be problematic because it can facilitate lithium plating and subsequent dendrite growth on the electrode surface leading to shorting and cell failure. The most commonly studied metal oxide based anode that is also used in commercial lithium ion batteries is lithium titanate (LTO). The main problem with LTO is that it has a much lower capacity (~170 mAh/g) [32] than graphite. However, its plateau potential is much higher (~1.5 V vs. Li^+/Li) [32] and would eliminate the possibility of lithium plating and dendrite formation. Although LTO can be paired with a high voltage cathode (≥4.8 V), its operating voltage in a full cell is typically less than 2.4 V [32]. The aluminum anode would have a much higher operating voltage (>3.5 V) [29] in a full cell set-up than LTO when coupled with a high voltage cathode. The novel electrochemical properties of an aluminum anode make it a unique anode candidate in the solid-state.

The increased energy demand from batteries has led to the investigation of novel high capacity anode/cathode materials as well as solid state electrolytes. This paper address two of the three research needs for the next generation of lithium ion batteries through the investigation of the solid-state electrolyte $LiBH_4$ paired with a high capacity anode candidate, aluminum. The goal of the work is to further expand this field to include new high capacity anodes that can be paired and successfully cycled with $LiBH_4$ or other composite metal hydride based solid state electrolytes.

2. Results and Discussion

Aluminum has a native oxide layer (~4 nm thick) on the surface of the material which must be traversed in order to achieve reversible lithiation/delithiation. Others have investigated the use of thermal evaporation to produce thin films [29], nanopillar aluminum arrays [33,34], or various particle sizes [35]. To avoid this native oxide layer and provide a high surface area aluminum to interact with Li^+, we utilized decomposed alane (AlH_3) as the source of aluminum [36,37]. To obtain the aluminum used for this work, a sample of alane was dehydrogenated under inert conditions to produce high surface area aluminum nanoparticles that are free of the native oxide layer on the surface (Figure 1).

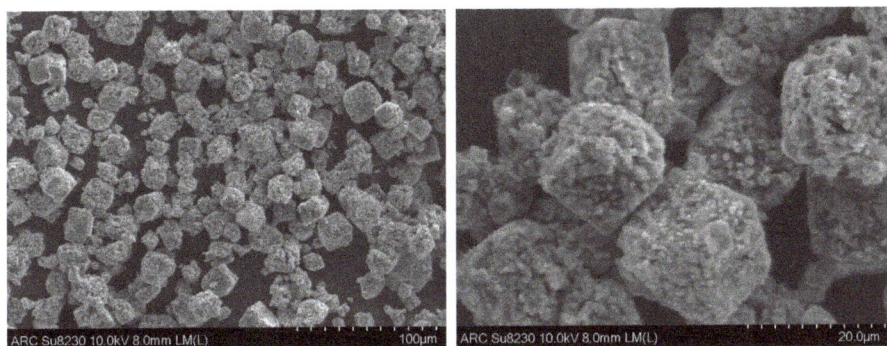

Figure 1. SEM of the aluminum obtained from the dehydrogenation of alane and utilized to prepare the composite anode material.

To determine if aluminum can be lithiated in the solid state, a composite electrode was prepared with aluminum as the active material (Al), $LiBH_4$ to facilitate Li^+ ion transport through the composite, carbon black (CB) to provide electrical conductivity, and polyvinylidene fluoride (PVDF) as a binder. The composite anode was pressed into a nickel foam (~2–4 mg) and then pelletized with $LiBH_4$ (~90 mg). Li foil was then attached to the opposite side and the pellet was sandwiched between two nickel disks and compressed with a spring inside of a 1/2 in. Swagelok union with two electrode posts (Figure 2). To enhance the ionic conductivity of $LiBH_4$ and allow for the even distribution of the solid electrolyte throughout the aluminum composite anode, it was Spex milled as previously described [38,39].

The cycling of the aluminum composite anode in the solid state with $LiBH_4$ as the solid-state electrolyte (Figure 3) is consistent with our previous study of aluminum (derived from AlH_3) based anodes in a liquid electrolyte (1.0 M $LiPF_6$ in EC/DMC) [40]. During the first cycle, a capacity of 895 mAh/g is achieved which is close to the theoretical capacity of a (1:1) LiAl alloy (993 mAh/g).

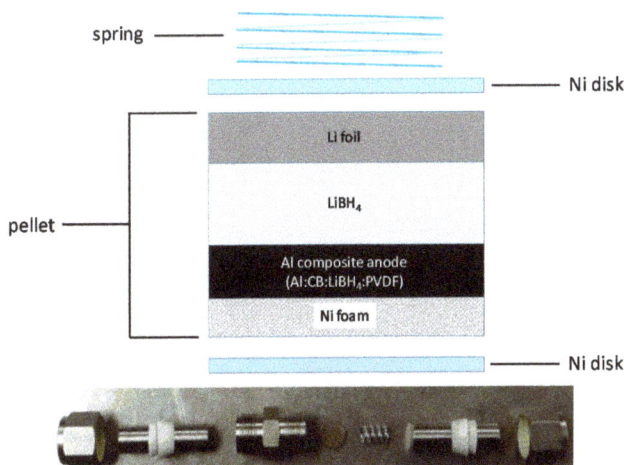

Figure 2. (Top) Schematic of the pellet inserted inside of the (**bottom**) Swagelok cell utilized for the electrochemical characterization.

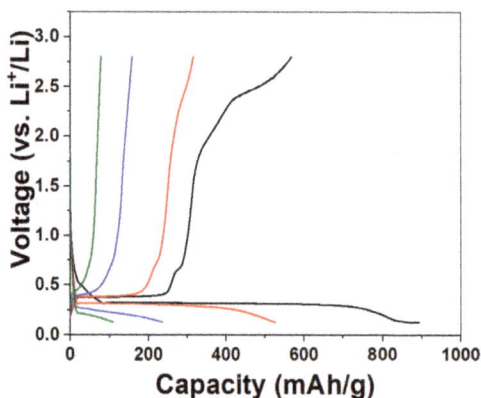

Figure 3. Lithiation/delithiation cycles of the Li/LiBH$_4$/Al composite cell. Cycling was performed at 0.1 C between 0.13 and 2.8 V (vs. Li$^+$/Li) at 135 °C. Cycle number: Black—1st, Red—2nd, Blue—5th, and Green—10th.

Subsequent cycling of the material results in a loss in capacity. This behavior is common to aluminum based anode materials and is likely attributed to the low lithium diffusion coefficient in LiAl which is 6×10^{-12} cm^{-2}·s^{-1} at 298 K [29]. This is likely the primary cause of low reversibility in the system due to the lithium being trapped within the LiAl alloy formed during the 1st and subsequent lithiation cycles. This same type of behavior has also been observed in the solution state and is due to lithium entrapment during the LiAl alloy formation [41]. In the same study, authors also report a phase transformation from α-LiAl to β-LiAl which occurs via a solid solution mediated crystallization to form the β-LiAl phase when an Al foil anode is utilized. Additional cycling studies were also performed to understand how the cycling rate effects the capacity retention (Figure 4). Cycling rates of 0.1 C, 0.5 C, and 1.0 C, based on the amount of aluminum within the anode, were also evaluated for the system. This indicates that the LiBH$_4$ can potentially support high charge/discharge rates.

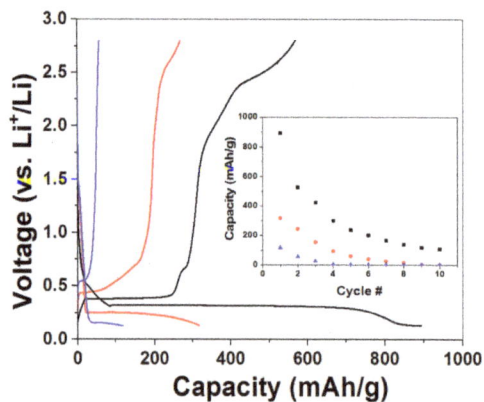

Figure 4. Effect of cycling rate on the first cycle capacity of the Li/LiBH$_4$/Al composite cell at 135 °C. Inset shows the capacity as a function of cycling rate over the first 10 cycles. Black—0.1 C, Red—0.5 C, and Blue—1.0 C.

CV was performed on the Li/LiBH$_4$/Al composite cell at 135 °C from 2.8 V to 0.13 V vs. Li$^+$/Li (Figure 5). The CV was consistent with the galvanostatic charging/discharging experiments over the potential window. The CV clearly shows the onset of lithiation at 0.32 V while the delithiation onset occurs at 0.38 V. With each subsequent cycle, the area of the oxidation and reduction peaks in the CV gradually decreases as observed in the cycling experiments.

To understand how the morphology of the Al composite electrode changed with cycling, SEM analysis was performed (Figure 6). SEM clearly showed the presence of spherical shaped particles (20–40 µm in diameter) and a relatively smooth surface of the electrode in the as prepared state. However, after the lithiation cycle of the aluminum, these spherical particles are no longer present. Additionally, the surface increases in roughness and the formation of voids within the electrode are clearly present. The formation of a rough surface and the creation of voids within the electrode could be due to the lattice expansion that occurs during the lithiation of aluminum to form the LiAl alloy. Going from metallic aluminum to the LiAl alloy results in a 97% expansion in the lattice [42]. However, this is significantly less than the lattice expansion during the lithiation of Si to $Li_{22}Si_5$ and Sn to $Li_{22}Sn_5$ which are 323% and 300% [43] respectively. The expansion of these 3 alloys are all significantly more than the graphite electrode currently utilized in commercial LIB, which expands only 9% upon lithiation [44].

Next, SEM-EDS was utilized to ascertain the identity of the spherical particles and obtain chemical information about the surface. Figure 7 shows the elemental distribution of B, Al, F, O, and C on the surface of the aluminum composite anode in the initial state. By looking at the boron distribution it is more concentrated around the spherical particles indicating that they are composed of $LiBH_4$. The aluminum particles are also clearly distinguishable. The distribution mapping of carbon from the CB and PVDF and the fluorine from PVDF are nearly identical. Oxygen is also present in the sample but it has a very low concentration where the aluminum particles are present and is consistent with our hypothesis that the aluminum is relatively free of an oxide layer.

SEM-EDS was then taken of the aluminum composite electrode after the first lithiation (Figure 8). The distribution of boron throughout the electrode is significantly affected by this process because it is now randomly distributed through the sample and not concentrated in a certain location/particle. This is also consistent with the disappearance of the spherical shapes that were identified as $LiBH_4$ in Figure 7. This unique morphological change has not been previously reported for a solid-state cell utilizing pure $LiBH_4$ as a solid electrolyte. This was unexpected because $LiBH_4$ does not melt until >275 °C. However, the operation temperature of the cell is above the orthorhombic to hexagonal phase transition that occurs at temperatures >115 °C, and could be responsible for the change in morphology. Further investigation of this morphological change and its possible impact on electrochemical systems is needed.

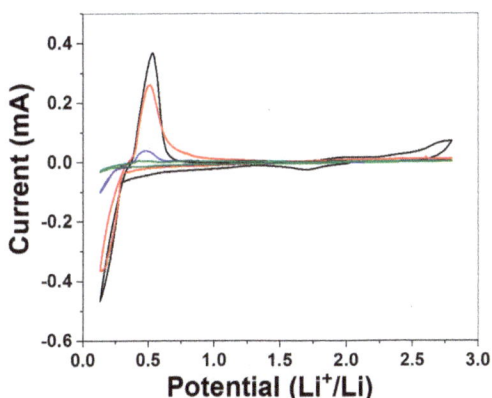

Figure 5. Cyclic voltammogram of the Li/$LiBH_4$/Al composite cell at 135 °C. Cycle number: Black—1st, Red—2nd, Blue—5th, and Green—10th.

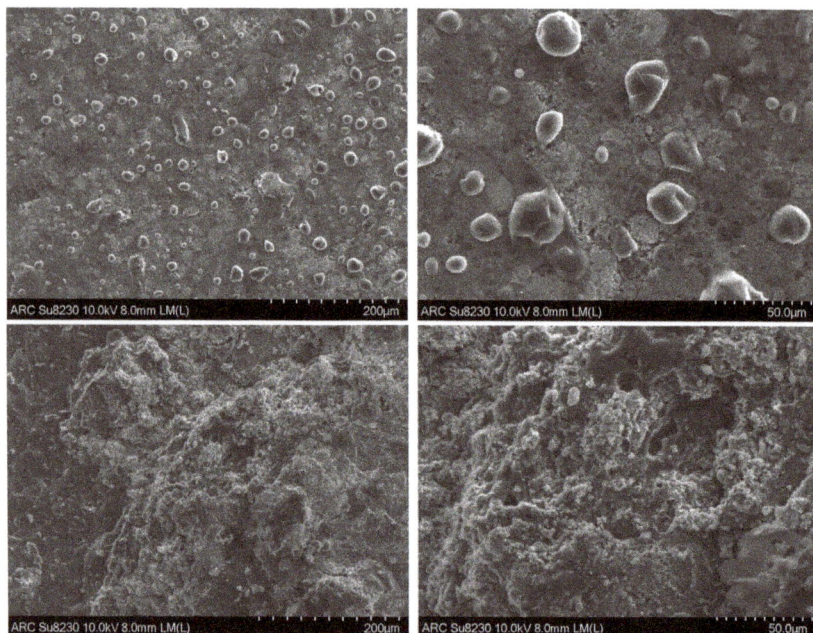

Figure 6. (**Top**, **left** and **right**) Wide field view and zoomed in view of the Al composite anode side of the pellet in the initial state. (**Bottom**, **left** and **right**) Wide field view and zoomed in view of the Al composite anode side of the pellet after the first lithiation.

The aluminum signal is still present in the sample; however, the edges of the aluminum particles are less sharp and not as defined which could be due to pulverization during the lithiation. The oxygen content of the sample is significantly increased in the sample; however, this could have occurred during the cycling of the sample in the Swagelok cell or during the brief exposure of the sample to air when it is introduced into the SEM for analysis.

XRD was utilized to confirm that the electrochemical plateaus observed during the galvanostatic charge/discharge cycles were attributed to the formation of a LiAl alloy (Figure 9). To do this a pellet was assembled as described in Figure 2, however, the Ni foam was excluded from the process so that diffraction pattern wouldn't be dominated by the high Z nickel. In the initial state (before lithiation) the presence of aluminum is confirmed in the XRD (denoted as *). After the first lithiation, the formation of the LiAl alloy (denoted as #) is then observed with the formation of 4 well defined and resolved peaks. There are still pure aluminum peaks present in the diffraction pattern and is attributed to the fact that the Li^+ must diffuse all of the way through the electrolyte and then the full thickness of electrode to achieve lithiation at the bottom of the pellet. It may be possible to demonstrate full conversion of aluminum to the LiAl alloy, but it would require a much slower lithiation rate and reduced thickness of the composite electrode. However, the purpose of this experiment was to spectroscopically verify the formation of the LiAl alloy via lithium migration through the $LiBH_4$ electrolyte. The unlabeled peaks in the spectrum are from the $LiBH_4$ used in the composite electrode and as the solid-state electrolyte.

To further characterize the interface between the electrodes and electrolyte electrochemical impedance spectroscopy (EIS) and Raman spectroscopy was performed (Figure 10). The EIS spectra for the cell ($Li/LiBH_4/Al$) at room temperature before cycling is shown in the inset. The Nyquist plot for this sample shows a very large resistance for the electrolyte and the charge transfer process occurring at the interface of the electrolyte/electrodes. Upon heating to 135 °C, there is a significant reduction in the resistance for the electrolyte and the charge transfer at the interface as expected before cycling. After the first cycle, there is a reduction in the resistance of the electrolyte. This is likely due to disappearance of the spherical LiBH$_4$ particles (observed in the SEM-EDS) resulting in better Li ion diffusion at the interface. After the 10th cycle, the resistance of the electrolyte and charge transfer increases and is likely due to the large volume expansion/contraction of the Al active material during the lithiation/delithiation process. This likely results in small gaps at the electrode-electrolyte interface which effectively closes the lithium ion transport into and out of the Al.

Raman spectroscopy was carried out on the anode side of the electrolyte pellet in the as prepared sample and after 5 cycles at room temperature. For these experiments, PDVF and CB were removed from the anode for clarity and to obtain suitable spectra due to the absorbance of scattered light from the black carbon material. The B–H vibration modes [45], shown in Figure 10, demonstrate no significant change in frequency due to the formation of additional Li–B–H species, such as Li$_2$B$_{12}$H$_{12}$, and are consistent with orthorhombic LiBH$_4$. This is expected since the hexagonal phase is only stabilized at high temperatures or with the incorporation of suitable additives [7].

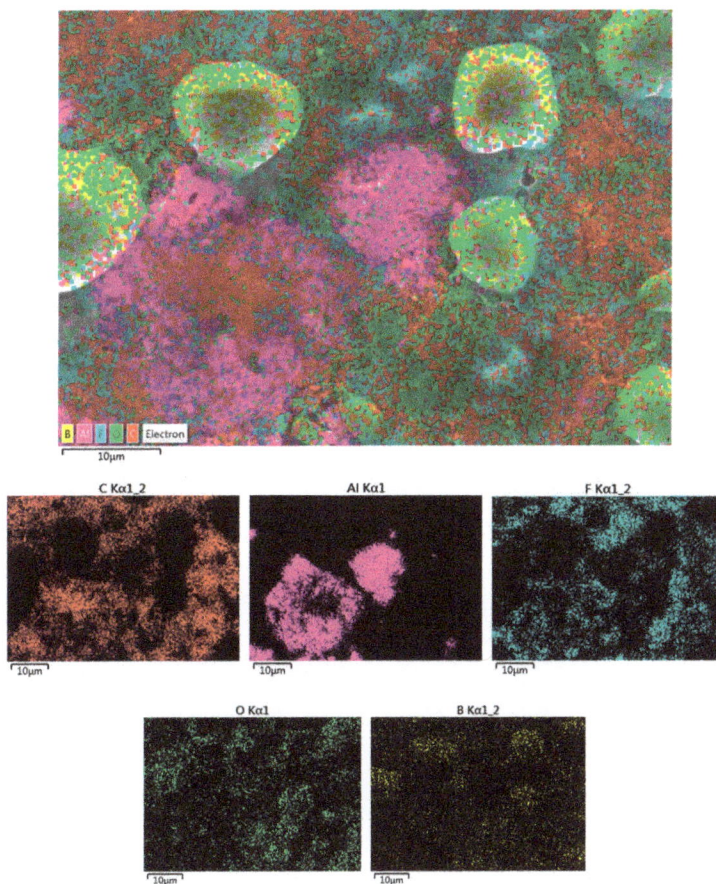

Figure 7. SEM-EDS of the aluminum composite anode in the initial state. The top image is the composite map of carbon, aluminum, fluorine, oxygen, and boron elemental distribution. The bottom images are the individual atomic distributions of the elements.

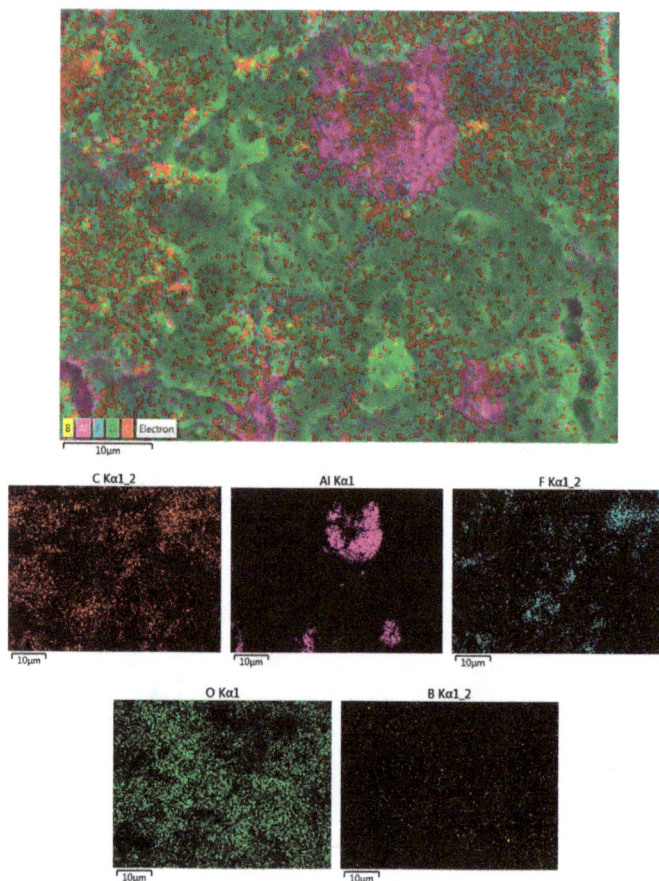

Figure 8. SEM-EDS of the aluminum composite anode after the first lithiation. The top image is the composite map of carbon, aluminum, fluorine, oxygen, and boron elemental distribution. The bottom images are the individual atomic distributions of the elements.

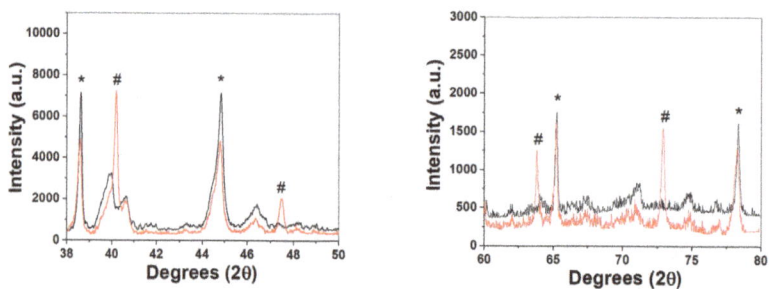

Figure 9. XRD of the aluminum anode composite before (black) and after the 1st lithiation cycle (red). The left and the right panels show the low 2θ and high 2θ respectively. Aluminum (*, pdf 00-0040787) and LiAl alloy (#, pdf 01-072-3428).

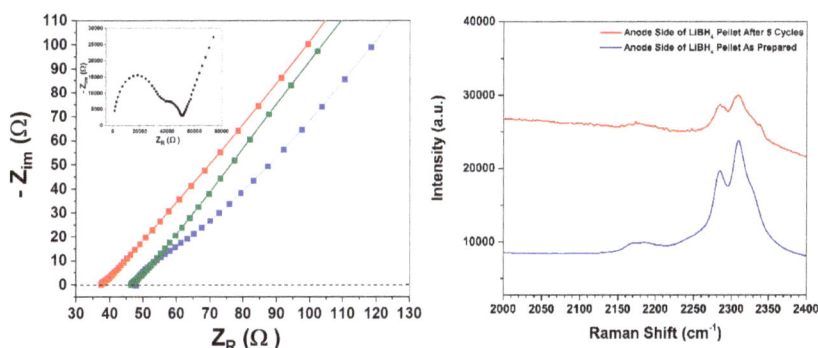

Figure 10. (**Left**) Electrochemical impedance spectroscopy (EIS) of the Li/LiBH$_4$/Al cell. The inset (black) shows the cell at room temperature. The other 3 EIS traces are for the cell at 135 °C before cycling (blue), after the 1st cycle (red), and after the 10th cycle (green). (**Right**) Raman spectra of the B–H stretching modes of LiBH$_4$ in the as prepared electrolyte pellet from the anode side (blue) and after cycling five times (red).

3. Materials and Methods

3.1. Materials

Chemicals were provided by the supplier and are listed by supplier as follows. Sigma Aldrich (St. Louis, MO, USA): LiBH$_4$ and lithium foil; MTI corp. (Richmond, CA, USA): TIMCAL graphite carbon super P (CB) and polvinylidene fluoride (PVdF). Aluminum (Al) utilized as the anode material was obtained from the dehydrogenation of AlH$_3$ (ATK) at 220 °C under a constant argon flow on a Schlenk line for 2 h.

3.2. Electrolyte and Anode Preparation

LiBH$_4$ was ball milled for a total of thirty minutes with a 30:1 ball to powder ratio utilizing a SPEX ball mill. After ball milling, the LiBH$_4$ material was ground up with the use of a mortar and pestle. The aluminum composite was prepared with equal parts of activated aluminum; LiBH$_4$; PVdF; and CB (mass ratio of 10:6:3:3 respectively). This mixture was then homogenized with the use of a mortar and pestle. The pellets were assembled by first obtaining a flattened nickel foam current collector and manually pressing the aluminum composite anode material (2.0–3.0 mg) into it. The foam was then placed at the bottom of a die 10 mm die set (International Crystal Laboratories) with the aluminum anode composite facing up. Next, approximately ~90 mg of LiBH$_4$ was added on top of the nickel foam/anode composite. The die set was then fully assembled and pressed in a hydraulic press at 2 tons of pressure. After the pellet was pressed, it was removed from the die set with the nickel foam/anode composite attached to one side of the LiBH$_4$ pellet.

3.3. Cell Assembly and Electrochemical Characterization

Electrochemical measurements were performed using a Swagelok cell. The cells contained two polished nickel disk current collectors. The nickel foam/anode composite/LiBH$_4$/Li foil pellet was placed between two solid nickel disks to maintain sufficient electrical connectivity and provide uniform pressure on the pellet while in the cell under the pressure of a spring. The inside of the Swagelok cell was lined with a Mylar film and Teflon ferrules that were used to prevent shorting. The measurements were taken on a Bio-Logic VMP3 multichannel potentiostat (Bio-Logic Ltd., Grenoble, France). In order to increase the conductivity of the LiBH$_4$ electrolyte and reduce the contact resistance at the electrode/electrolyte interfaces, the cells were electrochemically evaluated at

135 °C. The heating process was conducted with the use of heating tape submerged in a sand bath. The temperature was monitored with the use of a K-type thermocouple positioned against the wall of the Swagelok cell. The cells were then cycled at various rates (0.1 C, 0.5 C, and 1.0 C) based on the amount of aluminum in the composite anode, for conducting galvanostatic charging/discharging of the cells. Cyclic voltammograms were collected at a cycling rate of 0.100 mV/s with a voltage window of 0.13 V to 2.80 V. The EIS was measured in a frequency range of 1 MHz to 100 Hz at ±20 mV amplitude.

3.4. Ex-Situ XRD, SEM-EDS, and Raman Characterization

XRD was utilized to confirm the formation of the LiAl alloy after the first lithiation cycle. For this set-up, the anode composite was pressed into the $LiBH_4$ pellet without the use of the nickel foam current collector. This was done to eliminate the large signal from the nickel foam current collector during the XRD data collection. This allowed for the aluminum and the LiAl alloy to be readily detectable during the analysis. XRD was performed using a PANalytical X'pert Pro (PANalytical Ltd., Almelo, The Netherlands) with Cu Kα radiation, and the samples were protected with a Kapton film to minimize oxidation of the sample.

SEM was used to analyze the change in surface morphology that occurred as a result of lithiating the active material. This process was done with the use of a Hitachi Ultra-High Resolution Scanning Electron Microscope SU8230 series (Hitachi, Ltd., Tokyo, Japan) with an X-Max^n Silicon Drift Detector attachment. Further atomic composition studies (EDS) were performed and the data from the analysis was interpreted using Aztec software.

Raman spectra were obtained with 15 mW 532 nm excitation using a 5 mm focal length optical fiber probe focused onto the pellet surface. Spectra were recorded at ~5 cm^{-1} resolution using a Holospec transmission grating spectrometer (Andor Technology Ltd., Belfast, UK) with CCD detection.

4. Conclusions

In this study, we demonstrated the reversible electrochemical lithiation of a composite aluminum anode facilitated by a $LiBH_4$ solid state electrolyte. It was also demonstrated that the $LiBH_4$ solid state electrolyte can facilitate the reversible lithiation of aluminum at charge/discharge rates up to 1.0 C. Also, this study showed that metal hydrides (such as AlH_3) can serve as the source of oxide free metals for battery applications. XRD confirmed the formation of a LiAl alloy after the first lithiation cycle. SEM analysis identified unique morphological changes that occur within the electrode during the cycling of the material. It is likely that similar changes are also occurring in other cells that utilize a $LiBH_4$ based material as the solid electrolyte. This type of morphological change should also be considered along with other expansion/contraction processes in the high capacity anode materials in future studies. A better understanding of this mechanism will be needed for the further development of $LiBH_4$ based electrolytes. Although the cycle stability of the aluminum composite anode is poor, recent work has also investigated an Al/TiO_2 yolk-shell nano-architechture (ATO) that facilitates reversible lithiation/delithiation of aluminum for 500 cycles with excellent capacity retention [46]. Our future work will focus on this type of architecture in combination with the $LiBH_4$ based electrolyte as well as the closo-borane solid state electrolytes.

Acknowledgments: Work at SRNL was supported by the U.S. Department of Energy, Office of Science, Basic Energy Sciences, Materials Sciences and Engineering Division. Joseph A. Teprovich would like to thank the California State University—Northridge, College of College of Science & Mathematics start-up funding for support.

Author Contributions: Joseph A. Teprovich Jr. and Ragaiy Zidan conceived and designed the experimental approach. Jason A. Weeks, Spencer C. Tinkey and Patrick A. Ward performed the electrochemical and spectroscopic analysis of the materials. Robert Lascola performed the Raman analysis of the materials and interface. All of the authors wrote, commented, and contributed to the manuscript.

Conflicts of Interest: The authors declare no conflict of interest.

References

1. Yu, X.; Bates, J.B.; Jellison, G.E.; Hart, F.X. A Stable Thin-Film Lithium Electrolyte: Lithium Phosphorus Oxynitride. *J. Electrochem. Soc.* **1997**, *144*, 524–532. [CrossRef]

2. Murugan, R.; Thangadurai, V.; Weppner, W. Fast Lithium Ion Conduction in Garnet-Type $Li_7La_3Zr_2O_{12}$. *Angew. Chem. Int. Ed.* **2007**, *46*, 7778–7781. [CrossRef] [PubMed]

3. Liu, Z.; Fu, W.; Payzant, E.A.; Yu, X.; Wu, Z.; Dudney, N.J.; Kiggans, J.; Hong, K.; Rondinone, A.J.; Liang, C. Anomalous High Ionic Conductivity of Nanoporous β Li_3PS_4. *J. Am. Chem. Soc.* **2013**, *135*, 975–978. [CrossRef] [PubMed]

4. Matsuo, M.; Nakamori, Y.; Orimo, S.; Maekawa, H.; Takamura, H. Lithium superionic conduction in lithium borohydride accompanied by structural transition. *Appl. Phys. Lett.* **2007**, *91*, 224103. [CrossRef]

5. Gorbunov, V.E.; Gavrichev, K.S.; Zalukaev, V.L.; Sharpataya, G.A.; Bakum, S.I. Heat capacity and phase transition of lithium borohydride. *Zhurnal Neorganicheskoj Khimii* **1984**, *29*, 2333–2337.

6. Maekawa, H.; Matsuo, M.; Takamura, H.; Ando, M.; Noda, Y.; Karahashi, T.; Orimo, S. Halide-Stabilized $LiBH_4$, a Room-Temperature Lithium Fast-Ion Conductor. *J. Am. Chem. Soc.* **2009**, *131*, 894–895. [CrossRef] [PubMed]

7. Matsuo, M.; Takamura, H.; Maekawa, H.; Li, H.; Orimo, S. Stabilization of lithium superionic conduction phase and enhancement of conductivity of $LiBH_4$ by LiCl addition. *Appl. Phys. Lett.* **2009**, *94*, 084103. [CrossRef]

8. Skripov, A.V.; Soloninin, A.V.; Rude, L.H.; Jensen, T.R.; Filinchuk, Y. Nuclear Magnetic Resonance Studies of Reorientational Motion and Li Diffusion in $LiBH_4$–LiI Solid Solutions. *J. Phys. Chem. C* **2012**, *116*, 26177–26184. [CrossRef]

9. Choi, Y.S.; Lee, Y.; Oh, K.H.; Cho, Y.W. Interface-enhanced Li ion conduction in a $LiBH_4$–SiO_2 solid electrolyte. *Phys. Chem. Chem. Phys.* **2016**, *18*, 22540–22547. [CrossRef] [PubMed]

10. Shane, D.T.; Corey, R.L.; McIntosh, C.; Rayhel, L.H.; Bowman, R.C., Jr.; Vajo, J.J.; Gross, A.F.; Conradi, M.S. $LiBH_4$ in Carbon Aerogel Nanoscaffolds: An NMR Study of Atomic Motions. *J. Phys. Chem. C* **2010**, *114*, 4008–4014. [CrossRef]

11. Blanchard, D.; Nale, A.; Sveinbjornsson, D.; Eggenhuisen, T.M.; Verkuijlen, M.H.; Vegge, T.; Kentgens, A.P.; de Jongh, P.E. Nanoconfined $LiBH_4$ as a fast lithium ion conductor. *Adv. Funct. Mater.* **2015**, *25*, 184–192. [CrossRef]

12. Das, S.; Ngene, P.; Norby, P.; Vegge, T.; de Jongh, P.E.; Blanchard, D. All-Solid-State Lithium-Sulfur Battery Based on a Nanoconfined $LiBH_4$ Electrolyte. *J. Electrochem. Soc.* **2016**, *163*, A2029–A2034. [CrossRef]

13. Shane, D.T.; Corey, R.L.; Rayhel, L.H.; Wellons, M.; Teprovich, J.A., Jr.; Zidan, R.; Hwang, S.; Bowman, R.C., Jr.; Conradi, M.S. NMR Study of $LiBH_4$ with C_{60}. *J. Phys. Chem. C* **2010**, *114*, 19862–19866. [CrossRef]

14. Teprovich, J.A., Jr.; Colón-Mercado, H.R.; Ward, P.A.; Peters, B.; Giri, S.; Zhou, J.; Greenway, S.; Compton, R.N.; Jena, P.; Zidan, R. Experimental and Theoretical Analysis of Fast Lithium Ionic Conduction in a $LiBH_4$–C_{60} Nanocomposite. *J. Phys. Chem. C* **2014**, *118*, 21755–21761. [CrossRef]

15. Matsuo, M.; Remhof, A.; Martelli, P.; Caputo, R.; Ernst, M.; Miura, Y.; Sato, T.; Oguchi, H.; Maekawa, H.; Takamura, H.; et al. Complex Hydrides with $(BH_4)^-$ and $(NH_2)^-$ Anions as New Lithium Fast-Ion Conductors. *J. Am. Chem. Soc.* **2009**, *131*, 16389–16391. [CrossRef] [PubMed]

16. GharibDoust, S.P.; Brighi, M.; Sadikin, Y.; Ravnsbæk, D.B.; Černý, R.; Skibsted, J.; Jensen, T.R. Synthesis, Structure, and Li-Ion Conductivity of $LiLa(BH_4)_3X$, X = Cl, Br, I. *J. Phys. Chem. C* **2017**, *121*, 19010–19021. [CrossRef]

17. Skripov, A.V.; Soloninin, A.V.; Ley, M.B.; Jensen, T.R.; Filinchuk, Y. Nuclear Magnetic Resonance Studies of BH_4 Reorientations and Li Diffusion in $LiLa(BH_4)_3Cl$. *J. Phys. Chem. C* **2013**, *117*, 14965–14972. [CrossRef]

18. Roedern, E.; Lee, Y.; Ley, M.B.; Park, K.; Cho, Y.W.; Skibsted, J.; Jensen, T.R. Solid state synthesis, structural characterization and ionic conductivity of bimetallic alkali-metal yttrium borohydrides $MY(BH_4)_4$ (M = Li and Na). *J. Mater. Chem. A* **2016**, *4*, 8793–8802. [CrossRef]

19. Ley, M.B.; Ravnsbæk, D.B.; Filinchuk, Y.; Lee, Y.; Janot, R.; Cho, Y.W.; Skibsted, J.; Jensen, T.R. $LiCe(BH_4)_3Cl$, a New Lithium-Ion Conductor and Hydrogen Storage Material with Isolated Tetranuclear Anionic Clusters. *Chem. Mater.* **2012**, *24*, 1654–1663. [CrossRef]

20. Lee, Y.; Ley, M.B.; Jensen, T.R.; Cho, Y.W. Lithium Ion Disorder and Conduction Mechanism in $LiCe(BH_4)_3Cl$. *J. Phys. Chem. C* **2016**, *120*, 19035–19042. [CrossRef]

21. Yamauchi, A.; Sakuda, A.; Hayashi, A.; Tatsumisago, M. Preparation and ionic conductivities of (100 − *x*) (0.75Li$_2$S·0.25P$_2$S$_5$)·*x*LiBH$_4$ glass electrolytes. *J. Power Sources* **2013**, *244*, 707–710. [CrossRef]

22. Lee, H.; Hwang, S.; To, M.; Lee, Y.; Cho, Y.W. Discovery of Fluidic LiBH$_4$ on Scaffold Surfaces and Its Application for Fast Co-confinement of LiBH$_4$–Ca(BH$_4$)$_2$ into Mesopores. *J. Phys. Chem. C* **2015**, *119*, 9025–9035. [CrossRef]

23. Takahashi, K.; Hattori, K.; Yamazaki, T.; Takada, K.; Matsuo, M.; Orimo, S.; Maekawa, H.; Takamura, H. All-Solid-State Lithium Battery with LiBH$_4$ Solid Electrolyte. *J. Power Sources* **2013**, *226*, 61–64. [CrossRef]

24. Suzuki, S.; Kawaji, J.; Yoshida, K.; Unemoto, A.; Orimo, S. Development of complex hydride-based allsolid-state lithium ion battery applying low melting point electrolyte. *J. Power Sources* **2017**, *359*, 97–103. [CrossRef]

25. Unemoto, A.; Chen, C.L.; Wang, Z.; Matsuo, M.; Ikeshoji, T.; Orimo, S. Pseudo-binary electrolyte, LiBH$_4$–LiCl, for bulk-type all-solid-state lithium-sulfur battery. *Nanotechnology* **2015**, *26*, 254001. [CrossRef] [PubMed]

26. Lopez-Aranguren, P.; Berti, N.; Dao, A.H.; Zhang, J.; Cuevas, F.; Latroche, M.; Jordy, C. An all-solid state metal hydride—Sulfur lithium-ion battery. *J. Power Sources* **2017**, *357*, 56–60. [CrossRef]

27. Unemoto, A.; Ikeshoji, T.; Yasaku, S.; Matsuo, M.; Stavila, V.; Udovic, T.J.; Orimo, S. Stable Interface Formation between TiS$_2$ and LiBH$_4$ in Bulk-Type All-Solid-State Lithium Batteries. *Chem. Mater.* **2015**, *27*, 5407–5416. [CrossRef]

28. Unemoto, A.; Wu, H.; Udovic, T.J.; Matsuo, M.; Ikeshoji, T.; Orimo, S. Fast lithium-ionic conduction in a new complex hydride–sulphide crystalline phase. *Chem. Commun.* **2016**, *52*, 564–566. [CrossRef] [PubMed]

29. Hamon, Y.; Brousse, T.; Jousse, F.; Topart, P.; Buvat, P.; Schleich, D.M. Aluminum negative electrode in lithium ion batteries. *J. Power Sources* **2001**, *97–98*, 185–187. [CrossRef]

30. Buiel, E.; Dahn, J.R. Li-insertion in hard carbon anode materials for Li-ion batteries. *Electrochim. Acta* **1999**, *45*, 121–130. [CrossRef]

31. Hudak, N.; Huber, D. Size Effects in the Electrochemical Alloying and Cycling-of Electrodeposited Aluminum with Lithium. *J. Electrochem. Soc.* **2012**, *159*, A688–A695. [CrossRef]

32. Shenouda, A.Y.; Murali, K.R. Electrochemical properties of doped lithium titanate compounds and their performance in lithium rechargeable batteries. *J. Power Sources* **2008**, *176*, 332–339. [CrossRef]

33. Au, M.; McWhorter, S.; Ajo, H.; Adams, T.; Zhao, Y.; Gibbs, J. Free standing aluminum nanostructures as anodes for Li-ion rechargeable batteries. *J. Power Sources* **2010**, *195*, 3333–3337. [CrossRef]

34. Oltean, G.; Tai, C.; Edstrom, K.; Nyholm, L. On the origin of the capacity fading for aluminum negative electrodes in Li-ion batteries. *J. Power Sources* **2014**, *269*, 266–273. [CrossRef]

35. Lei, X.; Wang, C.; Yi, Z.; Liang, Y.; Sun, J. Effects of particle size on the electrochemical properties of aluminum powders as anode materials for lithium ion batteries. *J. Alloys Compd.* **2007**, *429*, 311–315. [CrossRef]

36. Teprovich, J.A., Jr.; Motyka, T.; Zidan, R. Hydrogen System Using Novel Additives to Catalyze Hydrogen Release from the Hydrolysis of Alane and Activated Aluminum. *Int. J. Hydrog. Energy* **2012**, *37*, 1594–1603. [CrossRef]

37. Martínez-Rodríguez, M.J.; García-Díaz, B.L.; Teprovich, J.A., Jr.; Knight, D.A.; Zidan, R. Advances in the Electrochemical Regeneration of Aluminum Hydride. *Appl. Phys. A* **2012**, *106*, 545–550. [CrossRef]

38. Sveinbjörnsson, D.; Myrdal, J.S.G.; Blanchard, D.; Bentzen, J.J.; Hirata, T.; Mogensen, M.B.; Norby, P.; Orimo, S.; Vegge, T. Effect of Heat Treatment on the Lithium Ion Conduction of the LiBH$_4$–LiI Solid Solution. *J. Phys. Chem. C* **2013**, *117*, 3249–3257. [CrossRef]

39. Teprovich, J.A., Jr.; Colon-Mercado, H.; Washington, A.L., II; Ward, P.A.; Hartman, H.; Greenway, S.; Missimer, D.M.; Velten, J.; Christian, J.H.; Zidan, R. Bi-functional Li$_2$B$_{12}$H$_{12}$ for energy storage and conversion applications: Solid-state electrolyte and luminescent down-conversion dye. *J. Mater. Chem. A* **2015**, *3*, 22853–22859. [CrossRef]

40. Teprovich, J.A., Jr.; Zhang, J.; Colón-Mercado, H.; Cuevas, F.; Peters, B.; Greenway, S.; Zidan, R.; Latroche, M. Li-Driven Electrochemical Conversion Reaction of AlH$_3$, LiAlH$_4$, and NaAlH$_4$. *J. Phys. Chem. C* **2015**, *119*, 4666–4674. [CrossRef]

41. Liu, D.X.; Co, A.C. Revealing Chemical Processes Involved in Electrochemical (De)Lithiation of Al with in Situ Neutron Depth Profiling and X-ray Diffraction. *J. Am. Chem. Soc.* **2016**, *138*, 231–238. [CrossRef] [PubMed]

42. Besenhard, J.O.; Hess, M.; Komenda, P. Dimensionally stable Li-alloy electrodes for secondary batteries. *Solid State Ion.* **1990**, *40–41*, 525–529. [CrossRef]

43. Winter, M.; Besenhard, J.O. Electrochemical lithiation of tin and tin-based intermetallics and composites. *Electrochim. Acta* **1999**, *45*, 31–50. [CrossRef]
44. Reynier, Y.; Yazami, R.; Fultz, B. XRD evidence of macroscopic composition inhomogeneities in the graphite–lithium electrode. *J. Power Sources* **2007**, *165*, 616–619. [CrossRef]
45. Gomes, S.; Hagemann, H.; Yvon, K. Lithium boro-hydride LiBH$_4$: II. Raman spectroscopy. *J. Alloys Compd.* **2002**, *346*, 206–210. [CrossRef]
46. Li, S.; Niu, J.; Zhao, Y.C.; So, K.P.; Wang, C.; Wang, C.A.; Li, J. High-rate aluminum yolk-shell nanoparticle anode for Li-ion battery with long cycle life and ultrahigh capacity. *Nat. Commun.* **2015**, *6*, 7872. [CrossRef] [PubMed]

inorganics

MDPI

Article

Lewis Base Complexes of Magnesium Borohydride: Enhanced Kinetics and Product Selectivity upon Hydrogen Release

Marina Chong [1], Tom Autrey [1,2,*] and Craig M. Jensen [2,*]

[1]   Pacific Northwest National Laboratory, 902 Battelle Blvd, Richland, WA 99352, USA; marinac628@gmail.com
[2]   Department of Chemistry, University of Hawaii at Manoa, 2545 McCarthy Mall, Honolulu, HI 96822, USA
*   Correspondence: tom.autrey@pnnl.gov (T.A.); Jensen@hawaii.edu (C.M.J.);
     Tel.: +1-509-375-3792 (T.A.); +1-808-956-2769 (C.M.J.)

Received: 3 November 2017; Accepted: 28 November 2017; Published: 6 December 2017

Abstract: Tetrahydofuran (THF) complexed to magnesium borohydride has been found to have a positive effect on both the reactivity and selectivity, enabling release of H_2 at <200 °C and forms $Mg(B_{10}H_{10})$ with high selectivity.

Keywords: hydrogen storage; Lewis base adducts; borohydride

1. Introduction

Over the past decade, there has been a significant international effort involving chemists, materials scientists and physicists to discover and demonstrate a solid-state hydrogen storage material that would enable a fuel cell electric vehicle 5 min refueling time and a 500 km driving range. Only a few of the thousands of materials investigated have garnered as much interest as $Mg(BH_4)_2$ [1–12]. The high gravimetric density of H_2, ca. 14.7 wt % H_2 and thermodynamics for H_2 release lie in the narrow window required for reversibility under moderate pressure and temperature. The dehydrogenation of the borohydride to MgB_2 has a calculated ΔH_0 of 38.6 kJ/(mol H_2) and ΔS of 111.5 J/(K·mol H_2), predicting a plateau pressure of 1 bar H_2 of 73 °C [13]. These thermodynamic properties together with the borohydride's high gravimetric hydrogen density, and demonstrated hydrogen cycling compatibility [1,9] suggest its application as a reversible hydrogen carrier for PEM fuel cell applications. Two critical challenges remaining are (i) the slow rates of hydrogen release and (ii) the thermodynamic stability of the major dehydrogenation product, magnesium dodecaborane, $Mg(B_{12}H_{12})$, occasionally referred to as the dead-end for reversibility.

At temperatures greater than 460 °C the borohydride releases ~14 wt % hydrogen, giving mixture of products, i.e., MgB_2, MgH_2, Mg and amorphous boron, depending on reaction conditions [6,10,11,14,15]. Hydrogenation of this product mixture at 400 bar H_2 and 270 °C results in the uptake of 6.1 wt % hydrogen [5]. NMR studies concluded that $MgB_{12}H_{12}$, forming at temperatures greater than 250 °C, is a thermodynamic endpoint, preventing re-hydrogenation to $Mg(BH_4)_2$ [4]. On the other hand, the reversal of MgB_2 to $Mg(BH_4)_2$ occurs, albeit, under extreme conditions of 950 bar H_2 and 400 °C [9]. This demonstrated that reversibility can be achieved, however, under conditions that are impractical for commercial hydrogen storage applications. Similarly, the lithium, sodium, and potassium salts of $B_{12}H_{12}{}^{2-}$ have been hydrogenated, in the presence of metal hydrides, to the corresponding borohydride under 1000 bar of H_2 at 500 °C [16]. Whether the pathway for the hydrogenation of MgB_2 to $Mg(BH_4)_2$ involves $MgB_{12}H_{12}$ remains an open question.

The use of additives to enhance kinetics of hydrogen release from $Mg(BH_4)_2$ has been the subject of several investigations [17–20]. An early study found that $TiCl_3$ lowered the onset temperature of hydrogen release from 262 to 88 °C [17]. More recently, significant reductions in the onset temperature of hydrogen release have been observed upon the addition of $NbCl_5$ and a Ti–Nb nanocomposite [18]; metal fluorides such as CaF_2, ZnF_2, TiF_3, and NbF_5 [18–20] and $ScCl_3$ [19]. Hydrogen release is also induced by mechanically milling $Mg(BH_4)_2$ with TiO_2 resulting in release of 2.4 wt % H_2 at 271 °C while undergoing reversible dehydrogenation to $Mg(B_3H_8)_2$ [20]. Alternatively, the thermal dehydrogenation of $Mg(BH_4)_2$ has been shown to be accelerated in eutectic mixtures with $LiBH_4$ [21–23]. Another study claimed that the addition of LiH to $Mg(BH_4)_2$ induced hydrogen evolution at temperatures as low as 150 °C and enabled the cycling of 3.6 wt % H_2 through 20 cycles at 180 °C [24].

2. Results

The high temperature and pressure required for reversibility led us to explore the decomposition pathways at lower temperatures. The decomposition of $Mg(BH_4)_2$ over a prolonged period (5 weeks) under 1 bar nitrogen at 200 °C yields $Mg(B_3H_8)_2$ as the major product [1]. While formation of the $B_3H_8{}^-$ anion has been recognized from thermal condensation studies of $BH_4{}^-$ in solution [25], this finding provided evidence that an analogous process may take place during solid state decomposition contrary to theoretical predictions. Furthermore, under 120 bar hydrogen pressure and 250 °C, the $Mg(B_3H_8)_2$ intermediate was completely converts back to $Mg(BH_4)_2$ after 48 h. The subsequent hydrogenation of independently synthesized $Mg(B_3H_8)_2 \cdot$ THF (THF = tetrahydofuran), where attempts to remove the solvent were unsuccessful, then demonstrated that quantitative re-hydrogenation to $Mg(BH_4)_2$ could be achieved under 50 bar H_2 and 5 h at 200 °C [26]. We concluded that the faster rate exhibited by the solvated sample resulted from a phase change induced by the coordination of the THF. Studies of borohydrides and boranes in the context of hydrogen storage, have typically focused on complete solvent removal. The presence of residual solvent is generally considered problematic and the various synthetic routes to $Mg(BH_4)_2$ often call for rigorous efforts to obtain a pure, solvent-free product. However, our findings suggested that the solvent coordination might have the beneficial effect of enhancing dehydrogenation kinetics. Only a handful of studies have explored the dehydrogenation of $Mg(BH_4)_2$ coordinated to a solvent, the majority of which have highlighted nitrogen donors [27–30].

Our observation of the kinetic enhancement of the hydrogenation of $Mg(B_3H_8)_2$ to $Mg(BH_4)_2$ prompted us to further explore how solvent coordination affects hydrogen release temperatures. We have examined the effect of dimethyl sulfide (DMS), diethyl ether (Et_2O), triethylamine (TEA), diglyme (Digly), dimethoxy ethane (DME) and THF, encompassing a range of Lewis basicity, on the decomposition of $Mg(BH_4)_2$. Alternative syntheses, complex polymorphism, predicted thermodynamic properties, and attempts to improve the hydrogen cycling capacity of $Mg(BH_4)_2$ have been widely explored and reviews of these activities were recently published [20,31]. However, the solid-state chemistry of the interconversion of the borane intermediates involved in these systems remains largely unexplored. Therefore, a unique aspect of this work has been the direct observation and characterization of the borane products and metastable reaction intermediates by MAS and solution phase ^{11}B NMR studies.

Table 1. Ligand ratios in synthesized solvates, determined by ^{1}H NMR.

Solvate	Mg:Ligand Ratio
$Mg(BH_4)_2 \cdot$ DMS [§]	1:0.34
$Mg(BH_4)_2 \cdot$ TEA	1:1.8
$Mg(BH_4)_2 \cdot Et_2O$	1:0.36
$Mg(BH_4)_2 \cdot$ Digly	1:1.18
$Mg(BH_4)_2 \cdot$ DME	1:2.2
$Mg(BH_4)_2 \cdot$ THF	1:2.8

[§] The dimethyl sulfide (DMS) solvate was obtained through the synthetic protocol as described by Zanella et al. [32]. The DMS is weakly bond to the magnesium cation and readily removed by heating.

The TEA, Et$_2$O, Digly, DME and THF solvates of Mg(BH$_4$)$_2$, were prepared by adding an excess of solvent to Mg(BH$_4$)$_2$ at room temperature. Subsequently the solvent was removed *en vacuo* to obtain a crystalline solid. The stoichiometry of the solvates was determined from the relative integrated intensities of the signals observed in the ^{1}H NMR spectra as summarized in Table 1. Where we could find crystal structure information for solvates of Mg(BH$_4$)$_2$, the stoichiometry of solvate to Mg cation determine by NMR in our work is slightly greater than reported for Mg(BH$_4$)$_2$·DME 1:1.5 and slightly lower for Mg(BH$_4$)$_2$·THF 1:3 [33].

Unsolvated Mg(BH$_4$)$_2$ and solvate powders were dehydrogenated via combinatorial screening equipment made by Unchained Labs® (Pleasanton, CA, USA), consisting of a 24 well plate design. Heating of the samples was conducted in a screening pressure reactor at 180 °C for 24 h under N$_2$ flow. Product ratios determined by ^{11}B NMR are shown in Table 2. Entry 1 shows the low reactivity of unsolvated Mg(BH$_4$)$_2$ at 180 °C with 93% BH$_4^-$ remaining. This result is typical of the slow kinetics of dehydrogenation for borohydride complexes at temperatures below 300 °C. Dehydrogenation of the TEA complex favored formation of B$_3$H$_8^-$, along with a trace amount of B$_{10}$H$_{10}^{2-}$. The ether additives showed higher levels of dehydrogenation at 180 °C. Another difference found with the ether complexes is the observation of B$_{10}$H$_{10}^{2-}$ as the major product, suggesting either a competing dehydrogenation path or that the presence of these ether ligands encourages further reactivity of the B$_3$H$_8^-$ to form more deeply dehydrogenated products. Of the ether solvates, DME and THF provided the highest conversion of BH$_4^-$ with B$_{10}$H$_{10}^{2-}$ as the major product. Only small amounts of B$_{12}$H$_{12}^{2-}$, demonstrating that the decomposition was ~10× more selective for B$_{10}$H$_{10}^{2-}$ than B$_{12}$H$_{12}^{2-}$, much higher than the ~1.5× selectivity exhibited by the Digly solvate. These findings motivated further exploration of the dehydrogenation reaction.

Table 2. Distribution of products of Mg(BH$_4$)$_2$ solvates determined from integration of ^{11}B NMR peaks in mol % after dehydrogenation at 180 °C, 24 h, 1 atm N$_2$. The balance of products consist of trace quantities of boric acid due to hydrolysis of unstable polyboranes.

Sample	B$_{10}$H$_{10}^{2-}$	B$_3$H$_8^-$	B$_{12}$H$_{12}^{2-}$	BH$_4^-$
Mg(BH$_4$)$_2$		3		93
Mg(BH$_4$)$_2$·TEA	2	6		89
Mg(BH$_4$)$_2$·Et$_2$O	4	4		88
Mg(BH$_4$)$_2$·Digl	5	2	3	82
Mg(BH$_4$)$_2$·DME	46	14	4	30
Mg(BH$_4$)$_2$·THF	31	12	3	39

3. Discussion

A recent study asserted that closo-boranes are secondary products formed upon aqueous workup of the low temperature dehydrogenation reactions [34]. To determine if formation of B$_{10}$H$_{10}^{2-}$ occurs directly in the solid state reaction, the ^{11}B VT MAS NMR spectrum of dehydrogenated (1 atm N$_2$ at 180 °C for 24 h) Mg(BH$_4$)$_2$·THF was obtained (Figure 1). At room temperature, the observed resonances were broad, typical of solid state spectra for quadrupolar nuclei. Heating the sample to 160 °C sharpened the BH$_4^-$ peak and the resonances for B$_{10}$H$_{10}^{2-}$ at −2 and −27 ppm could be resolved. The peaks at −2 and −27 ppm assigned to the basal and apical boron in B$_{10}$H$_{10}$ based on the 1:4 ratio integration ratio in the solid state spectrum at 160 °C. At 20 °C the peaks are barely perceptible from the baseline. The sample is subsequently dissolved in a mixture of THF/D$_2$O for solution NMR analysis. A solution phase spectrum of the dehydrogenated complex was obtained for comparison of product distribution and line width after dissolution in THF/D$_2$O (Figure 1c). The same high selectivity for B$_{10}$H$_{10}^{2-}$ is observed in both solution (−2, −30 ppm) and solid-state NMR with respective yields of 19% and 18%.

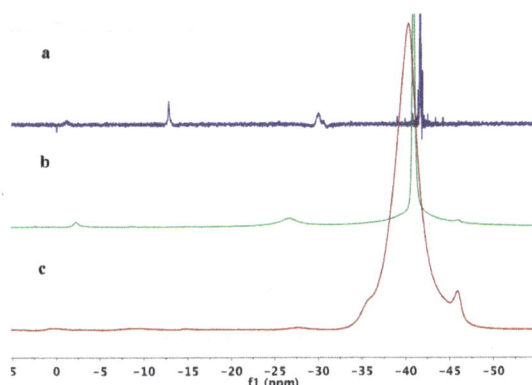

Figure 1. ^{11}B NMR spectra of Mg(BH$_4$)$_2$·THF (tetrahydofuran) dehydrogenated (**a**) solution phase dissolved in 1:2 THF:D$_2$O, (**b**) solid state collected at 160 °C and (**c**) solid state collected at 20 °C. Experimental set-up described in references [35,36].

In situ VT MAS ^{11}B NMR studies of the Mg(BH$_4$)$_2$·THF complex provides additional insight. As seen in Figure S1, the room temperature spectrum contains resonances for both unsolvated and solvated Mg(BH$_4$)$_2$ at −41 and −44 ppm respectively. See Figure S2 for a reference spectrum of unsolvated Mg(BH$_4$)$_2$. The downfield shift of the THF solvated BH$_4^-$ complex is comparable to the downfield shift reported for Mg(BH$_4$)$_2$·4NH$_3$ [37]. Upon heating the two peaks collapse into a single narrow peak at 90 °C. We interpret the narrow line width (FWHM = 32 Hz) as being indicative of a fluid phase. This is consistent with the observation of the melting of the THF solvate between 80–100 °C in a melting point apparatus and similar to the m.p. of 90 °C reported for Mg(BH$_4$)$_2$·2NH$_3$ [36].

A comparison of the IR spectra of the solvated and unsolvated Mg(BH$_4$)$_2$ complex is shown in Figure 2. The single prominent stretch observed in the B–H stretching region between 2300–2500 cm^{-1} [38] for Mg(BH$_4$)$_2$ is indicative of lack of directional bonding between the Mg cation and the tetrahedral environment of BH$_4^-$. The additional coordination of THF molecules results in the BH$_4^-$ also bonding in a mono or bidentate mode to the Mg cation. This lowering of symmetry leads to a spectrum with a number of overlapping bands occur between 2000–2500 cm^{-1}. The modified coordination may play a role in the dehydrogenation mechanism and energetics.

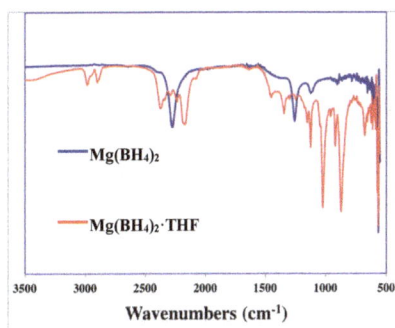

Figure 2. Attenuated Total Reflectance-Infrared spectra of unsolvated Mg(BH$_4$)$_2$ blue spectra with simple B–H stretching region and Mg(BH$_4$)$_2$·THF red spectrum with complex B–H stretching frequency.

The melting of the THF adduct is also likely to be a contributing factor to the enhanced kinetics. However, the onset of dehydrogenation occurs at temperatures above the melting point of the

THF complex. The THF may also reduce the activation energy of clustering to form more deeply dehydrogenated products by altering the coordination mode between Mg^{2+} and BH_4^- through donation of electron density or steric interactions. The high selectivity for $MgB_{10}H_{10}$ over $MgB_{12}H_{12}$ is surprising. Either THF influences the reaction pathway, i.e., lower the barrier for a branching point that pushes the reaction towards $MgB_{10}H_{10}$ formation or THF flips the thermodynamic stability of the closoboranes making $MgB_{10}H_{10}$ more stable than $MgB_{12}H_{12}$.

4. Materials and Methods

All sample preparation and storage was conducted either in a nitrogen glovebox or on a Schlenk line. Solvents were dried over molecular sieves and verified by NMR for purity before use.

4.1. Synthesis of Mg(BH₄)₂

Magnesium borohydride was synthesized following a method described by Zanella et al. Di-*n*-butylmagnesium (Sigma Aldrich, Milwaukee, WI, USA) was added dropwise to borane-dimethylsulfide complex (Sigma Aldrich) in toluene according to the reaction scheme:

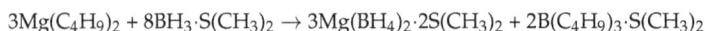

$$3Mg(C_4H_9)_2 + 8BH_3 \cdot S(CH_3)_2 \rightarrow 3Mg(BH_4)_2 \cdot 2S(CH_3)_2 + 2B(C_4H_9)_3 \cdot S(CH_3)_2$$

The mixture was allowed to stir at room temperature for a minimum of 3 h and subsequently filtered, washed with toluene, and dried *en vacuo* at room temperature for 6 h and then at 75 °C overnight. The product, a fine white powder, was found to consist of >95% α-Mg(BH₄)₂ by XRD analysis.

4.2. Synthesis of Solvent Adducts of Mg(BH₄)₂

The TEA, Et₂O, Digly, and THF solvates of Mg(BH₄)₂, were typically prepared by adding an excess of solvent to Mg(BH₄)₂ at room temperature and stirring for 30 min. Excess solvent was then removed *en vacuo* either at room temperature or up to 45 °C for higher boiling point solvents, for as long as needed to obtain a crystalline solid. The DMS adduct was obtained during the synthesis of Mg(BH₄)₂ as described above prior to removal of the DMS by heating.

4.3. Characterization of Mg(BH₄)₂ Adducts and Decomposition Products by Solution NMR

Powders were typically dissolved in a 1:2 mixture of THF:deuterium oxide (D₂O) and analyzed within 10 min on a Varian 300 MHz spectrometer with ^{11}B chemical shifts referenced to BF₃·Et₂O ($\delta = 0$ ppm) and ^1H referenced to TMS ($\delta = 0$ ppm). ^{11}B was measured at 96.23 MHz and ^1H was measured at 299.95 MHz. A relaxation delay of 15 s was used for all ^{11}B analyses with a 90° pulse width of 6 μs. An external standard was added to the quartz NMR tubes to determine the solubility of the powder in the THF/D₂O mixture. The standard consisted of an aqueous solution of sodium tetraphenylborate (NaBPh₄) sealed in a glass capillary. Calculation of percent composition of decomposed products was based on peak areas.

4.4. Solid State NMR

Sample powders were packed into 4 mm zirconium oxide rotors and spun at 12 kHz on a Varian 500 MHz spectrometer (Varian, Palo Alto, CA, USA) equipped with a HX 4 mm probe.

4.5. In Situ NMR

The characterization of Mg(BH₄)₂·THF during heating to 200 °C was conducted by variable temperature (VT) solid state magic angle spin (MAS) NMR in a Varian 500 MHz spectrometer 5 mm HXY probe. ^1H and ^{11}B shifts were referenced to tetramethylsilane at 0 ppm and lithium borohydride at −41.6 ppm and measured at 499.87 and 160.37 MHz respectively. ^1H and ^{11}B spectra were obtained with a 2 s and 5 s relaxation delay and 90° pulse width of 6 μs. The sample powder was packed in a

5 mm zirconia rotor under 1 atm N_2 with a Teflon spacer and then capped with a customized plastic bushing capable of withstanding pressures up to 200 bar. The details of the rotor design are given in detail elsewhere [32,33] and have been modified to accommodate 5 mm rotors. The rotors were spun at 5 kHz at room temperature and subsequently heated at a rate of about 6 °C/min and held at specific temperatures during the ramp at which ^{11}B and ^{1}H spectra were obtained. The duration of the analyses at the set temperatures was approximately 45 min.

5. Conclusions

In summary, characterization of the dehydrogenation products arising from $Mg(BH_4)_2 \cdot THF$ complex by solution and solid-state NMR shows that the dehydrogenation mechanism is highly selective for $B_{10}H_{10}{}^{2-}$ over $B_3H_8{}^-$ (theoretical H_2 release 8.1 wt % vs. 2.5 wt % in the absence of solvates) and $B_{12}H_{12}{}^{2-}$, a kinetic dead end. The dehydrogenation of $Mg(BH_4)_2$ at temperatures below 200 °C and potential for cycling between $Mg(BH_4)_2$ and $MgB_{10}H_{10}$ have significant implications for hydrogen storage applications. Further studies into optimizing the reaction through modification of ligand to Mg ratios are currently underway.

Supplementary Materials: The following are available online at www.mdpi.com/2304-6740/5/4/89/s1, Figure S1: In situ VT MAS ^{11}B NMR of $Mg(BH_4)_2 \cdot THF$ at room temperature and 90 °C. Figure S2: ^{11}B MAS spectra of unsolvated $Mg(BH_4)_2$ and solvated $Mg(BH_4)_2 \cdot THF$.

Acknowledgments: The authors gratefully acknowledge research support from the Hydrogen Materials—Advanced Research Consortium (HyMARC), established as part of the Energy Materials Network under the U.S. Department of Energy, Office of Energy Efficiency and Renewable Energy, Fuel Cell Technologies Office. We authors thank Junzhi Yang for preliminary experimental results on solvated magnesium borane complexes, Heather Job for assistance with the combinatorial decomposition experiments, Gary Edvenson for insightful discussion on borane cluster chemistry and David Hoyt and Sarah Burton from the Environmental Molecular Science Laboratory (EMSL) for assistance with the solid state NMR. EMSL is a DOE Office of Science User Facility sponsored by the Office of Biological and Environmental Research and located at Pacific Northwest National Laboratory (PNNL). PNNL a multi-program national laboratory operated by Battelle for the U.S. Department of Energy under Contract DE-AC05-76RL01830.

Author Contributions: Marina Chong, Tom Autrey and Craig Jensen conceived and designed the experiments; Marina Chong performed the experiments; Marina Chong, Tom Autrey and Craig Jensen contributed to analyzing the data and writing the paper.

Conflicts of Interest: The authors declare no conflict of interest.

References

1. Chong, M.; Karkamkar, A.; Autrey, T.; Orimo, S.; Jalisatgi, S.; Jensen, C.M. Reversible dehydrogenation of magnesium borohydride to magnesium triborane in the solid state under moderate conditions. *Chem. Commun.* **2011**, *47*, 1330–1332. [CrossRef]
2. Filinchuk, Y.; Richter, B.; Jensen, T.R.; Dmitriev, V.; Chernyshov, D.; Hagemann, H. Porous and dense magnesium borohydride frameworks: Synthesis, stability, and reversible absorption of guest species. *Angew. Chem. Int. Ed.* **2011**, *50*, 11162–11166. [CrossRef]
3. Hanada, N.; Chłopek, K.; Frommen, C.; Lohstroh, W.; Fichtner, M. Thermal decomposition of $Mg(BH_4)_2$ under He flow and H_2 pressure. *J. Mater. Chem.* **2008**, *18*, 2611–2614. [CrossRef]
4. Hwang, S.-J.; Bowman, R.C., Jr.; Reiter, J.W.; Rijssenbeek, J.; Soloveichik, G.; Zhao, J.-C.; Kabbour, H.; Ahn, C.C. NMR confirmation for formation of $[B_{12}H_{12}]^{2-}$ complexes during hydrogen desorption from metal borohydrides. *J. Phys. Chem. C* **2008**, *112*, 3164–3169. [CrossRef]
5. Li, H.-W.; Kikuchi, K.; Sato, T.; Nakamori, Y.; Ohba, N.; Aoki, M.; Miwa, K.; Towata, S.; Orimo, S. Synthesis and hydrogen storage properties of a single-phase magnesium borohydride $Mg(BH_4)_2$. *Mater. Trans.* **2008**, *49*, 2224–2228. [CrossRef]
6. Matsunaga, T.; Buchter, F.; Mauron, P.; Bielman, M.; Nakamori, Y.; Orimo, S.; Ohba, N.; Miwa, K.; Towata, S.; Züttel, A. Hydrogen storage properties of $Mg[BH_4]_2$. *J. Alloys Compd.* **2008**, *459*, 583–588. [CrossRef]

7.   Nakamori, Y.; Miwa, K.; Ninomiya, A.; Li, H.-W.; Ohba, N.; Towata, S.; Züttel, A.; Orimo, S. Correlation between thermodynamical stabilities of metal borohydrides and cation electronegativites: First-principles calculations and experiments. *Phys. Rev. B* **2006**, *74*, 045126. [CrossRef]

8.   Ozolins, V.; Majzoub, E.H.; Wolverton, C. First-principles prediction of a ground state crystal structure of magnesium borohydride. *Phys. Rev. Lett.* **2008**, *100*, 135501. [CrossRef]

9.   Severa, G.; Ronnebro, E.; Jensen, C.M. Direct hydrogenation of magnesium boride to magnesium borohydride: Demonstration of >11 weight percent reversible hydrogen storage. *Chem. Commun.* **2010**, *46*, 421–423. [CrossRef]

10. Van Setten, M.J.; de Wijs, G.A.; Fichtner, M.; Brocks, G.A. A Density Functional Study of alpha-Mg(BH$_4$)$_2$. *Chem. Mater.* **2008**, *20*, 4952–4956. [CrossRef]

11. Yan, Y.; Li, H.-W.; Nakamori, Y.; Ohba, H.; Miwa, K.; Towata, S.; Orimo, S. Differential scanning calorimetry measurements of magnesium borohydride Mg(BH$_4$)$_2$. *Mater. Trans.* **2008**, *49*, 2751–2752. [CrossRef]

12. Zavorotynska, O.; Deledda, S.; Hauback, B.C. Kinetics studies of the reversible partial decomposition reaction in Mg(BH$_4$)$_2$. *Int. J. Hydrog. Energy* **2016**, *41*, 9885–9892. [CrossRef]

13. Zhang, Y.; Majzoub, E.; Ozoliņš, V.; Wolverton, C. Theoretical prediction of metastable intermediates in the decomposition of Mg(BH$_4$)$_2$. *J. Phys. Chem. C* **2012**, *116*, 10522–10528. [CrossRef]

14. Chłopek, K.; Frommen, C.; Léon, A.; Zabara, O.; Fichtner, M. Synthesis and properties of magnesium tetrahydroborate, Mg(BH$_4$)$_2$. *J. Mater. Chem.* **2007**, *17*, 3496–3503. [CrossRef]

15. Li, H.-W. Dehydriding and rehydriding processes of well-crystallized Mg(BH$_4$)$_2$ accompanying with formation of intermediate compounds. *Acta Mater.* **2008**, *56*, 1342–1347. [CrossRef]

16. White, J.L.; Newhouse, R.J.; Zhang, J.Z.; Udovic, T.J.; Stavila, V. Understanding and mitigating the effects of stable dodecahydro-closo-dodecaborate intermediates on hydrogen-storage reactions. *J. Phys. Chem. C* **2016**, *120*, 25725–25731. [CrossRef]

17. Li, H.-W.; Kikuchi, K.; Nakamori, Y.; Miwa, K.; Towata, S.; Orimo, S. Effects of ball milling and additives on dehydriding behaviors of well-crystallized Mg(BH$_4$)$_2$. *Scr. Mater.* **2007**, *57*, 679–682. [CrossRef]

18. Bardají, E.G.; Hanada, N.; Zabara, O.; Fichtner, M. Effect of several metal chlorides on the thermal decomposition behaviour of α-Mg(BH$_4$)$_2$. *Int. J. Hydrog. Energy* **2011**, *36*, 12313–12318. [CrossRef]

19. Newhouse, R.J.; Stavila, V.; Hwang, S.-J.; Klebanoff, L.; Zhang, J.Z. Reversibility and improved hydrogen release of magnesium borohydride. *J. Phys. Chem. C* **2010**, *114*, 5224–5234. [CrossRef]

20. Saldan, I.; Frommen, C.; Llamas-Jansa, I.; Kalantzopoulos, G.N.; Hino, S.; Arstad, B.; Heyn, R.H.; Zavorotynska, O.; Deledda, S.; Sørby, M.H.; et al. Hydrogen storage properties of γ–Mg(BH$_4$)$_2$ modified by MoO$_3$ and TiO$_2$. *Int J. Hydrog. Energy* **2015**, *40*, 12286–12293. [CrossRef]

21. Bardají, E.G.; Zhao-Karger, Z.; Boucharat, N.; Nale, A.; van Setten, M.J.; Lohstroh, W.; Röhm, E.; Catti, M.; Fichtner, M. LiBH$_4$ –Mg(BH$_4$)$_2$: A physical mixture of metal borohydrides as hydrogen storage material. *J. Phys. Chem. C* **2011**, *115*, 6095–6101. [CrossRef]

22. Hagemann, H.; D'Anna, V.; Rapin, J.-P.; Černý, R.; Filinchuk, Y.; Kim, K.C.; Sholl, D.S.; Parker, S.T. New fundamental experimental studies on α-Mg(BH$_4$)$_2$ and other borohydrides. *J. Alloys Compd.* **2011**, *509*, S688–S690. [CrossRef]

23. Nale, A.; Catti, M.; Bardají, E.G.; Fichtner, M. On the decomposition of the 0.6LiBH$_4$–0.4Mg(BH$_4$)$_2$ eutectic mixture for hydrogen storage. *Int. J. Hydrog. Energy* **2011**, *36*, 13676–13682. [CrossRef]

24. Yang, J.; Fu, H.; Song, P.; Zheng, J.; Li, X. Reversible dehydrogenation of Mg(BH$_4$)$_2$–LiH composite under moderate conditions. *Int. J. Hydrog. Energy* **2012**, *37*, 6776–6783. [CrossRef]

25. Muetterties, E.L.; Knoth, W.H. *Polyhedral Boranes*; Marcel Dekker: New York, NY, USA, 1968.

26. Chong, M.; Matsuo, M.; Orimo, S.; Autrey, T.; Jensen, C.M. Selective reversible hydrogenation of Mg(B$_3$H$_8$)$_2$/MgH$_2$ to Mg(BH$_4$)$_2$: Pathway to reversible borane-based hydrogen storage? *Inorg. Chem.* **2015**, *54*, 4120–4125. [CrossRef]

27. Chen, J.; Chua, Y.S.; Wu, H.; Xiong, Z.; He, T.; Zhou, W.; Ju, X.; Yang, M.; Wu, G.; Chen, P. Synthesis, structures and dehydrogenation of magnesium borohydride–ethylenediamine composites. *Int. J. Hydrog. Energy* **2015**, *40*, 412–419. [CrossRef]

28. Yang, Y.; Liu, Y.; Zhang, Y.; Li, Y.; Gao, M.; Pan, H. Hydrogen storage properties and mechanisms of Mg(BH$_4$)$_2$·2NH$_3$–xMgH$_2$ combination systems. *J. Alloys Compd.* **2014**, *585*, 674–680. [CrossRef]

29. Zhao, S.; Xu, B.; Sun, N.; Sun, Z.; Zeng, Y.; Meng, L. Improvement in dehydrogenation performance of Mg(BH$_4$)$_2$·2NH$_3$ doped with transition metal: First-principles investigation. *Int. J. Hydrog. Energy* **2015**, *40*, 8721–8731. [CrossRef]

30. Soloveichik, G.; Her, J.H.; Stephens, P.W.; Gao, Y.; Rijssenbeek, J.; Andrus, M.; Zhao, J.-C. Ammine magnesium borohydride complex as a New material for hydrogen storage: Structure and properties of Mg(BH$_4$)$_2$·2NH$_3$. *Inorg. Chem.* **2008**, *47*, 4290–4298. [CrossRef]

31. Zavorotynska, O.; El-Kharbachi, A.; Deledda, S.; Hauback, B.C. Recent progress in magnesium borohydride Mg(BH$_4$)$_2$: Fundamentals and applications for energy storage. *Int. J. Hydrog. Energy* **2016**, *41*, 14387–14404. [CrossRef]

32. Zanella, P.; Crociani, L.; Masciocchi, N.; Giunchi, G. Facile high-yield synthesis of pure, crystalline Mg(BH$_4$)$_2$. *Inorg. Chem.* **2007**, *46*, 9039–9041. [CrossRef]

33. Wegner, W.; Jaroń, T.; Dobrowolski, M.A.; Dobrzycki, Ł.; Cyrański, M.K.; Grochala, W. Organic derivatives of Mg(BH$_4$)$_2$ as precursors towards MgB$_2$ and novel inorganic mixed-cation borohydrides. *Dalton Trans.* **2016**, *45*, 14370–14377. [CrossRef]

34. Yan, Y.; Remhof, A.; Rentsch, D.; Züttel, A. The role of MgB$_{12}$H$_{12}$ in the hydrogen desorption process of Mg(BH$_4$)$_2$. *Chem. Commun.* **2015**, *51*, 700–702. [CrossRef]

35. Hoyt, D.W.; Turcu, R.V.F.; Sears, J.A.; Rosso, K.; Burton, S.D.; Felmy, A.R.; Hu, J.-Z. High-pressure magic angle spinning nuclear magnetic resonance. *J. Magn. Res.* **2011**, *212*, 378–385. [CrossRef]

36. Turcu, R.V.F.; Hoyt, D.W.; Rosso, K.M.; Sears, J.A.; Loring, J.S.; Felmy, A.R.; Hu, J.-Z. Rotor design for high pressure magic angle spinning nuclear magnetic resonance. *J. Magn. Res.* **2013**, *226*, 64–69. [CrossRef]

37. Gao, L.; Guo, Y.H.; Li, Q.; Yu, X.B. The comparison in dehydrogenation properties and mechanism between MgCl$_2$(NH$_3$)/LiBH$_4$ and MgCl$_2$(NH$_3$)/NaBH$_4$ systems. *J. Phys. Chem. C* **2010**, *114*, 9534–9540. [CrossRef]

38. Marks, T.J.; Kolb, J.R. Covalent transition metal, lanthanide, and actinide tetrahydroborate complexes. *Chem. Rev.* **1977**, *77*, 263–293.

MDPI

St. Alban-Anlage 66

4052 Basel

Switzerland

Tel. +41 61 683 77 34

Fax +41 61 302 89 18

www.mdpi.com

Inorganics Editorial Office

E-mail: inorganics@mdpi.com

www.mdpi.com/journal/inorganics

www.ingramcontent.com/pod-product-compliance
Lightning Source LLC
Chambersburg PA
CBHW041215220326
41597CB00033BA/5935